T0371963

Quantum Optics
for Beginners

Quantum Optics for Beginners

Zbigniew Ficek
Mohamed Ridza Wahiddin

Pan Stanford Publishing

Published by

Pan Stanford Publishing Pte. Ltd.
Penthouse Level, Suntec Tower 3
8 Temasek Boulevard
Singapore 038988

Email: editorial@panstanford.com
Web: www.panstanford.com

British Library Cataloguing-in-Publication Data
A catalogue record for this book is available from the British Library.

Quantum Optics for Beginners

ISBN 978-981-4411-75-2 (Hardcover)
ISBN 978-981-4411-76-9 (eBook)

Printed in the USA

To our families

Latifah Mansoor, Sulaiman, Nasuha, Nadwiyah, and Renata

Contents

Preface

Quantum optics is the study of interactions between matter and the radiation field where quantum effects are important. Much of the fundamental interest in quantum optics is connected with its implications for the conceptual foundations of quantum mechanics. However, the major quantum optics problem is whether we have to quantize the electromagnetic field in order to get the correct picture of the interaction between matter and the field. The theoretical prediction and experimental verifications of photon antibunching and squeezing—the two nonclassical phenomena which do not exist in semiclassical theory—convinced researchers that the electromagnetic field should be quantized and stimulated considerable attention in other nonclassical effects such as quantum interference and entanglement.

This book is an extended and updated version of lecture notes published in 2004 as *Quantum Optics: Fundamentals and Applications* by the International Islamic University Malaysia Press, Kuala Lumpur, Malaysia. It is a compilation of the lectures given for postgraduate students at the University of Queensland, Brisbane, Australia, the University of Malaya and the International Islamic University, Malaysia in years 1995–2008. The chapters cover the background theory of various effects discussed from first principles, and as clearly as possible, to introduce students to the main ideas of quantum optics and to teach the mathematical methods and techniques used by researchers working in the fields of quantum and atom optics. Some of the key problems of quantum optics are also described, concentrating on the techniques, results, and interpretations. Although the chapters in the book do not provide a complete exploration of all the problems of quantum and atom optics, it is hoped that the problems explored will provide a useful starting point for those interested in learning more. The selected problems are not necessarily the most recent or advanced, but have been most influential in the directions of research in quantum and atom optics. Furthermore, the chapters contain numerous valuable derivations and calculations that are hard to find in scientific articles and textbooks on quantum optics. The goal of this book is to provide a compact logical exposition of the fundamentals of quantum optics and their application to atomic and quantum physics and to study quantum properties of matter and radiation. We are witnessing the development of new fields "atom optics" and "quantum information." As offsprings of quantum optics, they possess many resemblances to their parent field. These new fields can be approached and understood by using many of the same mathematical tools.

The chapters constitute the basic ideas and principles of quantum optics put in the order of the development of this subject, which is sophisticated enough to establish a firm basis for advanced study in this area. Current key problems of quantum optics, quantum information, and atom optics are included and treated in adequate depth to illustrate the basic concepts and also provide a nontrivial background in a diverse number of areas of current interest. Moreover, a number of exercises have been included at the end of each chapter. These exercises have been designed not only to help students learn how to apply the fundamental principles to many situations, but also to derive a number of important results not explicitly presented in the chapters.

Over the years we have collaborated with many colleagues and students, who directly or indirectly contributed to this work. We are particularly grateful to H. J. Carmichael, P. D. Drummond, G. J. Milburn, H. S. Freedhoff, B. J. Dalton, S. Swain, R. Tanaś, R. K. Bullough, S. S. Hassan, A. Messikh, M. R. Ferguson, T. Rudolph, U. Akram, and M. Salihi Abd Hadi. We are also indebted to students whose interesting questions and remarks have made the chapters more interesting and have helped purge them of typographical errors.

Zbigniew Ficek
Mohamed Ridza Wahiddin
Winter 2013

Chapter 1

General Description and Quantization of EM Fields

1.1 Introduction

We will begin our journey through the background of quantum optics with an elementary, but quantitative, classical theory of radiative fields. We will first briefly outline the electromagnetic (EM) theory of radiation, and describe how the EM radiation may be understood as a wave which can be represented by a set of harmonic oscillators. We shall describe how the free or non-interacting EM field may be understood as a collection of harmonic oscillators which is quantized in the standard manner, and whose energy cannot be zero as a consequence of the basic non-commutability of the canonical field variables. This chapter discusses the properties of plane EM waves and normalization of the EM field in one dimension. This is followed by a description of the Hamiltonian and the amplitudes of the EM field in terms of the annihilation and creation operators. Based on this chapter, it is possible to considerably simplify the formulation of the physical basis for the mathematical description of the major problems of quantum optics.

Quantum Optics for Beginners
Zbigniew Ficek and Mohamed Ridza Wahiddin

ISBN 978-981-4411-75-2 (Hardcover), 978-981-4411-76-9 (eBook)
www.panstanford.com

Why do we apply the quantum description of the EM field? The answer lies in the recent theoretical and experimental developments in quantum optics, which show that semiclassical radiation theory based on the quantum description of the radiation sources and classical description of EM fields does not always work. There are some optical phenomena, we will discuss about which during the course of this book, for which the field needs to be treated quantum mechanically. These phenomena were recognized as representing a radical departure from the traditional classical optics where the existing treatments turn out to be less than completely satisfactory. In other words, these phenomena are non-classical and do not exist in semiclassical radiation theory.

1.2 Maxwell's Equations for the EM Field

Let us consider the time-varying classical electric $\vec{E}$ and magnetic $\vec{B}$ fields that satisfy the Maxwell's equations [1]

$$\nabla \cdot \vec{E} = \rho_{\mathrm{f}}/\varepsilon_0, \tag{1.1}$$

$$\nabla \cdot \vec{B} = 0, \tag{1.2}$$

$$\nabla \times \vec{E} = -\frac{\partial}{\partial t}\vec{B}, \tag{1.3}$$

$$\nabla \times \vec{B} = \mu_0 \vec{J} + \frac{1}{c^2}\frac{\partial}{\partial t}\vec{E}, \tag{1.4}$$

where ρ_{f} is the density of free charges and $\vec{J}$ is the density of currents at a point where the electric and magnetic fields are evaluated. The parameters ε_0 and μ_0 are constants that determine the property of the vacuum and are called the electric permittivity and magnetic permeability, respectively. The parameter $c = 1/\sqrt{\varepsilon_0\mu_0}$ and its numerical value is equal to the speed of light in vacuum, $c = 3 \times 10^8$ [ms^{-1}].

In the Maxwell's equations, the fields $\vec{E}$ and $\vec{B}$ depend on $(\vec{r}, t)$, the charge and current densities also depend on $(\vec{r}, t)$. It is not explicitly stated in the above equations, but we shall remember about this dependence in the following calculations.

The fields $\vec{E}$ and $\vec{B}$ produced by the source charges ρ_{f} and currents $\vec{J}$ are found by solving the Maxwell's equations (1.1)–(1.4).

Note that the Maxwell's equations involve two fields that satisfy a system of four coupled differential equations. Generally, we do not find fields $\vec{E}$ and $\vec{B}$ by a direct integration of the Maxwell's equations. We rather first compute scalar and vector potentials from which the fields may be found.

Let us illustrate the concept of vector and scalar potentials in the solution of the Maxwell's equations. First, note that the field $\vec{B}$ always has zero divergence, $\nabla \cdot \vec{B} = 0$, and hence we can always write

$$\vec{B} = \nabla \times \vec{A}, \tag{1.5}$$

where $\vec{A}$ is the vector potential.

Since $\nabla \times \nabla\Phi \equiv 0$, where Φ is an arbitrary scalar function (scalar potential), we find from the Maxwell's equation (1.3) that the electric field can be written as[a]

$$\vec{E} = -\frac{\partial}{\partial t}\vec{A} - \nabla\Phi. \tag{1.6}$$

The electric field (1.6) depends on the specific choice of the potentials. However, the Maxwell's equations should be independent of the specific choice of the potentials.

Substituting Eq. (1.6) into Eq. (1.1), we get

$$\nabla \cdot \vec{E} = -\frac{\partial}{\partial t}\nabla \cdot \vec{A} - \nabla^2\Phi = 0. \tag{1.7}$$

Hence, the electric field (1.6) will satisfy the Maxwell's equation (1.1) when

$$-\frac{\partial}{\partial t}\nabla \cdot \vec{A} - \nabla^2\Phi = 0. \tag{1.8}$$

If we now substitute Eqs. (1.5) and (1.6) into Eq. (1.4), and expand the double curl $\nabla \times (\nabla \times \vec{A})$ to give $\nabla(\nabla \cdot \vec{A}) - \nabla^2\vec{A}$, we obtain a three-dimensional inhomogeneous wave equation for the vector potential

$$\nabla^2\vec{A} - \frac{1}{c^2}\frac{\partial^2}{\partial t^2}\vec{A} = \nabla\left(\nabla \cdot \vec{A} + \frac{1}{c^2}\frac{\partial}{\partial t}\Phi\right). \tag{1.9}$$

According to the Helmholtz theorem, a vector function is completely specified by its divergence and curl. Since Eq. (1.5) gives only the curl of $\vec{A}$, we can specify the divergence of $\vec{A}$ in any way we choose.

---

[a] In the static limit of $\partial\vec{A}/\partial t = 0$, the scalar function Φ reduces to the familiar electrostatic potential.

We can define new potentials

$$\vec{A}' = \vec{A} + \nabla\psi, \quad \Phi' = \Phi - \frac{\partial\psi}{\partial t}, \tag{1.10}$$

without changing the $\vec{E}$ and $\vec{B}$ fields, where ψ is an arbitrary scalar potential. This transformation is called a gauge transformation, and the invariance of the fields under such transformation is called gauge invariance.

Equation (1.8) implies that the electric field will satisfy the Maxwell's equations when

$$\nabla^2\Phi = -\frac{\partial}{\partial t}\nabla\cdot\vec{A}, \tag{1.11}$$

which is only for a specific choice of the potentials. However, the freedom of choosing $\vec{A}$ means that we can choose the potentials as

$$\nabla\cdot\vec{A} = 0, \qquad \Phi = 0. \tag{1.12}$$

This choice is called the *Coulomb gauge*, and this equation reduces Eq. (1.9) to

$$\nabla^2\vec{A} - \frac{1}{c^2}\frac{\partial^2}{\partial t^2}\vec{A} = 0, \tag{1.13}$$

which is much simpler than Eq. (1.9), and can be readily solved in terms of plane transverse waves.

1.3 Wave Equation

We have seen that the Maxwell's equations can be transferred, with the help of the Coulomb gauge, into a wave equation (1.13). The general solution of the wave equation is in the well-known form of an infinite set of plane waves[a]

$$\vec{A} = \sum_{\vec{k}s}\vec{A}_{\vec{k}s}\,\mathrm{e}^{-i\left(\omega_{\vec{k}s}t - \vec{k}\cdot\vec{r}\right)}, \tag{1.14}$$

where $\vec{k}s$ denotes the plane waves of the index of polarization s propagating in the $\vec{k}$ direction, $|\vec{k}| = \omega_{\vec{k}s}/c$, and $\vec{A}_{\vec{k}s}$ is the amplitude of the wave of frequency $\omega_{\vec{k}s}$.

[a]The solution (1.14) is readily verified by substitution into Eq. (1.13).

The Coulomb gauge condition, $\nabla \cdot \vec{A} = 0$, gives

$$\vec{k} \cdot \vec{A}_{\vec{k}s} = 0, \tag{1.15}$$

which is the transversal condition showing that the amplitude vectors of the field are orthogonal to the propagation direction. The amplitudes $\vec{A}_{\vec{k}s}$ being orthogonal to $\vec{k}$ can be specified in terms of components along two mutually orthogonal directions transverse to $\vec{k}$. Unit vectors along these directions, denoted by $\vec{e}_{\vec{k}s}$ $(s = 1, 2)$, obey the relations

$$\vec{e}_{\vec{k}i} \cdot \vec{e}_{\vec{k}j} = \delta_{ij}, \quad \vec{e}_{\vec{k}s} \cdot \vec{k} = 0, \quad \vec{e}_{\vec{k}1} \times \vec{e}_{\vec{k}2} = \vec{k}, \tag{1.16}$$

and $\vec{e}_{\vec{k}s}$ are usually called the unit vectors of the field polarization. In other words, they specify the polarization directions of the field.

Thus, we can represent the vector potential of the EM field in terms of plane waves

$$\begin{aligned} \vec{A} &= \sum_{\vec{k}s} \left[\vec{A}_{\vec{k}s}\, \mathrm{e}^{-i\left(\omega_{\vec{k}s} t - \vec{k}\cdot\vec{r}\right)} + \vec{A}^{*}_{\vec{k}s}\, \mathrm{e}^{i\left(\omega_{\vec{k}s} t - \vec{k}\cdot\vec{r}\right)} \right] \\ &= \sum_{\vec{k}s} \left[\vec{e}_{\vec{k}s} A_{\vec{k}s}\, \mathrm{e}^{-i\left(\omega_{\vec{k}s} t - \vec{k}\cdot\vec{r}\right)} + \vec{e}_{\vec{k}s} A^{*}_{\vec{k}s}\, \mathrm{e}^{i\left(\omega_{\vec{k}s} t - \vec{k}\cdot\vec{r}\right)} \right], \end{aligned} \tag{1.17}$$

propagating in the $\pm\vec{k}$ directions. Equation (1.17) allows us to calculate the transverse EM field vectors $\vec{E}$ and $\vec{B}$ at any space–time point through the relations (1.5) and (1.6).

1.4 Energy of the EM Wave

Consider an EM wave confined in a space of volume V. The energy of the three-dimensional EM field is given by the Hamiltonian

$$H_{\mathrm{F}} = \frac{1}{2} \int_V dV \left[\varepsilon_0 |\vec{E}|^2 + \frac{1}{\mu_0} |\vec{B}|^2 \right], \tag{1.18}$$

which is given by the integral of the energy density over the volume occupied by the field. Here, $\varepsilon_0 |\vec{E}^2|/2$ is the energy density of the electric field, and $|\vec{B}|^2/(2\mu_0)$ is the energy density of the magnetic field.

First, we will impose periodic boundary conditions on the field and normalize the energy to that contained in the finite volume V.

Next we will express the Hamiltonian in terms of the energy of discrete harmonic oscillators and quantize the Hamiltonian in the standard manner by associating the complex field amplitudes with creation and annihilation operators.

1.4.1 *Normalization of the EM Field*

In order to proceed further with the energy formula, Eq. (1.18), we have to formulate the normalization procedure of the EM field contained in a finite volume V. In general, the field is confined into three dimensions, but for simplicity and without loss of the generality, we will illustrate the normalization procedure in one dimension only.

Consider a plane-wave electric field confined between two perfectly reflecting walls, linearly polarized in the x-direction and propagating in one dimension, the z-direction, as illustrated in Fig. 1.1.

The field can be written as

$$\vec{E}(z, t) = \vec{i}\, E_x(z, t) = \vec{i}\, q(t) \sin(kz). \tag{1.19}$$

The walls of the field enclosure, located at $z = 0$ and $z = L$ are taken as perfectly reflecting surfaces, which implies that

$$E_x(0, t) = E_x(L, t) = 0, \tag{1.20}$$

and hence

$$\sin(kL) = 0. \tag{1.21}$$

From this result we see that the wave number k is given by

$$k = \frac{n\pi}{L}. \tag{1.22}$$

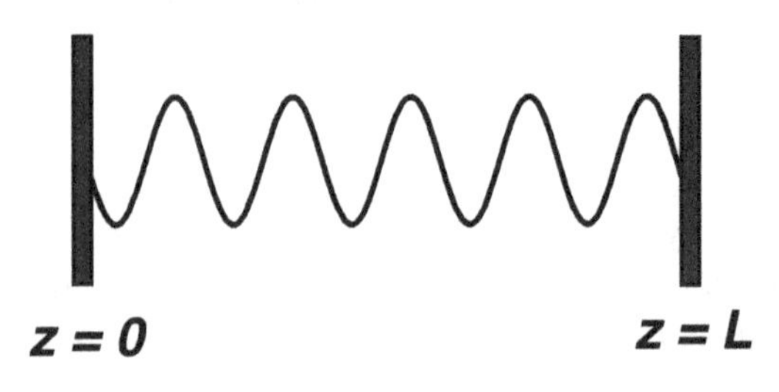

Figure 1.1 Plane wave electric field propagating in the z-direction and polarized in the x-direction.

Thus, the normalized EM field is represented by standing waves of the discrete wave (propagation) number k.

1.4.2 *Hamiltonian of the EM Wave*

In order to find the Hamiltonian (energy) of the EM field, we need both the electric $\vec{E}$ and magnetic $\vec{B}$ fields. The electric field is given by Eq. (1.19), and we will use the Maxwell's equations to find magnetic field.

The procedure of finding the magnetic field is as follows. According to the Maxwell's equations and the plane-wave representation of the EM field, the magnetic vector $\vec{B}$ of the field is perpendicular to $\vec{E}$ and oriented along the y-axis. Substituting Eq. (1.19) into the Maxwell's equation (1.4), we obtain

$$\nabla \times \vec{B} = \vec{i}\,\frac{1}{c^2}\dot{q}\,(t)\sin(kz). \qquad (1.23)$$

Since $B_x = B_z = 0$ and $B_y \neq 0$, the curl on the left-hand side contains only two non-zero terms. Thus, the equation takes the form

$$-\vec{i}\,\frac{\partial B_y}{\partial z} + \vec{k}\,\frac{\partial B_y}{\partial x} = \vec{i}\,\frac{1}{c^2}\dot{q}\,(t)\sin(kz). \qquad (1.24)$$

The coefficients on both sides of the equation at the same unit vectors should be equal. Hence, we find that

$$\frac{\partial B_y}{\partial x} = 0 \quad \text{and} \quad \frac{\partial B_y}{\partial z} = -\frac{1}{c^2}\dot{q}\,(t)\sin(kz). \qquad (1.25)$$

Integration of $\partial B_y/\partial z$ gives

$$B_y\,(z,t) = -\frac{1}{c^2}\dot{q}\,(t)\int dz\sin(kz) = \frac{1}{kc^2}\dot{q}\,(t)\cos(kz). \quad (1.26)$$

This equation gives the expression that determines the magnetic field of the one-dimensional EM wave propagating in the z-direction.

According to Eq. (1.18), the energy of the one-dimensional EM field is given by the Hamiltonian

$$\begin{aligned} H_{\mathrm{F}} &= \frac{1}{2}\int_0^L dz\left\{\varepsilon_0|\vec{E}|^2 + \frac{1}{\mu_0}|\vec{B}|^2\right\} \\ &= \frac{1}{2}\int_0^L dz\left\{\varepsilon_0 q^2\,(t)\sin^2(kz) + \frac{1}{k^2c^4\mu_0}\,(\dot{q}\,(t))^2\cos^2(kz)\right\}. \quad (1.27) \end{aligned}$$

Since

$$\int_0^L dz \sin^2(kz) = \int_0^L dz \cos^2(kz) = \frac{1}{2}L, \tag{1.28}$$

the Hamiltonian (1.27) reduces to

$$H_{\mathrm{F}} = \frac{1}{4}\varepsilon_0 q^2(t)\, L + \frac{1}{4}\frac{\varepsilon_0}{\omega^2} L (\dot{q}(t))^2 . \tag{1.29}$$

It is convenient to compare this energy with that of an harmonic oscillator given by the well-known formula

$$H_{\mathrm{osc}} = \frac{1}{2}m\omega^2 x^2 + \frac{1}{2}m(\dot{x})^2 . \tag{1.30}$$

In this case, we find that

$$q = \alpha x, \tag{1.31}$$

where

$$\alpha = \sqrt{\frac{2m\omega^2}{\varepsilon_0 L}}. \tag{1.32}$$

Hence, the electric and magnetic fields can be written in terms of the harmonic oscillator variables as

$$E_x(z,t) = \sqrt{\frac{2m\omega^2}{\varepsilon_0 L}}\, x(t)\sin(kz),$$

$$B_y(z,t) = \frac{1}{kc^2}\sqrt{\frac{2m\omega^2}{\varepsilon_0 L}}\, \dot{x}(t)\cos(kz). \tag{1.33}$$

Example 1.1 (EM field in terms of canonical variables) *An alternative representation of the EM field amplitudes is in terms of a pair of real canonical variables $q_k(t)$ and $p_k(t)$, defined as*

$$q_k(t) = \sqrt{\varepsilon_0}\left[A_k(t) + A_k^*(t)\right],$$

$$p_k(t) = -i\omega_k\sqrt{\varepsilon_0}\left[A_k(t) - A_k^*(t)\right]. \tag{1.34}$$

Since in harmonic motion $A_k(t) = A_k \exp(-i\omega_k t)$, the two canonical variables are related by

$$\frac{\partial}{\partial t}q_k(t) = p_k(t), \quad \frac{\partial}{\partial t}p_k(t) = -\omega_k^2 q_k(t). \tag{1.35}$$

Since p_k is obtained from the derivative of q_k and vice versa, q_k is obtained from the derivative of p_k, there is a phase shift between q_k

and p_k. For example, if q_k varies as a cosine function then p_k varies as a sine function.

Having the normalization procedure formulated in one dimension, we can now generalize the fields into three dimensions where the electric and magnetic fields, derived from Eqs. (1.5) and (1.6) with help of Eq. (1.17), can be written in the form

$$\vec{E} = i\sum_k \frac{\omega_k}{\sqrt{V}}\left[\vec{A}_k(t)\,\mathrm{e}^{i\vec{k}\cdot\vec{r}} - \text{c.c.}\right],$$

$$\vec{B} = i\sum_k \frac{1}{\sqrt{V}}\left[\vec{k}\times\vec{A}_k(t)\,\mathrm{e}^{i\vec{k}\cdot\vec{r}} - \text{c.c.}\right]. \qquad (1.36)$$

Here, c.c. stands for the complex conjugation of the first term in the bracket, $k \equiv (\vec{k}, s)$, and we have normalized the fields to the volume $\sqrt{V}$.

Substituting above equation into Eq. (1.18) and performing the integration with

$$\frac{1}{V}\int_V \mathrm{e}^{i\left(\vec{k}-\vec{k}'\right)\cdot\vec{r}}dV = \delta\left(\vec{k}-\vec{k}'\right), \qquad (1.37)$$

and a vector relation

$$\left(\vec{k}\times\vec{A}_k\right)\cdot\left(\vec{k}\times\vec{A}_k^*\right) = k^2\left|\vec{A}_k\right|^2, \qquad (1.38)$$

we obtain

$$H_\mathrm{F} = 2\varepsilon_0\sum_k \omega_k^2\left|\vec{A}_k\right|^2. \qquad (1.39)$$

Since in terms of the canonical variables

$$2\varepsilon_0\omega_k^2\left|\vec{A}_k\right|^2 = \frac{1}{2}\left[p_k^2(t) + \omega_k^2 q_k^2(t)\right], \qquad (1.40)$$

we get

$$H_\mathrm{F} = \frac{1}{2}\sum_k\left[p_k^2(t) + \omega_k^2 q_k^2(t)\right], \qquad (1.41)$$

which means that the EM field can be expressed as a set of harmonic oscillators, and the canonical variables q_k and p_k can be related to the position and momentum of the harmonic oscillator.

The energy can also be written as

$$H_\mathrm{F} = 2\varepsilon_0\sum_k \omega_k^2\left|A_k(t)\right|^2$$

$$= \varepsilon_0\sum_k \omega_k^2\left[A_k^*(t)\,A_k(t) + A_k(t)\,A_k^*(t)\right]. \qquad (1.42)$$

Introducing a new variable

$$a_k(t) = \sqrt{2\varepsilon_0\omega_k}A_k(t), \tag{1.43}$$

we finally can write the Hamiltonian of the classical EM field as

$$H_\mathrm{F} = \frac{1}{2}\sum_k \omega_k \left[a_k^*(t)a_k(t) + a_k(t)a_k^*(t)\right]. \tag{1.44}$$

Note the proportionality of the energy to frequency, the proportionality predicted in quantum physics, see, for example, [2]. However, so far we have used the classical description of the field.

1.5 Quantization of the EM Field

The most straightforward approach to the quantization of the EM field is to replace the classical canonical variables $q_k(t)$ and $p_k(t)$ by quantum mechanical operators $\hat{q}_k(t)$ and $\hat{p}_k(t)$ such that

$$\begin{aligned}
&[\hat{q}_k(t), \hat{q}_{k'}(t)] = 0, \quad [\hat{p}_k(t), \hat{p}_{k'}(t)] = 0, \\
&[\hat{q}_k(t), \hat{p}_{k'}(t)] = i\hbar\delta_{kk'},
\end{aligned} \tag{1.45}$$

that is, we associate with each mode k of the EM field a quantum mechanical harmonic oscillator.

The complex amplitudes a_k^* and a_k for the field mode k of the classical vector potential are quantized via the substitution

$$a_k \to \hat{a}_k, \quad a_k^* \to \hat{a}_k^\dagger. \tag{1.46}$$

The creation and annihilation operators are related to the quantum mechanical operators $\hat{q}_k$ and $\hat{p}_k$ as

$$\begin{aligned}
\hat{a}_k &= \frac{1}{\sqrt{2}}\left(\sqrt{\omega_k}\hat{q}_k + \frac{i}{\sqrt{\omega_k}}\hat{p}_k\right), \\
\hat{a}_k^\dagger &= \frac{1}{\sqrt{2}}\left(\sqrt{\omega_k}\hat{q}_k - \frac{i}{\sqrt{\omega_k}}\hat{p}_k\right).
\end{aligned} \tag{1.47}$$

Using the commutation relations (1.45), we readily find the commutation relation for the annihilation and creation operators

$$\left[\hat{a}_k(t), \hat{a}_{k'}^\dagger(t)\right] = \hbar\delta_{kk'}. \tag{1.48}$$

Hence, the Hamiltonian of the EM field takes the form

$$\hat{H}_\mathrm{F} = \sum_k \hbar\omega_k\left(\hat{a}_k^\dagger(t)\,\hat{a}_k(t) + \frac{1}{2}\right), \tag{1.49}$$

where we rescaled the operator $\hat{a}_k \rightarrow \hat{a}_k/\sqrt{\hbar}$ in order to have the commutation relation $[\hat{a}_k, \hat{a}_{k'}^\dagger] = \delta_{kk'}$.

In classical description of the EM field, the average energy is proportional to the intensity I of the field:

$$\langle H_\mathrm{F} \rangle = 2\varepsilon_0 \sum_k \omega_k^2 \langle |A_k(t)|^2 \rangle \sim I(t). \tag{1.50}$$

If the intensity $I(t) = 0$ then $\langle H_\mathrm{F} \rangle = 0$. Thus, in classical description of the field, the average energy can be equal to zero.

In quantum description of the field the average energy (the expectation value of the energy) is given by

$$\langle \hat{H}_\mathrm{F} \rangle = \sum_k \hbar\omega_k \left[\langle \hat{a}_k^\dagger \hat{a}_k \rangle + \frac{1}{2} \right] = \sum_k \hbar\omega_k \left[\langle \hat{n}_k \rangle + \frac{1}{2} \right], \tag{1.51}$$

where $\langle \hat{n}_k \rangle$ is the average number of photons in the kth mode of the field. In contrast to the classical energy, the average energy of a quantum field is different from zero even if $\langle \hat{n}_k \rangle = 0$.

The average energy depends on the state of the field, but is different from zero independent of the state of the field

$$\langle \psi | \hat{H}_\mathrm{F} | \psi \rangle = \sum_k \hbar\omega_k \left[\langle \psi | \hat{a}^\dagger \hat{a} | \psi \rangle + \frac{1}{2} \right]. \tag{1.52}$$

Since for any state $|\psi\rangle$ the expectation value $\langle \psi | \hat{a}^\dagger \hat{a} | \psi \rangle \geq 0$, we have that the average energy $\langle \psi | \hat{H}_\mathrm{F} | \psi \rangle > 0$.

1.6 Summary

We have seen that in quantum optics the EM field is represented as a set of independent quantized harmonic oscillators of energy

$$\hat{H}_\mathrm{F} = \sum_k \hbar\omega_k \left(\hat{a}_k^\dagger(t)\, \hat{a}_k(t) + \frac{1}{2} \right). \tag{1.53}$$

Moreover, we express the vector potential and the electric field in terms of plane waves whose amplitudes are quantized and determined by the creation $\hat{a}_k^\dagger$ and annihilation $\hat{a}_k$ operators associated with the mode k of the field as

$$\hat{\vec{A}}(\vec{r},t)=\sum_k\sqrt{\frac{\hbar}{2\omega_k\varepsilon_0 V}}\left[\vec{e}_k\hat{a}_k(t)\,\mathrm{e}^{i\vec{k}\cdot\vec{r}}+\mathrm{H.c.}\right],\qquad(1.54)$$

$$\begin{aligned}\hat{\vec{E}}(\vec{r},t)&=i\sum_k\sqrt{\frac{\hbar\omega_k}{2\varepsilon_0 V}}\left[\vec{e}_k\hat{a}_k(t)\,\mathrm{e}^{i\vec{k}\cdot\vec{r}}-\mathrm{H.c.}\right]\\&=\hat{\vec{E}}^{(+)}(\vec{r},t)+\hat{\vec{E}}^{(-)}(\vec{r},t),\end{aligned}\qquad(1.55)$$

where H.c. stands for the Hermitian conjugate of the first term in the bracket, and

$$\hat{\vec{E}}^{(+)}(\vec{r},t)=\left(\hat{\vec{E}}^{(-)}(\vec{r},t)\right)^{\dagger}=i\sum_k\sqrt{\frac{\hbar\omega_k}{2\varepsilon_0 V}}\,\vec{e}_k\hat{a}_k(t)\,\mathrm{e}^{i\vec{k}\cdot\vec{r}}.\qquad(1.56)$$

The constant $\sqrt{\hbar\omega_k/(2\varepsilon_0 V)}$ is called the quantum unit of electric strength.

Exercises

1.1 Explain the usefulness of the scalar and vector potentials in the solution of the Maxwell's equations.

1.2 Show that the $\vec{E}$ and $\vec{B}$ fields are invariant under the gauge transformation, Eq. (1.10).

1.3 Show that under the Coulomb gauge and in the presence of currents and charges, the wave equation for $\vec{A}$ involves only the transverse part of the current density.

1.4 Consider a source of electric $\vec{E}$ and magnetic $\vec{B}$ fields. If the fields are arranged so that $\vec{E}\perp\vec{B}$, should we expect to see an EM wave propagating in the direction determined by $\vec{E}\times\vec{B}$?

1.5 Show that in addition to the conditions (1.16), the unit polarization vectors $\vec{e}_{\vec{k}1}$, $\vec{e}_{\vec{k}2}$ and the unit propagation vector $\vec{k}/k$ form an orthonormal system

$$\sum_{s=1}^{2}(\vec{e}^{\,*}_{\vec{k}s})_i(\vec{e}_{\vec{k}s})_j+\frac{k_i k_j}{k^2}=\delta_{ij},\quad i,j=x,y,z,$$

where $(\vec{e}_{\vec{k}s})_i$ is the ith component of the unit polarization vector.

1.6 Show that the quantized electric field confined to a volume V can be expressed in terms of the annihilation and creation operators as

$$\hat{\vec{E}}(\vec{r}, t) = i\sum_k \sqrt{\frac{\hbar\omega_k}{2\varepsilon_0 V}} \left[\vec{e}_k \hat{a}_k(t)\, \mathrm{e}^{i\vec{k}\cdot\vec{r}} - \mathrm{H.c.}\right].$$

1.7 An electron moves in the xy-plane in a uniform magnetic field $\vec{B}$ propagating in the z-direction. The Hamiltonian of the electron is

$$H = \frac{1}{2m}\left(\vec{p} - e\vec{A}\right)^2,$$

where m is the mass of the electron, e is its charge, $\vec{p}$ is the linear momentum, and $\vec{A}$ is the vector potential of the field.

(a) Show that

$$H = \frac{1}{2m}\left[p_x^2 + p_y^2 + eB(yp_x - xp_y) + \frac{1}{4}e^2B^2(x^2+y^2)\right].$$

(b) Show that the operators

$$\hat{b} = \frac{1}{\sqrt{2eB\hbar}}\left(\frac{1}{2}eB\hat{x} + i\hat{p}_x + \frac{1}{2}ieB\hat{y} - \hat{p}_y\right),$$

$$\hat{b}^\dagger = \frac{1}{\sqrt{2eB\hbar}}\left(\frac{1}{2}eB\hat{x} - i\hat{p}_x - \frac{1}{2}ieB\hat{y} - \hat{p}_y\right),$$

have the same relation to the Hamiltonian as the annihilation and creation operators $\hat{a}$ and $\hat{a}^\dagger$ of the one-mode EM field, that is

$$\hat{b}\hat{b}^\dagger = \frac{H}{\hbar\omega_0} + \frac{1}{2}, \quad \hat{b}^\dagger\hat{b} = \frac{H}{\hbar\omega_0} - \frac{1}{2},$$

where $\omega_0 = eB/m$.

1.8 Show that for a single-mode EM field described by the annihilation and creation operators $\hat{a}$ and $\hat{a}^\dagger$:

$$\mathrm{e}^{i\hat{H}_{\mathrm{F}}t/\hbar}\hat{a}\mathrm{e}^{-i\hat{H}_{\mathrm{F}}t/\hbar} = \hat{a}\mathrm{e}^{-i\omega_0 t}, \quad \mathrm{e}^{i\hat{H}_{\mathrm{F}}t/\hbar}\hat{a}^\dagger\mathrm{e}^{-i\hat{H}_{\mathrm{F}}t/\hbar} = \hat{a}^\dagger\mathrm{e}^{i\omega_0 t},$$

where $\hat{H}_{\mathrm{F}} = \hbar\omega_0(\hat{a}^\dagger\hat{a} + 1/2)$.

1.9 Calculate the commutation relation between components of the quantized electric field to show that

$$\left[\hat{E}_i^{(+)}(\vec{r}, t), \hat{E}_j^{(-)}(\vec{r}', t)\right] = \frac{\hbar\omega_0}{2\varepsilon_0}\vec{\delta}_{ij}^{\mathrm{T}}(\vec{r} - \vec{r}'),$$

where ω_0 is the central frequency of the field and $\vec{\delta}_{ij}^{\mathrm{T}}(\vec{r} - \vec{r}')$ is the three-dimensional transverse Dirac delta function.

1.10 Consider the expression for the momentum of the EM field

$$\vec{p} = \varepsilon_0 \int dV \left(\vec{E} \times \vec{B}\right).$$

(a) Write the momentum $\vec{p}$ in terms of the creation and annihilation operators.

(b) Show that the momentum of a photon corresponding to the quantized EM plane wave of wave vector $\vec{k}$ is $\hbar\vec{k}$.

Chapter 2

Hamiltonians for Quantum Optics

2.1 Introduction

We often hear at seminars and presentations 'Show me the Hamiltonian of your problem and I will tell you what problem you are talking about'. Therefore, to understand what quantum optics is about, we start from the description of a standard Hamiltonian of the problems considered in quantum optics. We illustrate the method of derivation of the explicit form of the Hamiltonian for a simple system that is composed of two subsystems that interact (communicate) with each other. As we shall see, the explicit form of the Hamiltonian is essential to explicitly calculate energy levels of a given combined system and the temporal evolution of an arbitrary operator representing the system.

Hamiltonian for a standard quantum optics problem involving two systems that can interact with each other is composed of three terms

$$\hat{H} = \hat{H}_{\mathrm{S}} + \hat{H}_{\mathrm{F}} + \hat{H}_{\mathrm{int}}, \tag{2.1}$$

where the first term, $\hat{H}_{\mathrm{S}}$ describes a system Hamiltonian, the second term $\hat{H}_{\mathrm{F}}$ describes the field Hamiltonian and the third term $\hat{H}_{\mathrm{int}}$ is the Hamiltonian of the interaction between the system and

Quantum Optics for Beginners
Zbigniew Ficek and Mohamed Ridza Wahiddin

ISBN 978-981-4411-75-2 (Hardcover), 978-981-4411-76-9 (eBook)
www.panstanford.com

the field. Examples of systems considered in quantum optics are atoms, molecules, and solids. The field is usually taken as the free electromagnetic (EM) field represented as a set of independent quantized harmonic oscillators. Systems are usually represented as charge particles and the system–EM field interaction is considered as the interaction between the charge particles and the free EM field. The Hamiltonian (2.1) represents a closed system that is composed of two subsystems interacting with each other. We shall consider each term separately and illustrate a standard approach to obtain explicit forms of these terms.

2.2 Interaction Hamiltonian

Let us first consider the term representing the interaction between two subsystems. In quantum optics a free independent system is represented by charged particles and then the interaction Hamiltonian is the energy of the charges in the EM field. Following this observation, we now derive an explicit form of the interaction Hamiltonian involving charges in an external EM field.

From the EM theory, we know that energy of the charge particles of a volume density $\rho(\vec{r})$ located in an external field is given by the energy of the charges in the potential $\Phi(\vec{r})$ of the field

$$H_{\mathrm{int}} = \int d^3\vec{r}\, \rho(\vec{r})\, \Phi(\vec{r}), \tag{2.2}$$

where the integral is over the volume occupied by the charged particles.

We can expand the position-dependent potential $\Phi(\vec{r})$ into the Taylor series around a point $r_0 = 0$, and find

$$\Phi(\vec{r}) = \Phi(0) + \vec{r}\cdot\nabla\Phi(0) + \frac{1}{2}\sum_{ij} r_i r_j \frac{\partial^2\Phi}{\partial r_i \partial r_j}(0) + \cdots. \tag{2.3}$$

Since $\vec{E} = -\nabla\Phi(0)$, we can write the potential as

$$\Phi(\vec{r}) = \Phi(0) - \vec{r}\cdot\vec{E}(0) - \frac{1}{2}\sum_{ij} r_i r_j \frac{\partial E_j}{\partial r_i}(0) + \cdots. \tag{2.4}$$

Since for the free field $\nabla\cdot\vec{E} = 0$, we can add to the last term a factor

$$\frac{1}{6} r^2 \nabla\cdot\vec{E}(0) \tag{2.5}$$

and then the potential takes the form

$$\Phi(\vec{r}) = \Phi(0) - \vec{r} \cdot \vec{E}(0) - \frac{1}{6}\sum_{ij}\left(3r_i r_j - r^2\delta_{ij}\right)\frac{\partial E_j}{\partial r_i}(0) + \cdots. \tag{2.6}$$

Substituting this equation into Eq. (2.2) and performing the integration, we obtain

$$H_{\text{int}} = q\Phi(0) - \vec{\mu} \cdot \vec{E}(0) - \frac{1}{6}\sum_{ij} Q_{ij}\frac{\partial E_j}{\partial r_i}(0) + \cdots, \tag{2.7}$$

where

$$\vec{\mu} = q\vec{r} \tag{2.8}$$

is the *dipole moment*, and

$$Q_{ij} = \left(3r_i r_j - r^2\delta_{ij}\right) \tag{2.9}$$

is the *quadrupole moment* of the particles.

Equation (2.7) gives us a clear evidence how fields interact with systems that are represented by charges, dipole moments, etc. Thus, referring to Eq. (2.7), we conclude that

(1) The charge q interacts with potential $\Phi(0)$.
(2) The electric dipole moment $\vec{\mu}$ interacts with the field $\vec{E}$.
(3) The electric quadrupole moment Q interacts with the gradient of the field.

As we shall see latter, most of the models in quantum optics considers only the electric dipole interaction between systems and the EM field, that the interaction Hamiltonians are of the form $H_{\text{int}} = -\vec{\mu} \cdot \vec{E}(0)$. This choice of the interaction Hamiltonians is a consequence of the fact that in practice external fields such as lasers are often used to excite single-electron systems. Systems composed of a collection of free or bounded charges are difficult to be excited in a controlled way due to the presence of internal fields and forces between the charges.

2.3 Hamiltonian of an Atom

In quantum optics, a free independent system is represented by charges (electrons) and the system–field interaction is simply the

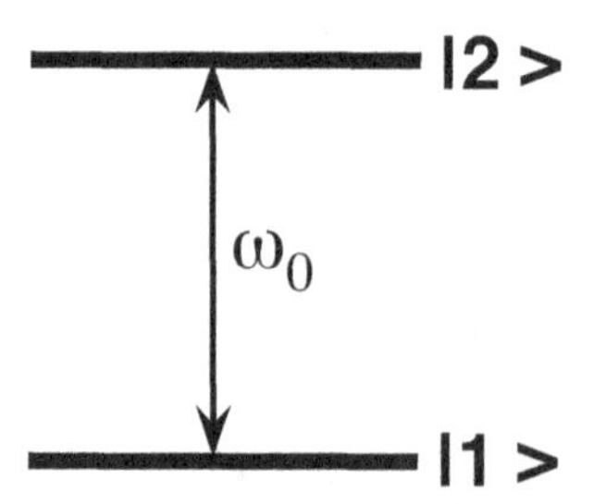

Figure 2.1 A schematic diagram of the simplest system in quantum optics: A two-level atom composed of a ground state $|1\rangle$ and an excited state $|2\rangle$ separated by frequency ω_0.

charge–field interaction. We know from the quantum mechanics that the motion of the electron in an atom is quantized and the electron can only be in some discrete energy states (the stationary energy levels). There is a non-zero probability that upon the interaction with an external field, the electron makes transitions between the quantized energy levels.

How do we model all of these behaviours of the electron? To answer this question, consider the simplest system used in quantum optics: a single-electron atom in which the electron can make transitions only between two energy states.

2.3.1 *A Two-Level System*

Figure 2.1 shows a schematic diagram of a two-level system, called a two-level atom, with the ground state $|1\rangle$ and the upper (excited) state $|2\rangle$. In fact, the electron can make transitions between many energy states, but we can limit the transitions to only between two states. In practice, it is done by a suitable choice of the frequency of an external field that will force the electron to oscillate only between these two selected states [3, 4]. Multi-level transitions involving more than two energy states are much more complex and we are not intend to consider them here, but the formalism presented here can be extended to multi-level cases.[a]

---

[a]For the derivation of the explicit form of the Hamiltonian of a multi-level atom see, for example, Z. Ficek and R. Tanaś, *Quantum-Limit Spectroscopy* (Springer, New York, 2014).

The energy of the electron in the selected states $|2\rangle$ and $|1\rangle$ is determined by the stationary Schrödinger equation

$$\hat{H}_{\mathrm{A}}|1\rangle = E_1|1\rangle, \qquad \hat{H}_{\mathrm{A}}|2\rangle = E_2|2\rangle, \tag{2.10}$$

where $\hat{H}_{\mathrm{A}}$ is the Hamiltonian of the atom whose explicit form is to be determine.

Note that the energy states of the two-level atom are orthonormal and satisfy the completeness relation, that is

$$\langle i|j\rangle = \delta_{ij} \quad \text{and} \quad \sum_{i=1}^{2}|i\rangle\langle i| = 1. \tag{2.11}$$

Let us first determine energies of the two states. If the atomic states are separated in energy by $\hbar\omega_0$, we can determine their energies relative to an average energy E_0 of the states

$$E_1 = E_0 - \frac{1}{2}\hbar\omega_0, \qquad E_2 = E_0 + \frac{1}{2}\hbar\omega_0, \tag{2.12}$$

we may choose $E_0 = 0$, which corresponds to the zero energy of the atom to be midway between the ground and excited states.

An obvious question arises: How to write the Hamiltonian of the electron which would contain information on the states in which the electron is?

This can be done by introducing the energy (population) difference operator, which in terms of the projection operators can be written as

$$S_z = \frac{1}{2}\left(|2\rangle\langle 2| - |1\rangle\langle 1|\right). \tag{2.13}$$

Since

$$\langle 1|\hat{H}_{\mathrm{A}}|1\rangle = -\frac{1}{2}\hbar\omega_0 \quad \text{and} \quad \langle 2|\hat{H}_{\mathrm{A}}|2\rangle = \frac{1}{2}\hbar\omega_0, \tag{2.14}$$

the Hamiltonian $\hat{H}_{\mathrm{A}}$ can be written in terms of the energy operator S_z as

$$\hat{H}_{\mathrm{A}} = \hbar\omega_0 S_z. \tag{2.15}$$

This equation is the explicit form of the Hamiltonian of a two-level atom, and shows that the energy of the electron in the two-level atom is determined by the operator S_z, the average value of which is between $\langle -\frac{1}{2} : \frac{1}{2}\rangle$.

2.3.2 *Spin Operators*

An electron interacting with an EM field jumps from the state $|1\rangle$ to $|2\rangle$, absorbing the energy, and from $|2\rangle$ to $|1\rangle$ emitting the energy. A jump (transition) can be represented by the spin operators S^+ and S^- (S^+: jump up) (S^-: jump down), as

$$\begin{aligned} S^+|1\rangle &= |2\rangle, \qquad S^-|2\rangle = |1\rangle, \\ S^+|2\rangle &= 0, \qquad S^-|1\rangle = 0. \end{aligned} \tag{2.16}$$

S^+ and S^- can be represented in terms of the projection operators of the two states involved as

$$S^+ = |2\rangle\langle 1|, \qquad S^- = |1\rangle\langle 2|, \tag{2.17}$$

and then it is easy to show that the spin operators satisfy the following properties

$$\begin{aligned} &\left(S^+\right)^2 = \left(S^-\right)^2 = 0, \\ &\left[S^+, S^-\right] = 2S_z, \\ &\left\{S^+, S^-\right\} = \left[S^+, S^-\right]_+ = 1. \end{aligned} \tag{2.18}$$

Since the spin operators S^+ and S^- are not Hermitian, it is convenient to introduce two Hermitian spin operators

$$S_x = \frac{1}{2}\left(S^+ + S^-\right), \quad S_y = \frac{1}{2i}\left(S^+ - S^-\right). \tag{2.19}$$

It follows from Eqs. (2.18) and (2.19) that the Hermitian spin operators obey the cyclic commutation relations

$$[S_\ell, S_m] = i\,\epsilon_{\ell mn}\,S_n, \qquad l, m, n = x, y, z, \tag{2.20}$$

where $\epsilon_{\ell mn}$ is the Levi–Civita tensor defined as

$$\epsilon_{\ell mn} = \begin{cases} 1 & lmn = xyz,\ yzx,\ zxy \quad \text{(even permutation of } xyz\text{)} \\ -1 & lmn = xzy,\ yxz,\ zyx \quad \text{(odd permutation of } xyz\text{)} \\ 0 & \qquad \text{when two or more indices are equal.} \end{cases} \tag{2.21}$$

On the basis of the states $|1\rangle$ and $|2\rangle$, the Hermitian spin operators are represented by matrices

$$\begin{aligned} S_x &= \frac{1}{2}\sigma_x = \frac{1}{2}\begin{pmatrix} 0 & 1 \\ 1 & 0 \end{pmatrix}, \qquad S_y = \frac{1}{2}\sigma_y = \frac{1}{2}\begin{pmatrix} 0 & i \\ -i & 0 \end{pmatrix}, \\ S_z &= \frac{1}{2}\sigma_z = \frac{1}{2}\begin{pmatrix} -1 & 0 \\ 0 & 1 \end{pmatrix}, \end{aligned} \tag{2.22}$$

where σ_x, σ_y and σ_z are the familiar Pauli spin matrices.

2.3.3 *Atomic Dipole Moment*

We have already shown that the operator S_z represents the energy of the electron. We can readily relate the spin raising and lowering operators, S^+ and S^- to the dipole moment $\vec{\mu}$ of the atom. To show this, we will use the completeness relation for the atomic states and multiply the dipole moment on both sides by unity in the form

$$1 = \sum_{i=1}^{2} |i\rangle\langle i|, \tag{2.23}$$

and obtain

$$\begin{aligned}\vec{\mu} &= \left(\sum_{i=1}^{2} |i\rangle\langle i|\right) \vec{\mu} \left(\sum_{j=1}^{2} |j\rangle\langle j|\right) \\ &= \vec{\mu}_{22} S^+ S^- + \vec{\mu}_{11} S^- S^+ + \vec{\mu}_{12} S^- + \vec{\mu}_{21} S^+, \end{aligned} \tag{2.24}$$

where $\vec{\mu}_{ij} = \langle i| \vec{\mu} |j\rangle$ are dipole matrix elements.

The diagonal matrix elements $\vec{\mu}_{11}$ and $\vec{\mu}_{22}$ determine dipole moments of the electron in the states $|1\rangle$ and $|2\rangle$, and are non-zero only in atoms with permanent dipole moments. The off-diagonal matrix elements $\vec{\mu}_{12}$ and $\vec{\mu}_{21}$ are transition dipole moments, which arise from a change of the size of the atom due to the transition of the electron from the state $|1\rangle$ to the state $|2\rangle$ and vice versa. The matrix element $\vec{\mu}_{12}$ can be real or complex and $\vec{\mu}_{21} = \vec{\mu}_{12}^*$. The transition dipole moments are real for $\Delta m = 0$ transitions in an atom, and are complex for $\Delta m = \pm 1$ transitions.

Let us consider a practical example to clarify the difference between the $\Delta m = 0$ and $\Delta m = \pm 1$ transition dipole moments. We wish to calculate the transition dipole moment between two energy states of atomic hydrogen.

Example 2.1 (Dipole moment of a $\Delta m = 0$ transition) *Consider, two energy states $\psi_{100} = |1\rangle$ and $\psi_{210} = |2\rangle$ of atomic hydrogen corresponding to a $\Delta m = 0$ transition*

$$\begin{aligned}\psi_{100} &= \sqrt{2}\mathcal{N} e^{-r/a_o}, \\ \psi_{210} &= \frac{1}{4}\mathcal{N} \frac{r}{a_o} e^{-r/2a_o} \cos\theta, \end{aligned} \tag{2.25}$$

where

$$\mathcal{N} = 1/\sqrt{2\pi a_o^3} \tag{2.26}$$

is the normalization constant and a_o is the Bohr radius.

The dipole matrix element $\vec{\mu}_{12}$ between the states (2.25) has the form

$$\vec{\mu}_{12} = \langle 1|\,\vec{\mu}\,|2\rangle = \int dV\,\psi_{100}^{*} e\vec{r}\,\psi_{210}. \tag{2.27}$$

Before integrating, we resolve $\vec{r}$ vector into components in the Cartesian coordinates, and next perform the integration in the spherical coordinates, where the dipole matrix element (2.27) can be written as

$$\vec{\mu}_{12} = e\int\int\int dr\,d\theta\,d\phi\,\sin^2\theta\;\psi_{100}^{*}\,(x\hat{x} + y\hat{y} + z\hat{z})\,\psi_{210}, \tag{2.28}$$

with the components

$$\begin{aligned} x &= r\sin\theta\cos\phi, \\ y &= r\sin\theta\sin\phi, \\ z &= r\cos\theta, \end{aligned} \tag{2.29}$$

and $\hat{x}$, $\hat{y}$, $\hat{z}$ are orthogonal unit vectors in the directions x, y and z, respectively.

Since ψ_{100} and ψ_{210} are independent of the azimuthal angle ϕ, the integrals over ϕ of the x and y components of the dipole moment are zero, but the integral over ϕ of the z component is non-zero, that is

$$\int_0^{2\pi} d\phi\cos\phi = \int_0^{2\pi} d\phi\sin\phi = 0, \qquad \int_0^{2\pi} d\phi = 2\pi. \tag{2.30}$$

Clearly, the dipole matrix element $\vec{\mu}_{12}$ of the $\Delta m = 0$ transition is real and oriented in the z-direction. Evaluating the remaining integrations over θ and r, we arrive to the following result

$$\vec{\mu}_{12} = \frac{128\sqrt{2}}{243} e a_o \hat{z}. \tag{2.31}$$

In terms of polarization, the dipole matrix element $\vec{\mu}_{12}$ of the $\Delta m = 0$ transition in a two-level atom is a vector linearly polarized in the z-direction.

In the following example, we calculate polarization and magnitude of the dipole moment of a $\Delta m = \pm 1$ transition.

Example 2.2 (Dipole moment of a $\Delta m = \pm 1$ transition) *In order to calculate the dipole moment of a $\Delta m = \pm 1$ transition, we choose two energy states $\psi_{100} = |1\rangle$ and $\psi_{211} = |2\rangle$. The state ψ_{100} is given in Eq. (2.25), and the state ψ_{211} is of the form*

$$\psi_{211} = \frac{1}{8\sqrt{\pi a_o^3}} \frac{r}{a_o} e^{-r/2a_o} e^{i\phi} \sin\theta. \tag{2.32}$$

Since

$$\int_0^{2\pi} d\phi\, e^{i\phi} = 0, \quad \int_0^{2\pi} d\phi\, e^{i\phi} \cos\phi \neq 0, \tag{2.33}$$

the z-component of the dipole moment is zero, whereas x, y components are non-zero. Hence

$$\vec{\mu}_{12} = -\frac{128}{243} e a_o \left(\hat{x} + i\hat{y}\right). \tag{2.34}$$

Thus, for a $\Delta m = \pm 1$ transition, the dipole matrix elements are complex numbers. In terms of polarization, the dipole matrix element $\vec{\mu}_{12}$ of a $\Delta m = \pm 1$ transition in a two-level atom is a vector circularly polarized in the xy-plane.

In the following we will consider atoms with zero permanent dipole moments ($\vec{\mu}_{11} = \vec{\mu}_{22} = 0$), and therefore we will write the dipole moment in terms of the dipole-lowering (S^-) and the dipole-raising (S^+) operators as

$$\vec{\mu} = \vec{\mu}_{12} S^- + \vec{\mu}_{21} S^+. \tag{2.35}$$

We can conclude that the dynamics of a two-level atom are completely described by the three spin operators S_x, S_y, S_z or S^+, S^-, S_z that obey the commutation relations (2.18) and (2.20).

2.4 Total Hamiltonian and the Rotating Wave Approximation

The Hamiltonian of the simplest system in quantum optics, a two-level atom interacting with the free EM field, is composed of three terms

$$\hat{H} = \hat{H}_0 + \hat{H}_{\mathrm{F}} + \hat{H}_{\mathrm{int}}, \tag{2.36}$$

where

$$\hat{H}_0 = \hbar\omega_0 S_z \tag{2.37}$$

is the Hamiltonian of the atom,

$$\hat{H}_{\mathrm{F}} = \sum_k \hbar\omega_k \left(\hat{a}_k^\dagger \hat{a}_k + \frac{1}{2}\right) \tag{2.38}$$

is the Hamiltonian of the field, and

$$\begin{aligned}\hat{H}_{\mathrm{int}} &= -\vec{\mu}\cdot\vec{E}(0,t) = -\vec{\mu}\cdot\left[\vec{E}^{(+)}(0,t) + \vec{E}^{(-)}(0,t)\right] \\ &= -i\sum_k \sqrt{\frac{\hbar\omega_k}{2\varepsilon_0 V}}\left[\vec{\mu}\cdot\vec{e}_k\hat{a}_k(t) - \mathrm{H.c.}\right] \\ &= -\frac{1}{2}i\hbar\sum_k \sqrt{\frac{2\omega_k}{\hbar\varepsilon_0 V}}\left[\vec{\mu}_{12}\cdot\vec{e}_k S^-\hat{a}_k(t) + \vec{\mu}_{21}\cdot\vec{e}_k S^+\hat{a}_k(t)\right. \\ &\quad \left. -\vec{\mu}_{12}^*\cdot\vec{e}_k^* S^+\hat{a}_k^\dagger(t) - \vec{\mu}_{21}^*\cdot\vec{e}_k^* S^-\hat{a}_k^\dagger(t)\right]\end{aligned} \tag{2.39}$$

is the interaction Hamiltonian between the atom and the field.

For $\Delta m = 0$ transitions, the transition dipole moment is real, and then the interaction Hamiltonian simplifies to

$$\hat{H}_{\mathrm{int}} = -\frac{1}{2}i\hbar\sum_k g_k\left[S^-\hat{a}_k(t) + S^+\hat{a}_k(t) - S^+\hat{a}_k^\dagger(t) - S^-\hat{a}_k^\dagger(t)\right], \tag{2.40}$$

where

$$g_k = (\vec{\mu}_{12}\cdot\vec{e}_k)\sqrt{\frac{2\omega_k}{\hbar\varepsilon_0 V}} \tag{2.41}$$

is the coupling constant (real) between the system and the EM field. It is often called the Rabi frequency of the atom–field interaction as it is proportional to the strength of the coupling between the atom and the field.

Note, that the interaction Hamiltonian $\hat{H}_{\mathrm{int}}$ contains both, the energy conserving terms $S^+\hat{a}_k$ and $S^-\hat{a}_k^\dagger$ as well as energy non-conserving terms $S^-\hat{a}_k$ and $S^+\hat{a}_k^\dagger$.

We can make the so-called rotating wave approximation (RWA), in which we ignore the energy non-conserving terms. More precisely, in the RWA approximation, we replace

$$\left(S^+ + S^-\right)\left[\hat{a}_k(t) - \hat{a}_k^\dagger(t)\right] \tag{2.42}$$

by

$$S^+\hat{a}_k(t) - S^-\hat{a}_k^\dagger(t), \tag{2.43}$$

that is, we exclude processes in which a photon is annihilated as the atom makes a downward transition (corresponding to $S^-\hat{a}_k(t)$), or a photon is created as the atom makes an upward transition (corresponding to $S^+\hat{a}_k^\dagger(t)$).

The RWA is a good approximation for long time processes, and is less valid for short time processes where the uncertainty of the energy is very large. As we will show later in Chapter 9, for weak couplings between a system and the field ($g_k \ll 1$) that are typical for the atom–vacuum field interaction, the non-RWA processes produce only a small frequency shift (the Bloch–Siegert shift). However, for strong couplings ($g_k \gg 1$), typical to the couplings inside optical cavities, they can have important dynamical consequences. For example, the Jaynes–Cummings model can exhibit chaotic dynamics, called "quantum chaos", that is, the question of how classical chaos might carry over into the corresponding quantum dynamics [5]. Some other interesting effects predicted when the RWA is not made include bifurcations in the phase space [6], a fine structure in the optical Stern–Gerlach effect [7], and entanglement between two atomic ensembles even if there is no initial excitation present in the system [8].

Exercises

2.1 Explain, why the spin operators S_x, S_y, S_z are often called spin-half (spin-$\frac{1}{2}$) operators?

2.2 Using the definition of the spin operators:

(a) Prove the commutation relations

$$\left[S^+, S^-\right] = 2S_z, \quad \left[S^\pm, S_z\right] = \mp S^\pm.$$

(b) Prove that the spin operators are unitary and that

$$S_x^2 = S_y^2 = S_z^2 = \frac{1}{4}.$$

(c) Prove that

$$e^{-i\pi(S_z+\frac{1}{2})}S^+e^{i\pi(S_z+\frac{1}{2})} = e^{-i\pi}S^+ = -S^+.$$

(d) Show that

$$S_x S_y = \frac{1}{2} i S_z, \quad S_y S_z = \frac{1}{2} i S_x, \quad S_z S_x = \frac{1}{2} i S_y.$$

2.3 For the spin operators S_x, S_y, S_z of a two-level atom with energy states $|1\rangle$ and $|2\rangle$, prove the following results

$$S_x|1\rangle = \frac{1}{2}|2\rangle, \quad S_y|1\rangle = -\frac{1}{2}i|2\rangle, \quad S_z|1\rangle = -\frac{1}{2}|1\rangle,$$

$$S_x|2\rangle = \frac{1}{2}|1\rangle, \quad S_y|2\rangle = \frac{1}{2}i|2\rangle, \quad S_z|2\rangle = \frac{1}{2}|2\rangle.$$

2.4 What is the physical consequence of the fact that the Hermitian spin operators S_x, S_y and S_z do not commute?

2.5 Show that the Pauli spin matrices satisfy the relation

$$\sigma_n \sigma_m = \delta_{nm} + i \in_{nmk} \sigma_k, \qquad n, m, k = x, y, z.$$

2.6 Consider the Pauli matrices representing the spin operators $\hat{\sigma}_x$, $\hat{\sigma}_y$ and $\hat{\sigma}_z$ of a two-level system in the basis of the states $|1\rangle$ and $|2\rangle$.

(a) Show that the operators $\hat{\sigma}_x$, $\hat{\sigma}_y$, $\hat{\sigma}_z$ each has eigenvalues $+1$, -1.

(b) Determine the normalised eigenvectors of each. Are $|1\rangle$ and $|2\rangle$ the eigenvectors of any of the matrices?

2.7 In the example on transition dipole moment between two energy states with $\Delta = 0$ we have chosen energy states ψ_{100} and ψ_{210} of atomic hydrogen.

(a) What is the transition dipole moment between states ψ_{100} and ψ_{200} of atomic hydrogen?

(b) How does the transition dipole moment depend on the parity of the energy states ψ_{nlm}?

2.8 Calculate dipole matrix element $\vec{\mu}_{12}$ of a $\Delta m = \pm 2$ transition between two states of the hydrogen atom. What is the polarization of the dipole moment?

2.9 Write the Hamiltonian (8.1) in the interacting picture to show that the energy non-conserving terms (counter rotating terms) contain time-dependent fast oscillating factors of the form $\exp[\pm i(\omega_0 + \omega_k)t]$, whereas the energy conserving terms contain slowly oscillating factors of the form $\exp[\pm i(\omega_0 - \omega_k)t]$.

2.10 Consider the Jaynes–Cummings model, which under the RWA is determined by the Hamiltonian

$$\hat{H} = \hbar\omega_0 S_z + \hbar\omega\left(\hat{a}^\dagger\hat{a} + \frac{1}{2}\right) - \frac{1}{2}i\hbar g\left(S^+\hat{a} - S^-\hat{a}^\dagger\right).$$

(a) Find the matrix representation of the Hamiltonian in the basis of the product states $|1\rangle|n\rangle$ and $|2\rangle|n-1\rangle$, where $|1\rangle$ and $|2\rangle$ are the energy states of a two-level atom and $|n\rangle$ is the n-photon energy state of the field.

(b) Find the eigenvalues and normalized eigenstates of the Hamiltonian of the Jaynes–Cummings model by the diagonalization of the matrix found in (a).

Chapter 3

Detection of the EM Field and Correlation Functions

3.1 Introduction

In this chapter, we will address one of the basic questions in quantum optics: How do we find an unknown state of the electromagnetic (EM) field, or in general, how do we find quantum state of a given system? The question is essentially about what are detectors and how an external field, that we want to detect, interacts with them. We may also see how the formulation of problems in quantum optics depends on the detection schemes.

In a laboratory, light fields are directly measured by photodetectors, devices in which an external field interacts with a photocathode composed of atoms (detectors) ionizing them. This process results in the emission of photoelectrons that form a photoelectric current, whose intensity or fluctuations are then measured. More precisely, the direct-detection experiments are sensitive to the intensity of the detected field and its fluctuations that are associated with statistical or spectral properties of the measured field. The direct measurement with photodetectors has a disadvantage that it destroys the incident field as the detector

Quantum Optics for Beginners
Zbigniew Ficek and Mohamed Ridza Wahiddin

ISBN 978-981-4411-75-2 (Hardcover), 978-981-4411-76-9 (eBook)
www.panstanford.com

absorbs all the field that falls on it and converts the field into a photocurrent. Moreover, the direct-detection experiments are not sensitive to the amplitude of the field and its fluctuations. Those require phase-sensitive detection schemes such as homodyne or heterodyne detectors.

3.2 Semiclassical Theory of Photodetection

Let us begin with a semiclassical theory of photodetection. This will give us some understanding of the process of detection of external fields and how it is formulated. Consequently, it will show which quantities are measured in the process of photodetection and what information they carry about the detected field.

In the semiclassical theory of photodetection, the probability to detect a classical field of intensity $I(\vec{R}, t)$ falling upon a photodetector in the time interval $(t, t + \Delta t)$ is defined as

$$P(\vec{R}, t)\Delta t = \eta \langle I(\vec{R}, t)\rangle \Delta t, \tag{3.1}$$

where η is the efficiency of the photodetector and $\langle I(\vec{R}, t)\rangle$ is the average intensity of the light evaluated at the location $\vec{R}$ of the photodetector. The average is taken over all possible instantaneous values of $I(\vec{R}, t)$, the ensemble average.

3.2.1 *First-Order Correlation Function*

In terms of the field amplitudes, $\langle I(\vec{R}, t)\rangle = \langle E^*(\vec{R}, t)E(\vec{R}, t)\rangle$, the probability can be written as

$$P(\vec{R}, t)\Delta t = \eta G^{(1)}(\vec{R}, t)\Delta t, \tag{3.2}$$

where

$$G^{(1)}(\vec{R}, t) = \langle E^*(\vec{R}, t)E(\vec{R}, t)\rangle \tag{3.3}$$

is called the *first-order correlation (coherence) function*.

From Eqs. (3.2) and (3.3) it is evident that an experimental measurement of the average intensity immediately provides information about the first-order correlation function of the detected field.

3.2.2 *Second-Order Correlation Function*

Consider two photodetectors located at points $\vec{R}_1$ and $\vec{R}_2$, respectively. The probability for the detector $\vec{R}_1$ to register a photodetection at time t_1 within Δt_1 is

$$P(\vec{R}_1, t_1)\Delta t_1 = \eta_1 I(\vec{R}_1, t_1)\Delta t_1, \tag{3.4}$$

and the probability for the detector $\vec{R}_2$ to register a photodetection at time t_2 within Δt_2 is

$$P(\vec{R}_2, t_2)\Delta t_2 = \eta_2 I(\vec{R}_2, t_2)\Delta t_2. \tag{3.5}$$

If the two detection processes are independent of each other, the joint probability of the two detections is defined as

$$P_2(\vec{R}_1, t_1; \vec{R}_2, t_2)\Delta t_1 \Delta t_2 = \eta_1\eta_2 \langle I(\vec{R}_1, t_1)\rangle\langle I(\vec{R}_2, t_2)\rangle\Delta t_1\Delta t_2. \tag{3.6}$$

We can use the joint probability to find whether the two detection processes are correlated or independent of each other. A correlation between the two detection processes can be determined from the joint probability of the form

$$P_2(\vec{R}_1, t_1; \vec{R}_2, t_2)\Delta t_1 \Delta t_2 = \eta_1\eta_2 \langle I(\vec{R}_1, t_1) I(\vec{R}_2, t_2)\rangle\Delta t_1\Delta t_2. \tag{3.7}$$

If the two detection processes are correlated

$$\langle I(\vec{R}_1, t_1) I(\vec{R}_2, t_2)\rangle \neq \langle I(\vec{R}_1, t_1)\rangle\langle I(\vec{R}_2, t_2)\rangle, \tag{3.8}$$

otherwise the detection processes are independent of each other.

We can write the joint probability in terms of the amplitudes of the EM field as

$$P_2(\vec{R}_1, t_1; \vec{R}_2, t_2)\Delta t_1 \Delta t_2 = \eta_1\eta_2 G^{(2)}(\vec{R}_1, t_1; \vec{R}_2, t_2)\Delta t_1\Delta t_2, \tag{3.9}$$

where

$$G^{(2)}(\vec{R}_1, t_1; \vec{R}_2, t_2) = \langle E^*(\vec{R}_1, t_1)E^*(\vec{R}_2, t_2)E(\vec{R}_2, t_2)E(\vec{R}_1, t_1)\rangle \tag{3.10}$$

is called the *second-order correlation (coherence) function.*

3.2.3 *Average Number of Photocounts*

It follows from Eqs. (3.7) and (3.9) that the joint probability of photodetection is proportional to the normally ordered correlation function of the fourth order in the field amplitude, or of the second order in the light intensity. Since the probability of a single photodetection is proportional to ηI, the average number of photocounts (classical particles) is given by

$$\langle n \rangle = \eta \langle I \rangle, \tag{3.11}$$

from which we get that the probability of detection of n photoelectron counts has a Poisson distribution

$$P_n = \frac{1}{n!} \langle (\eta I)^n \, \mathrm{e}^{-\eta I} \rangle, \tag{3.12}$$

where the average is the ensemble average over the intensity fluctuations.

Proof. Using the definition of statistical average, we obtain

$$\begin{aligned} \langle n \rangle &= \sum_n n P_n = \left\langle \sum_n \frac{n(\eta I)^n}{n!} \mathrm{e}^{-\eta I} \right\rangle \\ &= \left\langle (\eta I) \frac{d}{d(\eta I)} \left[\sum_n \frac{(\eta I)^n}{n!} \right] \mathrm{e}^{-\eta I} \right\rangle = \eta \langle I \rangle, \end{aligned} \tag{3.13}$$

as required. □

If the intensity of the detected light fluctuates during the detection time, the probability of photocounts in the interval t to $t + T$ is given by the ensemble average over the Poisson distribution

$$P_n(t, T) = \frac{1}{n!} \left\langle [\eta U(t, T)]^n \, \mathrm{e}^{-\eta U(t, T)} \right\rangle, \tag{3.14}$$

where

$$U(t, T) = \frac{1}{T} \int_t^{t+T} dt' I(t') \tag{3.15}$$

is the average intensity of the light in the interval t to $t + T$.

If the detection time is very short, so that the instantaneous intensity $I(t')$ can be substantially constant during the detection time T, the average intensity reduces to $U(t, T) = I$, and then the distribution (3.14) simplifies to that given in Eq. (3.13).

3.2.4 *Variance of the Number of Photocounts*

We can also calculate the variance of the number of counts n. Since

$$\langle n^2 \rangle = \sum_n n^2 P_n = \left\langle \left[(\eta I) \frac{d}{d(\eta I)} \right]^2 \left[\sum_n \frac{(\eta I)^n}{n!} \right] e^{-\eta I} \right\rangle$$
$$= \eta \langle I \rangle + \eta^2 \langle I^2 \rangle, \tag{3.16}$$

we find that the variance of the number of counts is

$$\langle (\Delta n)^2 \rangle = \langle n^2 \rangle - \langle n \rangle^2 = \langle n \rangle + \eta^2 \langle (\Delta I)^2 \rangle. \tag{3.17}$$

The formula (3.17) has a simple physical interpretation. The fluctuations of the photocounts are proportional to the fluctuations of the intensity of the incident field. For a non-fluctuating field, as it may be for a laser beam, the distribution of the photons obeys the Poisson distribution, $\langle (\Delta n)^2 \rangle = \langle n \rangle$, whereas for a fluctuating field, as it may be for a thermal field, $\langle (\Delta n)^2 \rangle > \langle n \rangle$. The first term in Eq. (3.17) is often called the *shot noise* associated with the random generation of discrete photoelectron counts in the detector. The second term is the noise in excess of the standard shot noise, and is often called the *wave noise*. Thus, for a non-fluctuating field only the shot noise is present.

The formula (3.17) also shows that the variance $\langle (\Delta n)^2 \rangle$ can never be smaller than $\langle n \rangle$. We will see in Chapter 4 that this is true for classical fields, but for certain quantum fields the variance $\langle (\Delta n)^2 \rangle$ can be smaller than $\langle n \rangle$.

3.3 Quantum Theory of Photodetection

We shall now give a quantum description of the detection theory [9, 10]. From quantum physics, we know that the probability of finding a system, located at $\vec{R}$ and described at time t by the total state $|\Psi(\vec{R}, t)\rangle$ in a particular state $|n\rangle$ is given by

$$P_n(\vec{R}, t) = |\langle n|\Psi(\vec{R}, t)\rangle|^2. \tag{3.18}$$

The total state vector at time t is related to the initial state $|\Psi(0)\rangle$ by the relation

$$|\Psi(\vec{R}, t)\rangle = |\Psi(0)\rangle + \frac{1}{i\hbar} \int_0^t dt_1 \hat{H}_{\text{int}}(t_1) |\Psi(0)\rangle, \tag{3.19}$$

which is a perturbation solution of the Schrödinger equation.

If the detection time is short, and we are interested only in transitions up (absorption of the external field), the probability of the transition takes the form

$$P_n(\vec{R}, t) = \frac{A^2t^2}{\hbar^2}|\langle n|\hat{E}^{(+)}(\vec{R}, t)|\Psi(0)\rangle|^2, \tag{3.20}$$

where A is a constant.

However, we cannot predict the final state of the atom as the electron in the detector can be in any state. Therefore, we have to sum the probabilities P_n over all possible final states, which gives

$$\begin{aligned} P(\vec{R}, t) &= \frac{A^2t^2}{\hbar^2}\sum_n |\langle n|\hat{E}^{(+)}(\vec{R}, t)|\Psi(0)\rangle|^2 \\ &= \frac{A^2t^2}{\hbar^2}\langle\Psi(0)|\hat{E}^{(-)}(\vec{R}, t)\hat{E}^{(+)}(\vec{R}, t)|\Psi(0)\rangle \\ &= \frac{A^2t^2}{\hbar^2}G^{(1)}(\vec{R}, t) = \frac{A^2t^2}{\hbar^2}I(\vec{R}, t), \end{aligned} \tag{3.21}$$

where

$$G^{(1)}(\vec{R}, t) = \langle\Psi(0)|\hat{E}^{(-)}(\vec{R}, t)\hat{E}^{(+)}(\vec{R}, t)|\Psi(0)\rangle \tag{3.22}$$

is the quantum first-order correlation (coherence) function.

Proceeding in a similar way, we can find the joint probability that a system described by the total state $|\Psi(\vec{R}, t)\rangle$ will be found in the state $|m\rangle$ at the time $t_2 > t_1$ if it was found in the state $|n\rangle$ at time t_1:

$$\begin{aligned} P(\vec{R}_1, t_1; \vec{R}_2, t_2) &= \frac{A^4t_1^2t_2^2}{\hbar^4}\sum_n\sum_m |\langle n|\hat{E}^{(+)}(\vec{R}_1, t_1)|\Psi(0)\rangle|^2 \\ &\quad\times|\langle m|\hat{E}^{(+)}(\vec{R}_2, t_2)|\Psi(0)\rangle|^2 \\ &= \frac{A^4t_1^2t_2^2}{\hbar^4}G^{(2)}(\vec{R}_1, t_1; \vec{R}_2, t_2), \end{aligned} \tag{3.23}$$

where

$$\begin{aligned} G^{(2)}(\vec{R}_1, t_1; \vec{R}_2, t_2) &= \langle\mathcal{T} : \hat{I}_1(\vec{R}_1, t_1)\hat{I}_2(\vec{R}_2, t_2) :\rangle \\ &= \langle\hat{E}^{(-)}(\vec{R}_1, t_1)\hat{E}^{(-)}(\vec{R}_2, t_2)\hat{E}^{(+)}(\vec{R}_2, t_2)\hat{E}^{(+)}(\vec{R}_1, t_1)\rangle \end{aligned} \tag{3.24}$$

is the quantum second-order correlation function, with $\mathcal{T}$ and :: denoting respectively the time and normal ordering of the field operators. In other words, the correlation function $G^{(2)}(\vec{R}_1, t_1; \vec{R}_2, t_2)$ is a relative measure of the joint probability that a photon is detected

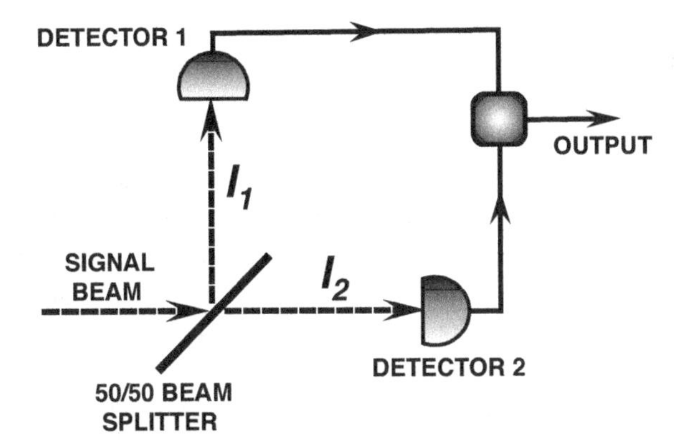

Figure 3.1 Experimental scheme for measurement of the second-order correlation function.

at a point $\vec{R}_2$ in time t_2 if one was detected at point $\vec{R}_1$ in time t_1. The time-ordering symbol $\mathcal{T}$ rearranges creation operators in forward time order and annihilation operators in backward time order, whereas the normal ordering symbol :: has the effect of rearranging the operators such that all the creation operators stand on the left of the annihilation operators.

Figure 3.1 shows an experimental scheme for measurement of the second-order correlation function. An incident signal beam is split on the 50/50 beam splitter into two beams I_1 and I_2 which are registered by two separate detectors. The signals from the detectors are then multiplied and averaged over all detected values to give the second order correlation function $\langle I_1 I_2 \rangle$. The correlation function is usually varied as a function of time difference between the two signals or difference in the position of the detectors.

In the laboratory, one can measure not only the first- and second-order correlation functions, that is, the real functions, but also complex functions such as average values of the field amplitude

$$\langle \hat{E}(\vec{R}, t) \rangle = \langle \Psi | \hat{E}(\vec{R}, t) | \Psi \rangle, \tag{3.25}$$

and the variance of the field amplitude

$$\langle \left(\Delta \hat{E}(\vec{R}, t) \right)^2 \rangle = \langle \Psi | \hat{E}^2(\vec{R}, t) | \Psi \rangle - \left(\langle \Psi | \hat{E}(\vec{R}, t) | \Psi \rangle \right)^2, \tag{3.26}$$

which are dependent on the phase of the field. Of course, as these are complex quantities, the average amplitude and its variance cannot be measured (detected) directly. The quantities depend on the phase of the field, which is a relative quantity, that is, it could be determined relative to some other well-known phase. It is done in practice by mixing the unknown measured light beam with the highly coherent light of the well-known amplitude and phase, usually derived from an intense narrow-band laser, and to study the intensity and fluctuations of the superposed beams. We will discuss later in Chapter 6 some experimental schemes for the non-direct measurement of the average value of the field amplitude and its fluctuations. The measurement schemes are different form that discussed above, referred as homodyne or heterodyne detection techniques.

In closing this section, we briefly comment about the need to measure higher order correlation functions. Although the foregoing detection schemes all measure the correlation functions only to the second order, the complete knowledge of a detected field requires measurements of the correlation functions to all orders of the field amplitude. However, the correlation functions of order higher than two have as yet played a negligible role in practice. The usefulness of the concept of higher order correlations still remains to be shown. For optical fields generated by lasers or thermal sources, the first- and second-order correlation functions contain all the required information about the field, and no correlation functions of order greater than two are required. A laser field is in general expressible in terms of the first-order correlation functions, whereas a thermal field is expressible in terms of products of second-order correlation functions.

Exercises

3.1 Using the time-dependent perturbation theory, show that to the first order in $\hat{H}_{\rm int}(t)$, the solution of the Schrödinger equation is of the form

$$|\Psi(\vec{R}, t)\rangle = |\Psi(0)\rangle + \frac{1}{i\hbar}\int_0^t dt_1\, \hat{H}_{\rm int}(t_1)\, |\Psi(0)\rangle.$$

3.2 Show that the correlation functions $G^{(1)}(\vec{R}, t) \equiv G^{(1)}$ and $G^{(2)}(\vec{R}_1, t_1; \vec{R}_2, t_2) \equiv G^{(2)}$ satisfy the wave equation

$$\left(\nabla^2 - \frac{1}{c^2}\frac{\partial^2}{\partial t^2}\right) G^{(i)} = 0, \quad i = 1, 2.$$

3.3 Using the definition of statistical average

$$\langle n^m \rangle = \sum_n n^m P_n = \left\langle \sum_n \frac{n^m (\eta I)^n}{n!} e^{-\eta I} \right\rangle,$$

calculate the average $\langle n(n-1)(n-2) \rangle$.

3.4 **(a)** Write the average intensity square $\langle \hat{I}^{\,2} \rangle$ of a quantum field in terms of the correlation functions $G^{(2)}$ and $G^{(1)}$.

(b) Find the average $\langle \hat{I}^{\,3} \rangle$ of a quantum field in terms of the correlation functions $G^{(3)}$, $G^{(2)}$ and $G^{(1)}$.

(c) What would be the values of $\langle \hat{I}^{\,2} \rangle$ and $\langle \hat{I}^{\,3} \rangle$ in terms of the correlation functions if the field was classical?

3.5 Write the quantum first- and second-order correlation functions in terms of the creation and annihilation operators.

3.6 Consider the joint probability of two simultaneous detections of a stationary or slowly varying field of intensity $I(t)$:

$$P_2(t)\Delta t_1 \Delta t_2 = \eta^2 \langle I(t) I(t) \rangle \Delta t_1 \Delta t_2.$$

(a) Show that the joint probability can be written as

$$P_2(t)\Delta t_1 \Delta t_2 = \eta^2 \langle I(t) \rangle^2 \left[1 + \lambda(t)\right] \Delta t_1 \Delta t_2,$$

where $\lambda(t)$ is the normalized intensity-fluctuations correlation function

$$\lambda(t) = \frac{\langle (\Delta I(t))^2 \rangle}{\langle I(t) \rangle^2}.$$

(b) Show that in terms of the number of photocounts n, the joint probability takes the form

$$P_2(t)\Delta t_1 \Delta t_2 = \langle n \rangle^2 \left[1 + \frac{\langle (\Delta n)^2 \rangle - \langle n \rangle}{\langle n \rangle^2}\right] \Delta t_1 \Delta t_2.$$

From this equation it then follows that when $\langle (\Delta n)^2 \rangle < \langle n \rangle$, the joint probability of the two photodetections is smaller than that of two independent detections. It is often said in the literature that these two photodetections are negatively correlated, since in this case $\lambda(t) < 0$. Conversely, when $\langle (\Delta n)^2 \rangle > \langle n \rangle$, these two photodetections are strongly correlated.

Chapter 4

Representations of the EM Field

4.1 Introduction

In this chapter, we will introduce different representations of the electromagnetic (EM) field. One may ask, why do we need different representations for the EM field? The answer is that usually we do not know the state of the EM field which in quantum optics is represented by the annihilation and creation operators. However, results of measurements of the field are given in terms of real variables, for example, intensity. The representations allow us to recognize the state of the field from values of the measured quantities such as average amplitude, intensity, and correlation functions. The nature of the state is present in terms of the interpretation of the apparently classical (measured) variables. We will discuss two basic types of representations often used in quantum optics, Fock state (photon number) representation and coherent states representation, the later one introduced by Glauber.[a] We also discuss properties of fields with thermal and Poisson distribution of photons. The photon number states are very often

[a] Roy Glauber was granted the Nobel prize in 2005 for his contribution to the quantum theory of optical coherence.

Quantum Optics for Beginners
Zbigniew Ficek and Mohamed Ridza Wahiddin

ISBN 978-981-4411-75-2 (Hardcover), 978-981-4411-76-9 (eBook)
www.panstanford.com

used as a basis for quantum optics problems, and despite of many difficulties have recently been generated experimentally [11–13].

4.2 Fock States Representation

In this representation, a state $|n\rangle$ of the EM field is characterized by the well-defined number of photons n. We will illustrate the concept of photon number states for a single-mode field and next we will generalize it to multi-mode fields.

4.2.1 *Single-Mode Number States*

Suppose, the EM field is composed of a single mode k, and we use the notation for the annihilation operator $\hat{a} = \hat{a}_k$.

Definition 4.1. Eigenstates $|n\rangle$ of the photon number operator $\hat{n} = \hat{a}^\dagger \hat{a}$ of the single-mode field are eigenstates of the Hamiltonian

$$\hat{H}_{\mathrm{F}} |n\rangle = \hbar\omega \left(\hat{a}^\dagger \hat{a} + \frac{1}{2} \right) |n\rangle = E_n |n\rangle , \tag{4.1}$$

and are called photon number states or Fock states.

The photon number states have the following properties

$$\begin{aligned} \hat{n} |n\rangle &= n |n\rangle , \\ \hat{H}_{\mathrm{F}} |0\rangle &= \frac{1}{2} \hbar\omega |0\rangle . \end{aligned} \tag{4.2}$$

The zero photon's state, that is, the eigenstate of $\hat{n}$ with the eigenvalue equal to zero, is called the vacuum state and has the energy $\frac{1}{2}\hbar\omega$.

The number state $|n\rangle$ can be generated by repeated application of the creation operator $\hat{a}^\dagger$ on the vacuum state

$$|n\rangle = \frac{\left(\hat{a}^\dagger\right)^n}{\sqrt{n!}} |0\rangle . \tag{4.3}$$

Example 4.1 (A property of the EM field in a photon number state) *Consider the amplitude of a single-mode EM field of the polarization in the x-direction and propagating in the z-direction*

$$\hat{E}_x = E \left(\hat{a} + \hat{a}^\dagger \right) \sin(kz). \tag{4.4}$$

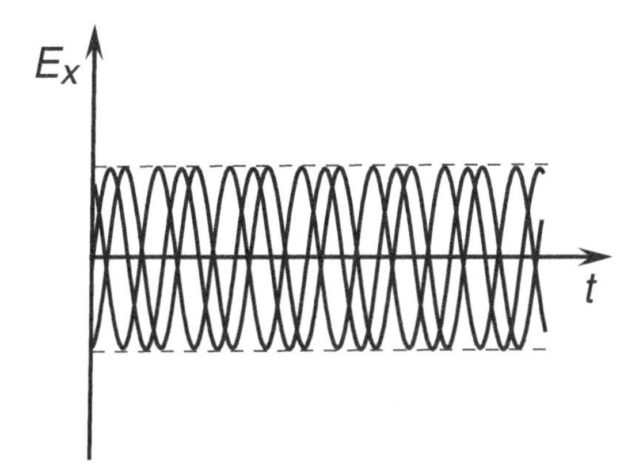

Figure 4.1 Illustration of a field of the definite amplitude but phase randomly distributed.

It is easy to show that the expectation value of the amplitude of the EM field in the photon number state $|n\rangle$ *is equal to zero. Since*

$$\langle n|\, \hat{E}_x \,|n\rangle = E\, \langle n|\hat{a} + \hat{a}^{\dagger}|n\rangle \sin(kz), \tag{4.5}$$

and

$$\hat{a}\,|n\rangle = \sqrt{n}\,|n-1\rangle\,, \quad \hat{a}^{\dagger}\,|n\rangle = \sqrt{n+1}\,|n+1\rangle\,, \tag{4.6}$$

with $\langle n|m\rangle = \delta_{nm}$, *we obtain*

$$\langle n|\, \hat{E}_x \,|n\rangle = 0. \tag{4.7}$$

Note that $\langle n|\hat{E}_x|n\rangle = 0$ independent of n, that is, $\langle n|\hat{E}_x|n\rangle = 0$ even if $n \to \infty$. This is unusual property of the quantum field as in the limit of $n \to \infty$ the properties of the photon-number state should convert into that corresponding to a classical state. However, we know that the classical field performs harmonic oscillations in time.

An alternative explanation of vanishing the expectation value of the field is that the field in the photon number state has a definite amplitude but phase randomly distributed over 2π, as it is schematically illustrated in Fig. 4.1. It therefore average to zero at any period of time.

Somehow surprising is that the expectation value of the square of the field amplitude $\hat{E}_x^2$ in the photon number state is different from zero, that is,

$$\langle n|\, \hat{E}_x^2 \,|n\rangle \neq 0. \tag{4.8}$$

Proof. Since

$$\langle n|\, \hat{E}_x^2 \,|n\rangle = E^2 \langle n|\, \hat{a}^\dagger \hat{a}^\dagger + \hat{a}\hat{a}^\dagger + \hat{a}^\dagger \hat{a} + \hat{a}\hat{a} \,|n\rangle \sin^2(kz), \quad (4.9)$$

we calculate expectation values of the four combinations of the field operators, and obtain

$$\begin{aligned} \langle n|\, \hat{a}^\dagger \hat{a}^\dagger \,|n\rangle &= \sqrt{(n+1)(n+2)}\, \langle n|n+2\rangle = 0, \\ \langle n|\, \hat{a}\hat{a} \,|n\rangle &= \sqrt{n(n-1)}\, \langle n|n-2\rangle = 0, \\ \langle n|\, \hat{a}\hat{a}^\dagger \,|n\rangle &= (n+1)\, \langle n|n\rangle = n+1, \\ \langle n|\, \hat{a}^\dagger \hat{a} \,|n\rangle &= n\, \langle n|n\rangle = n. \end{aligned} \quad (4.10)$$

Hence,

$$\langle n|\, \hat{E}_x^2 \,|n\rangle = 2E^2 \left(n + \frac{1}{2}\right) \sin^2(kz) \neq 0, \quad (4.11)$$

as required. □

In addition, since the average over all possible positions, $\overline{\sin^2(kz)} = 1/2$, we find after averaging Eq. (4.11) over z that

$$\langle n|\, \hat{E}_x^2 \,|n\rangle = E^2 \left(n + \frac{1}{2}\right). \quad (4.12)$$

The expectation value of the square of the field amplitude is different from zero even when $n = 0$. This simply shows that a vacuum field not only has a non-zero energy but also has non-zero fluctuations. We shall see later that these vacuum fluctuations lead to many interesting effects in quantum optics. An important challenge in quantum optics and quantum information science is a preparation or excitation of the field into a particular photon number state [12, 13]. It is a great practical difficulty to realize a single photon number state. In practice the only pure number state state commonly achieved is the vacuum state $|0\rangle$. Any attempt to excite a particular photon number state $|n\rangle$ with $n > 0$ leads also to a non-zero excitation of the neighbouring states $|n \pm m\rangle$.

In summary of this section, a field in the number state has a well-definite amplitude, but the phase is indefinite. The fluctuations of the field amplitude are different from zero even when $n = 0$. This feature is known as the vacuum fluctuations.

4.2.2 *Multi-Mode Number States*

We can generalize the concept of single-mode photon number states to the case of multi-mode fields. For a multi-mode field, we can define a state vector characterizing the entire field by taking the product of photon number states corresponding to different modes

$$|\{n\}\rangle = \prod_k |n_k\rangle \,, \tag{4.13}$$

where $\{n\}$ is to be interpreted as the set of all n. This simple generalization arises from the fact that the annihilation and creation operators corresponding to different modes commute.

The multi-mode Fock state $|\{n\}\rangle$ is an eigenstate of the photon number operator for the mode k

$$\hat{n}_k \,|\{n\}\rangle = n_k \,|\{n\}\rangle \,. \tag{4.14}$$

We can define the total number operator by summing $\hat{n}_k$ over all modes

$$\hat{n} = \sum_k \hat{n}_k, \tag{4.15}$$

and find that the Fock state $|\{n\}\rangle$ is also an eigenstate of $\hat{n}$ with an eigenvalue n, which is the total number of photons in the modes

$$\hat{n}\,|\{n\}\rangle = \left(\sum_k n_k\right)|\{n\}\rangle = n\,|\{n\}\rangle \,. \tag{4.16}$$

Since the energy of the field represented by a set of harmonic oscillators (modes) is a linear combination of the operators $\hat{n}_k$, see Eq. (1.49), it follows that the Fock states are also eigenstates of the Hamiltonian $\hat{H}$:

$$\hat{H}\,|\{n\}\rangle = \sum_k \left(\hat{n}_k + \frac{1}{2}\right)\hbar\omega_k \,|\{n\}\rangle = \sum_k \left(n_k + \frac{1}{2}\right)\hbar\omega_k \,|\{n\}\rangle \,. \tag{4.17}$$

Similar to the single-mode Fock states, we can create a multi-mode Fock state from the vacuum

$$|\{n\}\rangle = \prod_k \left\{\frac{\left(\hat{a}_k^{\dagger}\right)^n}{\sqrt{n!}}\right\}|0\rangle \,. \tag{4.18}$$

4.3 Correlation Functions for a Field in a Photon Number State

Having the photon number states defined, we shall now analyse characteristic properties of a single-mode field in the photon number state $|n\rangle$. For a single-mode case, the field operator takes a simple form

$$\hat{E} = \hat{E}^{(+)} + \hat{E}^{(-)} = \lambda\left(\hat{a} + \hat{a}^{\dagger}\right), \tag{4.19}$$

and with a little algebra, we find the following values for the correlation functions

$$\begin{aligned} \left\langle \hat{E} \right\rangle &= 0, \\ \left\langle \left(\Delta \hat{E}\right)^2 \right\rangle &= \lambda^2(2n+1), \\ \left\langle \hat{E}^{(-)}\hat{E}^{(+)} \right\rangle &= \lambda^2 \left\langle \hat{a}^{\dagger}\hat{a} \right\rangle = \lambda^2 n, \\ \left\langle \hat{E}^{(-)}\hat{E}^{(-)}\hat{E}^{(+)}\hat{E}^{(+)} \right\rangle &= \lambda^4 \left\langle \hat{a}^{\dagger}\hat{a}^{\dagger}\hat{a}\hat{a} \right\rangle = \lambda^4 n(n-1). \end{aligned} \tag{4.20}$$

In order to get a dipper insight into the properties of the field in the photon number state, we define the normalized second-order correlation function and variances of the so-called in-phase and out-of-phase components of the field amplitudes.

4.3.1 *Normalized Second-Order Intensity Correlation Function*

It is convenient to analyse second-order correlations in terms of the the normalized second-order correlation function that determines the correlations relative to the intensity of the field. The normalized second-order correlation function is defined as

$$g^{(2)}(\vec{R}_1, t_1; \vec{R}_2, t_2) = \frac{\langle \hat{E}^{(-)}(\vec{R}_1, t_1)\hat{E}^{(-)}(\vec{R}_2, t_2)\hat{E}^{(+)}(\vec{R}_2, t_2)\hat{E}^{(+)}(\vec{R}_1, t_1)\rangle}{\langle \hat{E}^{(-)}(\vec{R}_1, t_1)\hat{E}^{(+)}(\vec{R}_1, t_1)\rangle\langle \hat{E}^{(-)}(\vec{R}_2, t_2)\hat{E}^{(+)}(\vec{R}_2, t_2)\rangle}. \tag{4.21}$$

Let us confine to the case of $\vec{R}_1 = \vec{R}_2$ and $t_1 = t_2$. For a single-mode field, Eq. (4.19), and dropping the time and space arguments, the correlation function simplifies to

$$g^{(2)} = \frac{\langle \hat{a}^{\dagger}\hat{a}^{\dagger}\hat{a}\hat{a}\rangle}{\langle \hat{a}^{\dagger}\hat{a}\rangle\langle \hat{a}^{\dagger}\hat{a}\rangle} = \frac{\langle \hat{a}^{\dagger}(\hat{a}\hat{a}^{\dagger} - 1)\hat{a}\rangle}{\langle \hat{a}^{\dagger}\hat{a}\rangle^2} = \frac{\langle \hat{a}^{\dagger}\hat{a}\hat{a}^{\dagger}\hat{a}\rangle - \langle \hat{a}^{\dagger}\hat{a}\rangle}{\langle \hat{a}^{\dagger}\hat{a}\rangle^2}. \tag{4.22}$$

We can express the correlation function in terms of the photon number operator. Since $\hat{a}^{\dagger}\hat{a} = \hat{n}$, we obtain

$$g^{(2)} = \frac{\langle \hat{n}^2 \rangle - \langle \hat{n} \rangle}{\langle \hat{n} \rangle^2}. \tag{4.23}$$

Introducing the variance of the number of photons

$$\langle (\Delta \hat{n})^2 \rangle = \langle \hat{n}^2 \rangle - \langle \hat{n} \rangle^2, \tag{4.24}$$

we can finally write

$$g^{(2)} = 1 + \frac{\langle (\Delta \hat{n})^2 \rangle - \langle \hat{n} \rangle}{\langle \hat{n} \rangle^2}. \tag{4.25}$$

This formula shows that depending on the fluctuations of the number of photons, the correlation function $g^{(2)}$ can take the values $g^{(2)} < 1$, $g^{(2)} = 1$ or $g^{(2)} > 1$. Since the correlation function $g^{(2)}$ depends explicitly on the variance of the photon number. In quantum optics the function $g^{(2)}$ is often used to determine statistics of a given field.

Example 4.2 (Field in the photon number state) *When the field is in the photon number state $|n\rangle$, the variance and the average number of photons are*

$$\begin{aligned} \langle (\Delta \hat{n})^2 \rangle &= 0, \\ \langle \hat{n} \rangle &= n. \end{aligned} \tag{4.26}$$

Hence, the normalized second-order correlation function takes the form

$$g^{(2)} = 1 - \frac{1}{n}. \tag{4.27}$$

Thus, for a single-photon field, $n = 1$, the correlation function reduces to

$$g^{(2)} = 0, \tag{4.28}$$

whereas for a multi-photon field with $n \to \infty$

$$g^{(2)} \to 1. \tag{4.29}$$

A value of the correlation function $g^{(2)} < 1$ expresses the very interesting fact that photons in the field are separated from each other and move individually rather than in groups. This phenomenon is called ungrouping effect or *photon anticorrelation* [14, 15]. The photons are anticorrelated in the sense that the probability of appearance of a pair of photons at any particular time is very small. When photons are separated in time, they may exhibit the phenomenon of *photon antibunching*—that is, the correlation function may satisfy the inequality $g^{(2)}(t, \tau) > g^{(2)}(t, 0)$. Physically, the inequality means that the probability of detecting two photons separated by time τ is more likely than the probability of detecting two photons at the same time.

4.3.2 *Two-Level Atom as a Source of Antibunched Light*

As an example of a source of antibunched light, we consider a two-level atom that is represented by the spin operators $S^{\pm}$. The normalized two-time second-order correlation function of the field detected by a single photodetector and expressed in terms of the atomic spin operators can be written as

$$g^{(2)}(t, \tau) \equiv g^{(2)}(\vec{R}, t; \vec{R}, t+\tau)$$
$$= \frac{\langle S^+(t)S^+(t+\tau)S^-(t+\tau)S^-(t)\rangle}{\langle S^+(t)S^-(t)\rangle\langle S^+(t+\tau)S^-(t+\tau)\rangle}. \qquad (4.30)$$

Since $(S^{\pm})^2 \equiv 0$, we see that at $\tau = 0$ the correlation function $g^{(2)}(t, 0) = 0$, indicating that a two-level atom is an ideal source of antibunched light. Physically, the vanishing of $g^{(2)}(t, 0)$ for a single two-level atom implies that just after a photon is emitted, the atom is in the ground state, and cannot emit again until is re-excited so that a photon may be emitted again.

Using sodium atoms as a source of light, Dagenais, Kimble and Mandel [16] demonstrated experimentally photon anticorrelation, $g^{(2)}(0) < 1$, and photon antibunching, $g^{(2)}(t, \tau) > g^{(2)}(t, 0)$ from direct measurement of the second-order correlation function of the emitted resonance fluorescence. Their experiment provided evidence that light is composed of particles. Photon antibunching has also been observed in similar experiments involving trapped atoms [17] and a cavity quantum electrodynamic (QED) system [18].

We will show later that $g^{(2)}(t, \tau) = 1$ corresponds to a coherent state of the field, while $g^{(2)}(t, \tau) < 1$ corresponds to a non-classical state of the field (quantum field with no analogue in classical physics). Thus, the photon number state $|n\rangle$ is a non-classical state of the field with the fluctuations of the number of photons suppressed below the quantum (coherent) level.

4.3.3 *Fluctuations of the Field Amplitudes*

We can consider not only the photon number fluctuations, but also fluctuations in the electric field amplitude. The electric field $\hat{E}$ can be expressed in terms of the so-called in-phase and out-off phase quadrature (phase) components defined as

$$\hat{E}_{\text{in}} = \frac{1}{2}\left(\hat{a} + \hat{a}^{\dagger}\right), \quad \hat{E}_{\text{out}} = \frac{1}{2i}\left(\hat{a} - \hat{a}^{\dagger}\right). \tag{4.31}$$

The quadrature components obey the commutation relation

$$\left[\hat{E}_{\text{in}}, \hat{E}_{\text{out}}\right] = \frac{i}{2}, \tag{4.32}$$

from which we find that they satisfy the Heisenberg uncertainty relation

$$\sqrt{\left\langle \left(\Delta \hat{E}_{\text{in}}\right)^2 \right\rangle \left\langle \left(\Delta \hat{E}_{\text{out}}\right)^2 \right\rangle} \geq \frac{1}{4}, \tag{4.33}$$

where the factor $1/4$ on the right-hand side determines the vacuum level of the fluctuations.

For a field in the Fock state $|n\rangle$, the variances of the two quadrature components are equal and linearly increase with n:

$$\left\langle \left(\Delta \hat{E}_{\text{in}}\right)^2 \right\rangle = \left\langle \left(\Delta \hat{E}_{\text{out}}\right)^2 \right\rangle = \frac{1}{2}\left(n + \frac{1}{2}\right). \tag{4.34}$$

In summary of this section, a field in the Fock state $|n\rangle$ is characterized by the well-defined number of photons (intensity), but the amplitude of the field exhibits large fluctuations which increase with increasing number of photons n.

4.4 Probability Distributions of Photons

In many problems in quantum optics, we face with a difficulty of a lack of knowledge of the state of the field. However, in some situations that we will explore here, we know or at least we can infer the probability distribution of photons in the field. As we shall see, this is enough to determine the statistics of the field.

Let us introduce this idea with a simple question: What is the expectation value $\langle \hat{n} \rangle = \langle \psi | \hat{n} | \psi \rangle$ if, for example, $|\psi\rangle$ is a thermal state of the field?

We do not know the explicit form of the state, but if the probability distribution of photons is known, the expectation values $\langle n \rangle, \langle n^2 \rangle, \ldots$ still can be explicitly calculated. Namely, we find the expectation values from the definition of the statistical moments

$$\langle n^m \rangle = \sum_n n^m P_n, \qquad m = 1, 2, \ldots, \tag{4.35}$$

where P_n is the probability distribution function.

We illustrate the procedure for EM fields that are determined by the two well-known probability distribution functions, the thermal and Poisson distribution functions.

4.4.1 *Thermal Distribution*

The probability distribution function for a thermal field at temperature T is given by the Boltzmann distribution

$$P_n = \frac{e^{-nx}}{\sum\limits_{n=0}^{\infty} e^{-nx}}, \tag{4.36}$$

where

$$x = \frac{\hbar\omega}{k_B T}, \tag{4.37}$$

with k_B is the Boltzmann constant, and T is the absolute temperature.

Example 4.3 (P_n in terms of the average number of photons) *For some problems, it is convenient to have the distribution function P_n in terms of the average number of photons as*

$$P_n = \frac{\langle n \rangle^n}{(1 + \langle n \rangle)^{n+1}}. \tag{4.38}$$

To show it, consider first the sum $\sum_{n=0}^{\infty} \exp(-nx)$ appearing in the denominator of Eq. (4.36). The sum is a particular example of a *geometric series. Since* $|\exp(-nx)| < 1$*, the sum tends to the limit*

$$\sum_{n=0}^{\infty} e^{-nx} = \frac{1}{1-e^{-x}}. \tag{4.39}$$

Hence, we can write the probability distribution function (4.36) as

$$\begin{aligned} P_n &= e^{-nx}\left(1-e^{-x}\right) = e^{-nx}\frac{e^x-1}{e^x} = \frac{(e^x-1)^n\,(e^x-1)}{(e^x-1)^n\,(e^x)^{n+1}} \\ &= \frac{(e^x-1)^{n+1}}{(e^x-1)^n\,(e^x)^{n+1}} = \frac{\left(\frac{1}{e^x-1}\right)^n}{\left(1+\frac{1}{e^x-1}\right)^{n+1}}. \end{aligned} \tag{4.40}$$

We furthermore have that the expectation number of photons can be written as

$$\langle n \rangle = \sum_n nP_n = \left(1-e^{-x}\right)\sum_n ne^{-nx}. \tag{4.41}$$

In order to evaluate the sum in Eq. (4.41), we introduce a notation

$$z = \sum_n e^{-nx} = \frac{1}{1-e^{-x}}, \tag{4.42}$$

and find that

$$\begin{aligned} \sum_n ne^{-nx} &= -\frac{d}{dx}\sum_n e^{-nx} \\ &= -\frac{d}{dx}\left(\frac{1}{1-e^{-x}}\right) = \frac{-e^{-x}}{(1-e^{-x})^2}. \end{aligned} \tag{4.43}$$

Therefore,

$$\langle n \rangle = \frac{1}{e^x-1}, \tag{4.44}$$

and the expression for the distribution function becomes[a]

$$P_n = \frac{\langle n \rangle^n}{(1+\langle n \rangle)^{n+1}}, \tag{4.46}$$

as required.

---

[a] An interesting observation: According to Eq. (4.44), by measuring the average number of photons one can extract a temperature of the source

$$T = \frac{\hbar\omega}{k_B \ln\left(\frac{\langle n \rangle + 1}{\langle n \rangle}\right)}. \tag{4.45}$$

This formula is often used in cold atom optics to determine temperature of cooled trapped atoms.

Consider now the fluctuations of a thermal field. Since we know the distribution function P_n, we can calculate $\langle(\Delta n)^2\rangle$, the variance of the number of photons in the thermal field. Following the same procedure as in the above proof of Eq. (4.38), we find that

$$\langle n^2\rangle = \sum_n n^2 P_n = \frac{e^x + 1}{(e^x - 1)^2} = \frac{1}{e^x - 1} + \frac{2}{(e^x - 1)^2}$$
$$= \langle n\rangle + 2\langle n\rangle^2, \tag{4.47}$$

from which we find that the variance is composed of two terms

$$\left\langle(\Delta n)^2\right\rangle = \langle n\rangle + \langle n\rangle^2. \tag{4.48}$$

With this result, we find from Eqs. (4.25) and (4.48) that the normalized second-order correlations for a field with the thermal distribution of photons is equal to

$$g^{(2)} = 2. \tag{4.49}$$

This particular value of the correlation function means that in a thermal field correlations between the photons are large. In other words, the photons group together (move in large groups). We call this effect *photon bunching*.

Example 4.4 (Continuous versus discrete *n*) *Assume that n is a continuous rather than a discrete variable. The continuous variable approach makes the calculation of $\langle(\Delta n)^2\rangle$ appropriate to classical field. In this case when calculating $\langle n^2\rangle$, we replace the sum over n by integration over n, and obtain*

$$\langle n^2\rangle = \frac{\int_0^\infty dn\, n^2 e^{-nx}}{\int_0^\infty dn\, e^{-nx}} = \frac{(1/x)''}{(1/x)} = 2\langle n\rangle^2, \tag{4.50}$$

where $''$ denotes second derivative of $1/x$ with respect to x.

Hence, the variance of the number of photons is given by

$$\langle(\Delta n)^2\rangle = \langle n\rangle^2. \tag{4.51}$$

Comparing Eqs. (4.48) and (4.51), we see that the classical (continuous energy) and quantum (discrete energy) results differ by $\langle n\rangle$. The quantum result shows that radiation possess both a wave character, which gives the $\langle n\rangle^2$, and a particle character, which gives the $\langle n\rangle$ term.

4.4.2 *Poisson Distribution*

We now consider correlation functions for a field with a Poisson distribution of photons

$$P_n = \frac{\langle n\rangle^n}{n!}\mathrm{e}^{-\langle n\rangle}, \tag{4.52}$$

where $\langle n\rangle$ is the average number of photons in the field.

For example, light emitted from a perfectly stabilized laser working well above threshold can be described by the Poisson distribution function, see Chapter 18.

In this case, the first statistical moment ($m = 1$) is given by

$$\langle n\rangle = \sum_n \frac{n\langle n\rangle^n}{n!}\mathrm{e}^{-\langle n\rangle} = \langle n\rangle\,\mathrm{e}^{-\langle n\rangle}\sum_n \frac{\langle n\rangle^{n-1}}{(n-1)!} = \langle n\rangle\,, \tag{4.53}$$

which confirms the Poisson distribution of photons.

Similarly, we can find the higher statistical moments and the variance of the number of photons. First, calculate $\langle n^2\rangle$. Using the definition of the statistical moments, Eq. (4.35), we find

$$\langle n^2\rangle = \sum_n \frac{n^2\mathrm{e}^{-\langle n\rangle}\langle n\rangle^n}{n!} = \langle n\rangle\,\mathrm{e}^{-\langle n\rangle}\sum_n \frac{n\langle n\rangle^{n-1}}{(n-1)!}. \tag{4.54}$$

To proceed further with the sum over n, we change the variable by substituting $n - 1 = k$, and obtain

$$\langle n^2\rangle = \langle n\rangle\,\mathrm{e}^{-\langle n\rangle}\left\{\sum_k \frac{k\langle n\rangle^k}{k!} + \sum_k \frac{\langle n\rangle^k}{k!}\right\}. \tag{4.55}$$

The two sums over k are easy to evaluate, and finally we obtain

$$\langle n^2\rangle = \langle n\rangle^2 + \langle n\rangle\,. \tag{4.56}$$

Thus, the variance of the number of photons in a field with the Poisson distribution of photons is given by

$$\langle(\Delta n)^2\rangle = \langle n\rangle, \tag{4.57}$$

and with this result, we find from Eq. (4.25) that in this case the normalized second-order correlation function is

$$g^{(2)} = 1. \tag{4.58}$$

The value of the correlation function $g^{(2)} = 1$ means that photons in the coherent field are independent of each other. It is clearly seen

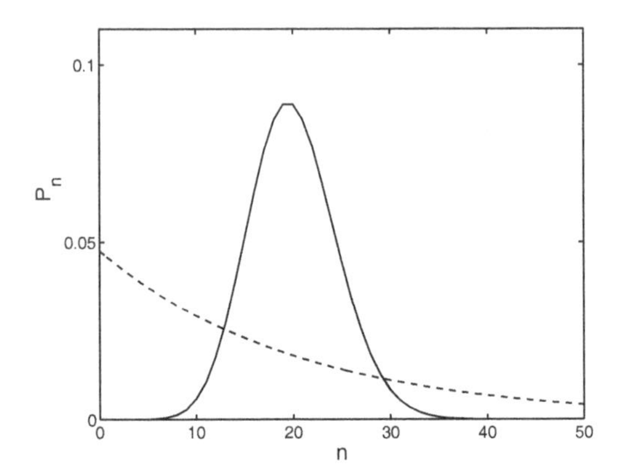

Figure 4.2 The Poisson distribution (solid line) and the thermal distribution (dashed line) as a function of n for $\langle n \rangle = 20$.

from the definition of $g^{(2)}$, Eq. (4.21), that $g^{(2)} = 1$ when the second-order correlation function factorizes into a product

$$\langle \hat{E}^{(-)} \hat{E}^{(-)} \hat{E}^{(+)} \hat{E}^{(+)} \rangle = \langle \hat{E}^{(-)} \hat{E}^{(+)} \rangle \langle \hat{E}^{(-)} \hat{E}^{(+)} \rangle. \tag{4.59}$$

Figure 4.2 shows the thermal and Poisson distributions of photons. It is seen that for the thermal field, photons are randomly distributed over a large range of n as the fluctuations are large. For the Poisson distribution, photons group around the average $\langle n \rangle$ and are distributed over the average in a range determined by the variance $\langle (\Delta n)^2 \rangle$.

Fluctuations in the photon number are often described in terms of the Mandel Q-parameter defined by [19]

$$Q = \frac{\langle (\Delta n)^2 \rangle - \langle n \rangle}{\langle n \rangle}. \tag{4.60}$$

Positive values of the Q-parameter indicate super-Poissonian statistics, whereas negative values indicate sub-Poissonian statistics, and the value of $Q = 0$ corresponds to a Poissonian statistics.[a]

According to Eqs. (4.25) and (4.60), the Mandel Q-parameter can be expressed in terms of the normalized second-order correlation

[a] In the laser theory, the signature of threshold behaviour is determined by the so-called Fano factor defined as $F = \langle (\Delta n)^2 \rangle / \langle n \rangle$. Below and at the threshold $F > 1$ and $F = 1$ above the threshold indicating coherent nature of the laser field. The Fano factor is related to the Mandel Q parameter as $F = Q + 1$.

function as

$$Q = \langle n \rangle \left(g^{(2)} - 1 \right). \qquad (4.61)$$

Thus, photon antibunching corresponds to sub-Poissonian statistics of photons, whereas photon bunching corresponds to super-Poissonian statistics.

4.5 Coherent States of the EM Field

The number states (Fock states) visualize a field with defined amplitude but with phase randomly distributed. Then a question arises: Are there any states which, in the limit of large amplitude, reproduce a state of a classical field of stable amplitude and fixed phase?

The closest quantum states to classical states are coherent states that are defined as eigenstates of the annihilation operator

$$\hat{a} \left| \alpha \right\rangle = \alpha \left| \alpha \right\rangle. \qquad (4.62)$$

Since the operator $\hat{a}$ is non-Hermitian, we cannot use the coherent states as eigenstates of any observable. However, the states correspond to measurable features.

To establish the form of the coherent state we begin by expanding $|\alpha\rangle$ in terms of the Fock states, which act as an appropriate basis due to their orthogonality

$$|\alpha\rangle = \sum_{n=0}^{\infty} |n\rangle\langle n|\alpha\rangle = \sum_{n} c_n |n\rangle, \qquad (4.63)$$

where $c_n = \langle n|\alpha\rangle$ is the transformation from the coherent state to the number state representation. The $|\langle n|\alpha\rangle|^2$ is the probability that the n photons in the field being in the coherent state $|\alpha\rangle$.

In order to determine the coefficients c_n we write

$$\hat{a} |\alpha\rangle = \sum_{n} c_n \hat{a} |n\rangle = \sum_{n-1}^{\infty} c_n \sqrt{n} |n-1\rangle. \qquad (4.64)$$

On the other hand

$$\hat{a} |\alpha\rangle = \alpha |\alpha\rangle = \sum_{n=0}^{\infty} \alpha c_n |n\rangle. \qquad (4.65)$$

Hence, comparing Eqs. (4.64) and (4.65), we obtain a recurrence relation for the coefficients c_n:

$$c_{n+1}\sqrt{n+1} = \alpha c_n. \tag{4.66}$$

By iterations, we find that

$$c_n = \frac{\alpha^n}{\sqrt{n!}} c_0. \tag{4.67}$$

Therefore, the coherent state can be written as a superposition of the photon number states as

$$|\alpha\rangle = \sum_{n=0}^{\infty} \frac{\alpha^n}{\sqrt{n!}}\, c_0\, |n\rangle\,. \tag{4.68}$$

We choose c_0 such that the $|\alpha\rangle$ will be normalized, $\langle\alpha|\alpha\rangle = 1$. Then, using Eq. (4.68), we find

$$|c_0|^2 = \mathrm{e}^{-|\alpha|^2}, \tag{4.69}$$

and finally, we obtain

$$|\alpha\rangle = \mathrm{e}^{-\frac{1}{2}|\alpha|^2} \sum_{n=0}^{\infty} \frac{\alpha^n}{\sqrt{n!}}\, |n\rangle\,. \tag{4.70}$$

Thus, the probability of finding n photons in the field being in a coherent state is given by

$$P_n = |\langle n|\alpha\rangle|^2 = \frac{|\alpha|^{2n}}{n!}\mathrm{e}^{-|\alpha|^2}, \tag{4.71}$$

which is a Poisson distribution.

With the distribution (4.71), we find that

$$\begin{cases} \langle n\rangle = \langle\alpha|\, n\, |\alpha\rangle = |\alpha|^2\,, \\ g^{(2)} = 1, \\ \left\langle \left(\Delta \hat{E}_{\mathrm{in}}\right)^2 \right\rangle = \left\langle \left(\Delta \hat{E}_{\mathrm{out}}\right)^2 \right\rangle = \frac{1}{4}. \end{cases} \tag{4.72}$$

The coherent states, first introduced by Glauber [20], have a close analogy with classical states of definite amplitudes which is apparent from the definition $\hat{a}\,|\alpha\rangle = \alpha\,|\alpha\rangle$. For this reason, coherent

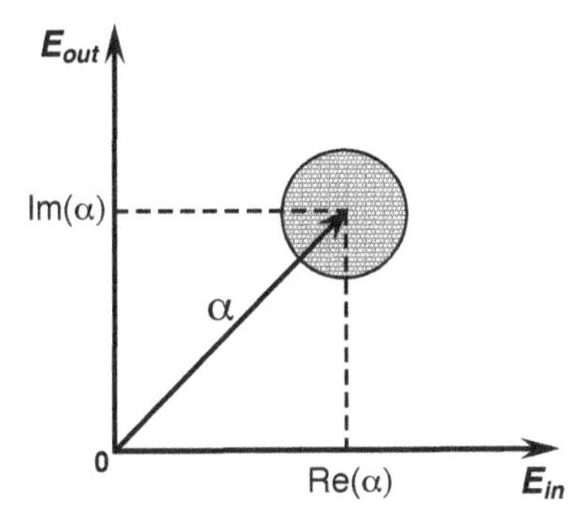

Figure 4.3 Representation of the coherent state $|\alpha\rangle$ in terms of the quadrature amplitudes E_{in} and E_{out}.

states are often called the "most classical" states. It is interesting to note that the fluctuations of the field amplitudes in the coherent state are equal to the vacuum level fluctuations independent of α. This is in contrast to the field in the Fock state, where the fluctuations depend on the number of photons n, see Eq. (4.34). This property makes the field in a coherent state very useful in experimental physics, as the noise produced by the field is always on the same vacuum level independent of the intensity of the field. Laser light is a good approximation to the ideal coherent field which exhibits Poisson fluctuations.

Figure 4.3 shows the mean values and variances of the quadrature amplitudes E_{in} and E_{out} of the field in a coherent state $|\alpha\rangle$. The state is represented by a circle of radius $1/4$ displaced from the origin by α. The coordinates of the centre of the circle are $(\langle E_{\text{in}}\rangle, \langle E_{\text{out}}\rangle) = (\text{Re}(\alpha), \text{Im}(\alpha))$.

Since the repeated application of the annihilation operator on a coherent state does not change the state, the photons can be continuously absorbed without changing the state of the field. Thus, the field remains in the coherent state during the interaction with another system.

What about the phase of the field? Before we answer the question, let us introduce some useful representations of the coherent states.

4.5.1 *Displacement Operator*

Since the photon number state $|n\rangle$ can be obtained from the vacuum state $|0\rangle$ by successive application of the creation operator

$$|n\rangle = \frac{\left(\hat{a}^{\dagger}\right)^{n}}{\sqrt{n!}}\,|0\rangle\,, \tag{4.73}$$

we can express a coherent state in terms of the vacuum state as

$$\begin{aligned}|\alpha\rangle &= \mathrm{e}^{-\frac{1}{2}|\alpha|^{2}}\sum_{n}\frac{\alpha^{n}}{\sqrt{n!}}\frac{\left(\hat{a}^{\dagger}\right)^{n}}{\sqrt{n!}}\,|0\rangle \\ &= \mathrm{e}^{-\frac{1}{2}|\alpha|^{2}}\sum_{n}\frac{\left(\alpha\hat{a}^{\dagger}\right)^{n}}{n!}\,|0\rangle = \mathrm{e}^{-\frac{1}{2}|\alpha|^{2}}\mathrm{e}^{\alpha\hat{a}^{\dagger}}\,|0\rangle\,. \end{aligned} \tag{4.74}$$

Note that

$$\mathrm{e}^{-\alpha^{*}\hat{a}}\,|0\rangle = |0\rangle\,, \tag{4.75}$$

which allows us to write the coherent state as

$$|\alpha\rangle = \mathrm{e}^{-\frac{1}{2}|\alpha|^{2}}\mathrm{e}^{\alpha\hat{a}^{\dagger}}\mathrm{e}^{-\alpha^{*}\hat{a}}\,|0\rangle\,. \tag{4.76}$$

We can apply the Campbell–Baker–Hausdorff operator identity

$$\mathrm{e}^{\hat{A}+\hat{B}} = \mathrm{e}^{\hat{A}}\mathrm{e}^{\hat{B}}\mathrm{e}^{-\frac{1}{2}[\hat{A},\hat{B}]}, \tag{4.77}$$

which is valid for two operators satisfying the commutation relations

$$[\hat{A},[\hat{A},\hat{B}]] = [\hat{B},[\hat{A},\hat{B}]] = 0. \tag{4.78}$$

The condition (4.78) is obviously satisfied for any pair of operators $\hat{A}$, $\hat{B}$ whose commutator $[\hat{A},\hat{B}]$ is a *c*-number. If we put $\hat{A} = \alpha\hat{a}^{\dagger}$ and $\hat{B} = -\alpha^{*}\hat{a}$, we obtain

$$|\alpha\rangle = \mathrm{e}^{\alpha\hat{a}^{\dagger}-\alpha^{*}\hat{a}}\,|0\rangle = \hat{D}\left(\alpha\right)|0\rangle\,, \tag{4.79}$$

where $\hat{D}\left(\alpha\right)$ is called the *displacement operator*. Thus, a coherent state is obtained by applying the displacement operator on the vacuum state. The coherent state is therefore the displaced form of the harmonic oscillator ground state.

4.5.2 *Properties of the Displacement Operator*

The displacement operator has the following properties:

$$\hat{D}^{\dagger}(\alpha) = \hat{D}^{-1}(\alpha) = \hat{D}(-\alpha),$$
$$\hat{D}^{\dagger}(\alpha)\,\hat{D}(\alpha) = \hat{D}(\alpha)\,\hat{D}^{\dagger}(\alpha) = 1, \quad \text{(Unitary operator)}. \tag{4.80}$$

Further

$$\hat{a}\hat{D}(\alpha) = \hat{D}(\alpha)\,\hat{a} + \alpha\hat{D}(\alpha). \tag{4.81}$$

Therefore, the operator $\hat{D}(\alpha)$ represents a displacement operator in the sense that

$$\hat{D}^{\dagger}(\alpha)\,\hat{a}\hat{D}(\alpha) = \hat{a} + \alpha,$$
$$\hat{D}^{\dagger}(\alpha)\,\hat{a}^{\dagger}\hat{D}(\alpha) = \hat{a}^{\dagger} + \alpha^{*}. \tag{4.82}$$

The action of successively applied displacement operators is also additive up to phase factors

$$\hat{D}(\alpha)\,\hat{D}(\beta) = \hat{D}(\alpha+\beta)\exp\left[\frac{1}{2}\left(\alpha\beta^{*} - \beta\alpha^{*}\right)\right]. \tag{4.83}$$

Note that the extra phase on the right-hand side makes the displacement operators non-commutative in general.

4.5.3 *Representation in Terms of Coherent States*

The number state $|n\rangle$ may be represented in terms of coherent states $|\alpha\rangle$. Multiplying the expansion

$$|\alpha\rangle = \mathrm{e}^{-\frac{1}{2}|\alpha|^{2}} \sum_{n} \frac{\alpha^{n}}{\sqrt{n!}}\,|n\rangle \tag{4.84}$$

by a factor

$$\frac{1}{\pi}\frac{\alpha^{*n}}{\sqrt{n!}}\mathrm{e}^{-\frac{1}{2}|\alpha|^{2}}, \tag{4.85}$$

and integrating overall α, we obtain

$$|n\rangle = \frac{1}{\pi}\int d^{2}\alpha\,\mathrm{e}^{-\frac{1}{2}|\alpha|^{2}}\frac{\alpha^{*n}}{\sqrt{n!}}\,|\alpha\rangle, \tag{4.86}$$

where $d^{2}\alpha = d\,(\mathrm{Re}\alpha)\,d\,(\mathrm{Im}\alpha) = r\,dr\,d\psi$ denotes a double integration over the whole complex α-space.

From the completeness of the Fock state $\left(\sum_n |n\rangle\langle n| = 1\right)$, we readily obtain the completeness relation for coherent states

$$\frac{1}{\pi}\int d^2\alpha|\alpha\rangle\langle\alpha| = 1, \tag{4.87}$$

where we have used the integral property

$$\int d^2\alpha\, \alpha^n\alpha^{*m}e^{-s|\alpha|^2} = \frac{\pi n!}{s^{n+1}}\delta_{nm}. \tag{4.88}$$

Thus, coherent states are complete.

However, the coherent states are not orthogonal. To show this, we calculate a product of two coherent states

$$\begin{aligned}\langle\beta|\alpha\rangle &= e^{-\frac{1}{2}|\alpha|^2-\frac{1}{2}|\beta|^2}\sum_{n,m}\frac{\beta^{*n}}{\sqrt{n!}}\frac{\alpha^m}{\sqrt{m!}}\langle n|m\rangle \\ &= e^{-\frac{1}{2}|\alpha|^2-\frac{1}{2}|\beta|^2}\sum_{n}\frac{(\alpha\beta^*)^n}{n!} \\ &= e^{-\frac{1}{2}\left(|\alpha|^2+|\beta|^2\right)+\alpha\beta^*}.\end{aligned} \tag{4.89}$$

Then, we find

$$|\langle\beta|\alpha\rangle|^2 = e^{-|\alpha-\beta|^2} \neq \delta_{\alpha\beta}, \tag{4.90}$$

which shows that the two coherent states are not orthogonal.

The coherent states become approximately orthogonal when the difference $|\alpha-\beta|^2 \to \infty$. Since the coherent states are not orthogonal they are sometimes called *over-complete*.

Exercises

4.1 The state $|n\rangle$ is an eigenstate of the photon number operator $\hat{n} = \hat{a}^\dagger\hat{a}$ with eigenvalue n. Show that

(a) the state $\hat{a}|n\rangle$ is also an eigenstate of $\hat{n}$ with eigenvalue $n-1$.

(b) the state $\hat{a}^\dagger|n\rangle$ is also an eigenstate of $\hat{n}$ with eigenvalue $n+1$.

4.2 Show that $\mathrm{Tr}\left[\hat{D}(\alpha)\right] = \pi\delta(\alpha)$, where $\delta(\alpha)$ is the two-dimensional Dirac delta function.

4.3 Using $\mathrm{Tr}\left[\hat{D}(\alpha)\right] = \pi\delta(\alpha)$, show that the $\delta(\alpha)$-function can be represented by a Fourier integral in the complex form

$$\delta(\alpha) = \frac{1}{\pi^2}\int d^2\beta\ \exp(\alpha\beta^* - \alpha^*\beta).$$

4.4 Show that

$$\hat{D}(\alpha)\,\hat{D}(\beta) = \hat{D}(\alpha+\beta)\exp\left[\frac{1}{2}(\alpha\beta^* - \beta\alpha^*)\right],$$

$$\hat{D}^{-1}(\alpha)\,\hat{a}\,\hat{D}(\alpha) = \hat{a} + \alpha,$$

$$\hat{D}^{-1}(\alpha)\,\hat{a}^\dagger\,\hat{D}(\alpha) = \hat{a}^\dagger + \alpha^*.$$

4.5 Prove the identity

$$\int d^2\alpha\,\alpha^n\alpha^{*m}\mathrm{e}^{-s|\alpha|^2} = \frac{\pi n!}{s^{n+1}}\delta_{nm}.$$

4.6 Prove that

$$\hat{D}(\alpha) \equiv \mathrm{e}^{\alpha\hat{a}^\dagger - \alpha^*\hat{a}} = \mathrm{e}^{-\frac{1}{2}|\alpha|^2}\mathrm{e}^{\alpha\hat{a}^\dagger}\mathrm{e}^{-\alpha^*\hat{a}} = \mathrm{e}^{\frac{1}{2}|\alpha|^2}\mathrm{e}^{-\alpha^*\hat{a}}\mathrm{e}^{\alpha\hat{a}^\dagger}.$$

4.7 Show that the expectation value of the displacement operator $\hat{D}(\alpha)$ for a chaotic (thermal) field is given by

$$\langle\hat{D}(\alpha)\rangle = \exp\left[-|\alpha|^2\left(\langle\hat{n}\rangle + \frac{1}{2}\right)\right],$$

where $\langle\hat{n}\rangle$ is the mean number of photons in the field. A thermal field is characterized by the following correlation functions

$$\langle\hat{a}\rangle = \langle\hat{a}^\dagger\rangle = 0,$$
$$\langle\hat{a}^2\rangle = \langle(\hat{a}^\dagger)^2\rangle = 0,$$
$$\langle\hat{a}^\dagger\hat{a}\rangle = \langle\hat{n}\rangle,$$
$$\langle\hat{a}\hat{a}^\dagger\rangle = 1 + \langle\hat{n}\rangle.$$

4.8 Assume that the field is in a superposition state

$$|\Psi\rangle = C_1|\alpha\rangle + iC_2|\beta\rangle,$$

where $|\alpha\rangle$, $|\beta\rangle$ are two different coherent states, $\langle\Psi|\Psi\rangle = 1$ and all four numbers C_1, C_2, α, β are real. Consider the Mandel Q parameter defined as

$$Q = \frac{\langle(\Delta\hat{n})^2\rangle - \langle\hat{n}\rangle}{\langle\hat{n}\rangle}.$$

The parameter Q determines the statistics of the field. For a Poissonian statistics $Q = 0$, for a sub-Poissonian statistics $Q < 0$, and $Q > 0$ for a super-Poissonian statistics.

(a) Under which conditions the statistics of the field being in the superposition state $|\Psi\rangle$ is sub-Poissonian?

(b) Under which conditions $Q = 0$?

4.9 The probability of finding n photons in the mode being in a coherent state $|\alpha\rangle$ is given by the Poisson distribution

$$P_n = \frac{\langle n\rangle^n}{n!} \mathrm{e}^{-\langle n\rangle}.$$

Use the Sterling's formula for the factorial to show that this distribution can be approximated by

$$P_n \approx \frac{1}{\sqrt{2\pi \langle n\rangle}} \exp\left[-\frac{(n - \langle n\rangle)^2}{2\langle n\rangle}\right].$$

In what limit is this a good approximation?

4.10 Calculate the variances of the position and momentum operators of a harmonic oscillator in a coherent state $|\alpha\rangle$ to prove that irrespectively of α the coherent states are minimum uncertainty states.

4.11 Show that in the coordinates of the quadrature phase components, $\hat{E}_{\mathrm{in}}$ and $\hat{E}_{\mathrm{out}}$, the average amplitudes $\langle \hat{E}_{\mathrm{in}}\rangle$, $\langle \hat{E}_{\mathrm{out}}\rangle$ and the associated variances in a coherent state $|\alpha\rangle$ can be represented by a circle of radius $1/4$ centred on the complex amplitude vector α.

Chapter 5

Photon Phase Operator

5.1 Introduction

We have already shown that the coherent states are very close to the classical state of the well-defined amplitude and phase. A question not answered yet is: Does there exist an analogous quantum mechanical observable for the phase that is given by some Hermitian operator? We know from classical optics that phase of an EM field is an observable quantity. Hence, in quantum physics it should be associated with a Hermitian operator. In order to set up a quantitative description of the phase operator, we will introduce some of the ideas presented in the literature which led to the formulation of the Hermitian phase operator.

5.2 Exponential Phase Operator

In classical optics, an electric field $\vec{E}(\vec{r}, t)$ is often written as

$$\vec{E}(\vec{r}, t) = \vec{E}_0(\vec{r}, t)e^{i\psi}, \tag{5.1}$$

where $\vec{E}_0(\vec{r}, t)$ is the amplitude of the field, whose modulus square is the intensity of the field, and the argument ψ is the phase.

Quantum Optics for Beginners
Zbigniew Ficek and Mohamed Ridza Wahiddin

ISBN 978-981-4411-75-2 (Hardcover), 978-981-4411-76-9 (eBook)
www.panstanford.com

Both intensity and the phase can be measured simultaneously with arbitrary accuracy.

In quantum optics the situation is completely different. The field amplitude becomes an operator acting in a Hilbert space of field states. In addition, the polar decomposition of the field amplitude, which is trivial for classical fields, becomes far from being trivial for quantum fields because of the problem with proper definition of the Hermitian phase operator.

The early studies of the quantum field (photon) phase were concerned with the exponential form of the phase operator. We start from the concept of the exponential phase operator, introduced by Dirac [21], to develop interesting concepts of the Hermitian phase operator.

It is well known from the complex analysis that an arbitrary complex number can be written as

$$z = |r|\, e^{i\psi}, \tag{5.2}$$

where $|r|$ is the modulus and ψ (real) is the argument of the complex number.

By analogy, Dirac[a] proposed to decompose the annihilation (non-Hermitian) operator

$$\hat{a} = \hat{g} e^{i\hat{\psi}}, \tag{5.3}$$

where $\hat{\psi}$ can be treated as a phase operator.

An obvious question arises: Is this decomposition valid? Suppose, the decomposition is valid. Then the Hermitian conjugate of $\hat{a}$ is given by

$$\hat{a}^{\dagger} = e^{-i\hat{\psi}} \hat{g}, \tag{5.4}$$

and the product of $\hat{a}$ and $\hat{a}^{\dagger}$ gives

$$\hat{a}\hat{a}^{\dagger} = \hat{n} + 1 = \hat{g}^2, \tag{5.5}$$

from which we have

$$\hat{g} = (\hat{n} + 1)^{1/2}. \tag{5.6}$$

Using the above results, we can write

$$e^{i\hat{\psi}} = \frac{1}{(\hat{n}+1)^{1/2}} \hat{a}, \qquad e^{-i\hat{\psi}} = \hat{a}^{\dagger} \frac{1}{(\hat{n}+1)^{1/2}}. \tag{5.7}$$

[a] Paul Dirac was granted the Nobel prize in 1933 for the discovery of new productive forms of atomic theory.

Hence, the properties of the phase operator can be calculated using the properties of the annihilation, creation and photon number operators.

We find from Eq. (5.7) that the product of the two exponents gives

$$\begin{aligned} e^{i\hat{\psi}}e^{-i\hat{\psi}} &= \frac{1}{(\hat{n}+1)^{1/2}}\hat{a}\hat{a}^{\dagger}\frac{1}{(\hat{n}+1)^{1/2}} \\ &= \frac{(\hat{n}+1)}{(\hat{n}+1)^{1/2}(\hat{n}+1)^{1/2}} = 1. \end{aligned} \tag{5.8}$$

However, the product

$$e^{-i\hat{\psi}}e^{i\hat{\psi}} = \hat{a}^{\dagger}\frac{1}{\hat{n}+1}\hat{a} \neq 1, \tag{5.9}$$

and it gives zero when it operates either to the left or to the right on the vacuum state.

Evidently, $e^{i\hat{\psi}}$ is not unitary and hence $\hat{\psi}$ is not Hermitian.

5.3 Susskind–Glogower Phase Operator

Susskind and Glogower [22] argued that a consistent way out of the difficulty of the exponent phase operator is to introduce $\widehat{\cos\psi}$ and $\widehat{\sin\psi}$ operators that are combinations of the exponent phase operator

$$\widehat{\cos\psi} = \frac{1}{2}\left[e^{i\hat{\psi}} + e^{-i\hat{\psi}^{\dagger}}\right], \tag{5.10}$$

$$\widehat{\sin\psi} = \frac{1}{2i}\left[e^{i\hat{\psi}} - e^{-i\hat{\psi}^{\dagger}}\right], \tag{5.11}$$

where, as before, $e^{i\hat{\psi}} = \left(1/(\hat{n}+1)^{1/2}\right)\hat{a}$.

It is obvious that $\widehat{\cos\psi}$ and $\widehat{\sin\psi}$ are Hermitian, despite the fact that $e^{i\hat{\psi}}$ is not unitary.

Proof. Matrix elements of a Hermitian operator $\hat{A}$ should satisfy the following relation

$$\langle i|\hat{A}|j\rangle = \langle j|\hat{A}|i\rangle^{*}. \tag{5.12}$$

We will check whether the operators $\widehat{\cos\psi}$ and $\widehat{\sin\psi}$ satisfy this relation. First, we consider

$$e^{i\hat{\psi}}|n\rangle = (\hat{n}+1)^{-1/2}\hat{a}|n\rangle = (\hat{n}+1)^{-1/2}\sqrt{n}|n-1\rangle. \tag{5.13}$$

However, there is a square root of $(\hat{n}+1)$ in the denominator. In order to have this operator in its basic form, the function $(\hat{n}+1)^{-1/2}$ has to be expanded into a power series with respect to $\hat{n}$.

Let $\hat{n} = x$, then

$$(1+x)^{-1/2} = 1 - \frac{1}{2}x + \frac{1}{2}\frac{3}{4}x^2 - \frac{1}{2}\frac{3}{4}\frac{5}{6}x^3 + \cdots \tag{5.14}$$

Hence

$$\begin{aligned}(\hat{n}+1)^{-1/2}|n-1\rangle &= \left\{1 - \frac{1}{2}\hat{n} + \frac{1}{2}\frac{3}{4}\hat{n}^2 - \cdots\right\}|n-1\rangle \\ &= \left\{1 - \frac{1}{2}(n-1)\right. \\ &\quad \left. + \frac{1}{2}\frac{3}{4}(n-1)^2 - \cdots\right\}|n-1\rangle,\end{aligned} \tag{5.15}$$

where we have used the relation

$$(\hat{n})^k|n-1\rangle = (n-1)^k|n-1\rangle. \tag{5.16}$$

Thus,

$$\begin{aligned}(\hat{n}+1)^{-1/2}|n-1\rangle &= [1+(n-1)]^{-1/2}|n-1\rangle \\ &= \frac{1}{\sqrt{n}}|n-1\rangle.\end{aligned} \tag{5.17}$$

Using this result in Eq. (5.13), we obtain

$$\mathrm{e}^{i\hat{\psi}}|n\rangle = |n-1\rangle. \tag{5.18}$$

Similarly, we can show that

$$\mathrm{e}^{-i\hat{\psi}^\dagger}|n\rangle = |n+1\rangle. \tag{5.19}$$

Using these relations, we find that the non-zero matrix elements of the $\widehat{\cos\psi}$ and $\widehat{\sin\psi}$ operators are

$$\langle n-1|\widehat{\cos\psi}|n\rangle = \langle n|\widehat{\cos\psi}|n-1\rangle = \frac{1}{2}, \tag{5.20}$$

$$\langle n-1|\widehat{\sin\psi}|n\rangle = -\langle n|\widehat{\sin\psi}|n-1\rangle = \frac{1}{2i}. \tag{5.21}$$

From this we can derive that

$$\begin{aligned}\langle i|\widehat{\cos\psi}|j\rangle &= \langle j|\widehat{\cos\psi}|i\rangle^*, \\ \langle i|\widehat{\sin\psi}|j\rangle &= \langle j|\widehat{\sin\psi}|i\rangle^*,\end{aligned} \tag{5.22}$$

as required. □

We therefore may conclude that $\widehat{\cos\psi}$ and $\widehat{\sin\psi}$ are Hermitian and can represent the observable phase properties of the EM field.

However, $\widehat{\cos\psi}$ and $\widehat{\sin\psi}$ do not commute

$$\left[\widehat{\cos\psi}, \widehat{\sin\psi}\right] = \frac{1}{2i}\left\{\hat{a}^{\dagger}\frac{1}{(\hat{n}+1)}\hat{a} - 1\right\}, \tag{5.23}$$

and therefore do not determine the same phase operator [23].

Proof. Calculate a commutator

$$\begin{aligned}\left[\widehat{\cos\psi}, \widehat{\sin\psi}\right]|n\rangle &= \left[\frac{1}{2}\left(\mathrm{e}^{i\hat{\psi}} + \mathrm{e}^{-i\hat{\psi}^{\dagger}}\right), \frac{1}{2i}\left(\mathrm{e}^{i\hat{\psi}} - \mathrm{e}^{-i\hat{\psi}^{\dagger}}\right)\right]|n\rangle \\ &= \frac{1}{2i}\left\{\mathrm{e}^{-i\hat{\psi}^{\dagger}}\mathrm{e}^{i\hat{\psi}} - \mathrm{e}^{i\hat{\psi}}\mathrm{e}^{-i\hat{\psi}^{\dagger}}\right\}|n\rangle. \end{aligned} \tag{5.24}$$

We have shown before that

$$\mathrm{e}^{i\hat{\psi}} = \frac{1}{(\hat{n}+1)^{1/2}}\hat{a}, \tag{5.25}$$

$$\mathrm{e}^{-i\hat{\psi}^{\dagger}} = \hat{a}^{\dagger}\frac{1}{(\hat{n}+1)^{1/2}}. \tag{5.26}$$

Hence,

$$\mathrm{e}^{-i\hat{\psi}^{\dagger}}\mathrm{e}^{i\hat{\psi}}|n\rangle = \hat{a}^{\dagger}(\hat{n}+1)^{-1}\hat{a}|n\rangle, \tag{5.27}$$

and

$$\mathrm{e}^{i\hat{\psi}}\mathrm{e}^{-i\hat{\psi}^{\dagger}}|n\rangle = |n\rangle. \tag{5.28}$$

Collecting the above results, we finally obtain

$$\left[\widehat{\cos\psi}, \widehat{\sin\psi}\right]|n\rangle = \frac{1}{2i}\left\{\hat{a}^{\dagger}(\hat{n}+1)^{-1}\hat{a} - 1\right\}|n\rangle, \tag{5.29}$$

as required. □

We can calculate the matrix elements of the commutator $\left[\widehat{\cos\psi}, \widehat{\sin\psi}\right]$

$$\begin{aligned}\langle m|\left[\widehat{\cos\psi}, \widehat{\sin\psi}\right]|n\rangle &= \frac{1}{2i}\langle m|\hat{a}^{\dagger}(\hat{n}+1)^{-1}\hat{a} - 1|n\rangle \\ &= \frac{1}{2i}\left[\langle m|\mathrm{e}^{-i\hat{\psi}^{\dagger}}\mathrm{e}^{i\hat{\psi}}|n\rangle - \delta_{mn}\right]. \end{aligned} \tag{5.30}$$

Since

$$\begin{aligned}\mathrm{e}^{i\hat{\psi}}|n\rangle &= |n-1\rangle = 0, \qquad \text{for} \qquad n = 0, \\ \mathrm{e}^{-i\hat{\psi}^{\dagger}}|n\rangle &= |n+1\rangle, \end{aligned} \tag{5.31}$$

we therefore obtain

$$\langle m| \left[\widehat{\cos\psi}, \widehat{\sin\psi}\right] |n\rangle = 0, \qquad n \neq 0,$$

$$\langle m| \left[\widehat{\cos\psi}, \widehat{\sin\psi}\right] |n\rangle = -\frac{1}{2i}\delta_{m0}, \qquad n = 0. \tag{5.32}$$

Only one of the infinite number of the matrix elements of the commutator is different from zero, the diagonal ground state matrix element, $m = n = 0$.

Can we find states of the EM field which are simultaneous eigenstates of the photon number and the photon phase? To answer this question, we will calculate two commutators $\left[\hat{n}, \widehat{\cos\psi}\right]$ and $\left[\hat{n}, \widehat{\sin\psi}\right]$, using the photon number representation

$$\left[\hat{n}, \widehat{\cos\psi}\right] |n\rangle = \frac{1}{2}\left\{\left[\hat{n}, \mathrm{e}^{i\hat{\psi}}\right] + \left[\hat{n}, \mathrm{e}^{-i\hat{\psi}^{\dagger}}\right]\right\} |n\rangle, \tag{5.33}$$

$$\left[\hat{n}, \widehat{\sin\psi}\right] |n\rangle = \frac{1}{2i}\left\{\left[\hat{n}, \mathrm{e}^{i\hat{\psi}}\right] - \left[\hat{n}, \mathrm{e}^{-i\hat{\psi}^{\dagger}}\right]\right\} |n\rangle. \tag{5.34}$$

First, we calculate the commutator

$$\begin{aligned}\left[\hat{n}, \mathrm{e}^{\hat{\psi}}\right] |n\rangle &= \hat{n}\,|n-1\rangle - n\mathrm{e}^{i\hat{\psi}}\,|n\rangle \\ &= (n-1)\,|n-1\rangle - n\,|n-1\rangle = -\,|n-1\rangle \\ &= -\mathrm{e}^{i\hat{\psi}}\,|n\rangle.\end{aligned} \tag{5.35}$$

Similarly, we find

$$\left[\hat{n}, \mathrm{e}^{-\hat{\psi}^{\dagger}}\right] |n\rangle = \mathrm{e}^{-i\hat{\psi}^{\dagger}}\,|n\rangle \tag{5.36}$$

and finally, we combine the results given in Eqs. (5.33) and (5.34), and obtain

$$\left[\hat{n}, \widehat{\cos\psi}\right] = -i\widehat{\sin\psi}, \tag{5.37}$$

$$\left[\hat{n}, \widehat{\sin\psi}\right] = i\widehat{\cos\psi}. \tag{5.38}$$

The above commutation relations show that the number and phase operators do not commute, and therefore, it is not possible to set up states of the radiation field which are simultaneous eigenstates of the two operators. The amplitude of an EM field, associated with $\hat{n}$, and the phase, associated with $\widehat{\cos\psi}$ or $\widehat{\sin\psi}$, cannot be both precisely specified.

The commutation relation (5.37) immediately leads to the uncertainty relation

$$(\Delta n)\,(\Delta\cos\psi) \geq \frac{1}{2}|\langle\widehat{\sin\psi}\rangle|, \tag{5.39}$$

where

$$(\Delta n)^2 = \langle\hat{n}^2\rangle - \langle\hat{n}\rangle^2 \tag{5.40}$$

and

$$(\Delta\cos\psi)^2 = \langle(\widehat{\cos\psi})^2\rangle - \langle\widehat{\cos\psi}\rangle^2. \tag{5.41}$$

In a Fock state

$$\Delta n = 0,$$
$$\Delta\cos\psi = \frac{1}{\sqrt{2}}. \tag{5.42}$$

The above result shows that the EM field which corresponds to the state $|n\rangle$ has a definite amplitude but the phase has an arbitrary value.

We have introduced two operators, $\widehat{\cos\psi}$ and $\widehat{\sin\psi}$, to represent the phase properties of the radiation field. In classical physics, the phase is a single quantity and it seems unnecessary to represent it by two different operators in quantum optics. Since the phase operators do not commute, it is impossible to form states which are simultaneously eigenstates of $\widehat{\cos\psi}$ and $\widehat{\sin\psi}$.

5.4 Unitary Exponential Phase Operator

To preserve the concept of exponential phase operators and to remove the non-unitary properties, we can define the exponential phase operator

$$\hat{V} = e^{i\hat{\psi}} = \sum_n |n\rangle\,\langle n|\,e^{i\hat{\psi}} = \sum_n |n\rangle\,\langle n+1|\,, \tag{5.43}$$

with

$$\hat{V}^\dagger = e^{-i\hat{\psi}^\dagger} = \sum_n e^{-i\hat{\psi}^\dagger}\,|n\rangle\,\langle n| = \sum_n |n+1\rangle\,\langle n|\,. \tag{5.44}$$

Hence,

$$\hat{V}\hat{V}^\dagger = \sum_{n,m} |n\rangle\,\langle n+1|\,m+1\rangle\,\langle n| = \sum_n |n\rangle\,\langle n| = 1, \tag{5.45}$$

and

$$\hat{V}^\dagger \hat{V} = \sum_{n,m} |n+1\rangle \langle n|\, m\rangle \langle m+1| = \sum_{n} |n+1\rangle \langle n+1|$$
$$= 1 - |0\rangle \langle 0|\,. \tag{5.46}$$

This is still a non-unitary operator.

However, we can remove the non-unitarity by extending the lower limit of the sum over n to $-\infty$, and obtain

$$\hat{V}\hat{V}^\dagger = \sum_{n,m=-\infty}^{\infty} |n\rangle\langle n+1|m+1\rangle\langle n| = 1, \tag{5.47}$$

and

$$\hat{V}^\dagger \hat{V} = \sum_{n=-\infty}^{\infty} |n+1\rangle\langle n+1| = 1. \tag{5.48}$$

Clearly, in this case the operator $\hat{V}$ is an unitary operator and then $\hat{\psi}$ is Hermitian.

If $\hat{\psi}$ is a Hermitian operator then there exists an eigenstate such that (postulate)

$$\hat{V}\,|\Phi\rangle = \mathrm{e}^{i\psi}\,|\Phi\rangle\,. \tag{5.49}$$

We can expand the state $|\Phi\rangle$ in terms of the number states $|n\rangle$ as

$$|\Phi\rangle = \sum_{n} a_n\,|n\rangle\,, \tag{5.50}$$

which gives

$$\hat{V}\,|\Phi\rangle = \sum_{n} a_n \hat{V}\,|n\rangle = \sum_{n,m} a_n |m\rangle\langle m+1|n\rangle = \sum_{n} a_{n+1}\,|n\rangle\,. \tag{5.51}$$

The right-hand side of Eq. (5.49) is equal to

$$\mathrm{e}^{i\psi}\,|\Phi\rangle = \sum_{n} a_n \mathrm{e}^{i\psi}\,|n\rangle\,. \tag{5.52}$$

Hence, comparing Eqs. (5.51) and (5.52), we obtain a recurrence relation for the coefficients a_n:

$$a_{n+1} = a_n \mathrm{e}^{i\psi}. \tag{5.53}$$

By iteration, we find that

$$a_n = a_0 \mathrm{e}^{in\psi}. \tag{5.54}$$

Therefore, the state $|\Phi\rangle$ can be written as

$$|\Phi\rangle = a_0 \sum_n e^{in\psi} |n\rangle . \tag{5.55}$$

From the normalization condition

$$\int d\Phi |\Phi\rangle\langle\Phi| = 1, \tag{5.56}$$

we find that $a_0 = 1/\sqrt{2\pi}$, and then the state $|\Phi\rangle$ takes the form

$$|\Phi\rangle = \frac{1}{\sqrt{2\pi}} \sum_{n=-\infty}^{\infty} e^{in\psi} |n\rangle . \tag{5.57}$$

Example 5.1 (Phase properties of the field in a coherent state.) *We wish to calculate the uncertainty relation* $\Delta n \Delta \cos\psi$ *for the field in a coherent state* $|\alpha\rangle$.

First, we calculate the expectation value of the phase operator $\widehat{\cos\psi}$ *in the coherent state* $|\alpha\rangle$:

$$\begin{aligned}\langle\alpha|\widehat{\cos\psi}|\alpha\rangle &= \frac{1}{2}(\alpha + \alpha^*) e^{-|\alpha|^2} \sum_n \frac{|\alpha|^{2n}}{n!} \frac{1}{\sqrt{n+1}} \\ &= |\alpha| \cos\theta e^{-|\alpha|^2} \sum_n \frac{|\alpha|^{2n}}{n!} \frac{1}{\sqrt{n+1}},\end{aligned} \tag{5.58}$$

where $\alpha = |\alpha| e^{i\theta}$.

Similarly, we find

$$\begin{aligned}\langle\alpha|(\widehat{\cos\psi})^2|\alpha\rangle &= \frac{1}{2} - \frac{1}{4}e^{-|\alpha|^2} + |\alpha|^2\left(\cos^2\theta - \frac{1}{2}\right)e^{-|\alpha|^2} \\ &\times \sum_n \frac{|\alpha|^{2n}}{n!\sqrt{(n+1)(n+2)}}.\end{aligned} \tag{5.59}$$

Unfortunately, it is not possible to evaluate the summations in the above equations analytically. There are, however, some simplifications in the limit of $|\alpha| \gg 1$, *where*

$$\sum_n \frac{|\alpha|^{2n}}{n!\sqrt{n+1}} \simeq \frac{1}{|\alpha|} e^{|\alpha|^2}\left(1 - \frac{1}{8|\alpha|^2} + \cdots\right) \tag{5.60}$$

and

$$\sum_n \frac{|\alpha|^{2n}}{n!\sqrt{(n+1)(n+2)}} = \frac{1}{|\alpha|^2} e^{|\alpha|^2}\left(1 - \frac{1}{2|\alpha|^2} + \cdots\right). \tag{5.61}$$

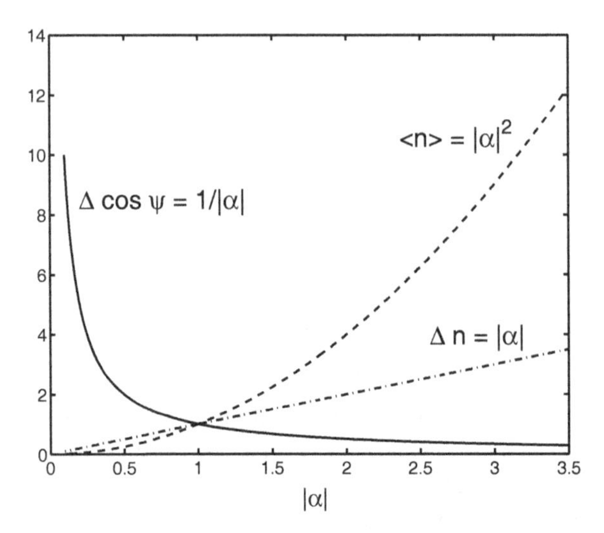

Figure 5.1 The dependence of the fluctuations of the phase $\Delta \cos \psi$ (solid line), the average number of photons (dashed line), and the fluctuations of the photon number (dashed-dotted line) on the amplitude of the coherent state.

In this case, we find that

$$\Delta \cos \psi = \frac{1}{2|\alpha|} \sin\theta, \tag{5.62}$$

which can be written as

$$\Delta n \Delta \cos \psi = \frac{1}{2} \sin\theta, \qquad |\alpha| \gg 1, \tag{5.63}$$

where $\Delta n = |\alpha|$ are the fluctuations of the number of photons.

This result shows that in the limit of $|\alpha| \gg 1$, the coherent state $|\alpha\rangle$ is the minimum uncertainty state for the photon number and the phase $(\widehat{\cos \psi})$ operators.

In Fig. 5.1, we plot $\Delta \cos \psi = |\alpha|^{-1}$, the average number of photons $\langle n \rangle = |\alpha|^2$, and the fluctuations $\Delta n = |\alpha|$.

Although the fluctuations Δn increase linearly with $|\alpha|$, the fractional uncertainty $\Delta n / \langle n \rangle = |\alpha|^{-1}$, which determines the fluctuations of the amplitude, decreases with increasing $|\alpha|$. In addition, the fluctuations of the phase $\Delta \cos \psi$ vary like $|\alpha|^{-1}$. Therefore, for $|\alpha| \gg 1$ the field is well defined in both the amplitude and phase.

5.5 Pegg–Barnett Phase Operator

The failure of $\widehat{\cos\psi}$ and $\widehat{\sin\psi}$ to commute, and the absence of a ready, intuitive interpretation of their eigenvalues, have led Pegg and Barnett to construct a Hermitian phase operator and a phase state $|\psi\rangle$ on a finite-dimensional Hilbert space [24]. The phase state $|\psi\rangle$, defined by the expansion in terms of the Fock states as

$$|\psi\rangle = \frac{1}{\sqrt{s+1}} \sum_{n=0}^{s} \mathrm{e}^{in\psi} |n\rangle , \tag{5.64}$$

behaves in some ways as a state of definite phase ψ when s is large, that the state $|\psi\rangle$ can be the common eigenstate of $\widehat{\cos\psi}$ and $\widehat{\sin\psi}$ in a certain limiting sense. In other words, the phase state is represented by the $(s+1)$-dimensional Hilbert space expanded on the complete basis of the photon number states $|n\rangle$, with the expansion coefficients weighted by the factor $\exp(in\psi)$.

However, the states $|\psi\rangle$ are not orthonormal unless we assume that the phase is a discrete quantity

$$\psi \to \psi_m = \psi_0 + \frac{2\pi m}{s+1}, \qquad m = 0, 1, 2, \ldots, s. \tag{5.65}$$

Proof. Consider a scalar product of two phase states

$$\langle \psi_{m'} | \, \psi_m \rangle = \frac{1}{s+1} \sum_{n,p}^{s} \mathrm{e}^{i(n\psi_m - p\psi_{m'})} \langle p|n\rangle = \frac{1}{s+1} \sum_{n}^{s} \mathrm{e}^{in(\psi_m - \psi_{m'})}. \tag{5.66}$$

If we choose

$$\psi_m = \psi_0 + \frac{2\pi m}{s+1}, \tag{5.67}$$

we obtain

$$\frac{1}{s+1} \sum_{n=0}^{s} \mathrm{e}^{\frac{2\pi n}{s+1}(m-m')} = \frac{1}{s+1} \frac{1 - \mathrm{e}^{2\pi i(m-m')}}{\left[1 - \mathrm{e}^{2\pi i(m-m')/(s+1)}\right]} = \delta_{m,m'}, \tag{5.68}$$

as required. □

It is natural to try and introduce a Hermitian operator in terms of projectors of the type $|\psi\rangle\langle\psi|$. Hence, we can say that the state $|\psi_m\rangle$ is the eigenstate of the Hermitian (Pegg–Barnett) phase operator

$$\hat{\psi} = \sum_{m=0}^{s} \psi_m |\psi_m\rangle\langle\psi_m|, \tag{5.69}$$

with the corresponding eigenvalues ψ_m, that is,

$$\hat{\psi} |\psi_m\rangle = \psi_m |\psi_m\rangle . \tag{5.70}$$

The eigenvalues ψ_m are restricted to lie within a phase window between ψ_0 and $\psi_0 + 2\pi s/(s+1)$.

In order to overcome the limit problem of finite s, Pegg and Barnett have proposed to work with finite s and letting $s \to \infty$ after expectation values have been calculated.

Example 5.2 (Difficulties of the Hermitian phase operator) *Let us consider an example that illustrates a difficulty with the concept of the Hermitian phase operator.*

If the state $|\psi\rangle$ is the eigenstate of $\hat{\psi}$ then $|\psi\rangle$ should also be the eigenstate of $\widehat{\cos\psi}$ and $\widehat{\sin\psi}$. We calculate $\widehat{\cos\psi}\,|\psi\rangle$, and find

$$\begin{aligned}\widehat{\cos\psi}\,|\psi\rangle &= \cos\psi\,|\psi\rangle \\ &\quad + \frac{1}{2}\lim_{s\to\infty}\frac{1}{\sqrt{s+1}}\left\{e^{is\psi}\,|s+1\rangle\right. \\ &\quad \left. - \;e^{i(s+1)\psi}\,|s\rangle - e^{-i\psi}\,|0\rangle\right\} . \end{aligned} \tag{5.71}$$

The state $|\psi\rangle$ thus fails to be a strict eigenstate of $\widehat{\cos\psi}$ because of the contribution of the second term. However, the magnitude of the second term tends to zero in the limit of $s \to \infty$.

Hence,

$$\widehat{\cos\psi}\,|\psi\rangle = \cos\psi\,|\psi\rangle \qquad \text{for} \quad s \to \infty. \tag{5.72}$$

In a similar way, we can show that

$$\begin{aligned}\langle\psi|\,\widehat{\sin\psi}\,|\psi\rangle &= \sin\psi, \\ \langle\psi|\,(\widehat{\cos\psi})^2\,|\psi\rangle &= \cos^2\psi, \\ \langle\psi|\,(\widehat{\sin\psi})^2\,|\psi\rangle &= \sin^2\psi. \end{aligned} \tag{5.73}$$

Therefore, the uncertainties tend to

$$(\Delta\cos\psi) = (\Delta\sin\psi) = 0 \qquad \text{for} \quad s \to \infty. \tag{5.74}$$

These results show that in the limit of $s \to \infty$, the phase state $|\psi\rangle$ behaves as a common eigenstate of the operators $\widehat{\cos\psi}$ and $\widehat{\sin\psi}$. Thus, ψ is the observable phase angle.

It is interesting to note that the number state $|n\rangle$ can be expanded in terms of the phase state basis $|\psi_m\rangle$ as

$$|n\rangle = \sum_{m=0}^{s} |\psi_m\rangle\langle\psi_m|n\rangle = \frac{1}{\sqrt{s+1}}\sum_{n=0}^{s} \mathrm{e}^{-in\psi_n}\,|\psi_n\rangle . \tag{5.75}$$

It is seen from Eqs. (5.64) and (5.75) that a system in a number state is equally likely to be found in any state $|\psi_m\rangle$, and a system in a phase state is equally likely to be found in any number state $|n\rangle$.

However, there is a problem when one calculates the number of photons present in the field being in a phase state $|\psi\rangle$. We explore the problem in the following example.

Example 5.3 (Expectation value in the Hermitian phase state) *Let us calculate the expectation values* $\langle\psi|\,\hat{n}\,|\psi\rangle$ *and* $\langle\psi|\,\hat{n}^2\,|\psi\rangle$*, for which we find*

$$\langle\psi|\,\hat{n}\,|\psi\rangle = \lim_{s\to\infty}\frac{1}{2}s \to \infty,$$
$$\langle\psi|\,\hat{n}^2\,|\psi\rangle = \lim_{s\to\infty}\frac{s\,(2s+1)}{6} \to \infty. \tag{5.76}$$

The photon number expectation values are thus infinite and so is the uncertainty Δn.

However, the ratio of the uncertainty Δn *to the expectation number of photons is finite*

$$\frac{\Delta n}{\langle\hat{n}\rangle} = \frac{\sqrt{\langle\hat{n}^2\rangle - \langle\hat{n}\rangle^2}}{\langle\hat{n}\rangle} = \lim_{s\to\infty}\frac{1}{\sqrt{3}}\frac{\sqrt{s(s+2)}}{s} = \frac{1}{\sqrt{3}}. \tag{5.77}$$

Thus, we may conclude that $|\psi\rangle$ is not a physical state, as it is rather impossible to excite a field to state $|\psi\rangle$ in a practical experiment. This property reflects one of the difficulties encountered in the search for a physical phase operator.

Despite the difficulties, the Hermitian phase operator or quantum phase became the subject of a great research activity in 1990s and in early 2000s. Intensive studies were done on aspects such as the phase probability [25] and quantum phase properties of various linear and nonlinear processes [26].

Exercises

5.1 Show that for a coherent state $\alpha = |\alpha|\exp(i\theta)$, with $|\alpha| \gg 1$, the mean values

$$\langle\alpha|\widehat{\cos\psi}|\alpha\rangle \approx \cos\theta \quad \text{and} \quad \langle\alpha|\widehat{\sin\psi}|\alpha\rangle \approx \sin\theta.$$

This shows that in the limit of $|\alpha| \gg 1$, the coherent state behaves as a classical wave with phase θ.

5.2 Suppose, there exists a Hermitian operator $\hat{A}$ canonically conjugated with photon number operator $\hat{n}$ such that

$$[\hat{A}, \hat{n}] = i.$$

(a) Show that the exponential operator $\exp(i\hat{A})$ is unitary.
(b) Use the Baker–Hausdorff formula to show that

$$e^{i\hat{A}}\hat{n}e^{-i\hat{A}} = \hat{n} - 1.$$

5.3 Calculate the expectation value of the operator $\hat{A}^{\dagger}\hat{A}$ in the coherent state $|\alpha\rangle$, where $\hat{A} = (\hat{n}+1)^{-1/2}\,\hat{a}$, and $\hat{n} = \hat{a}^{\dagger}\hat{a}$.

5.4 Calculate $\widehat{\cos\psi}\,|\psi\rangle$, where $\widehat{\cos\psi}$ is given in Eq. (5.10), and $|\psi\rangle$ is the photon phase state

$$|\psi\rangle = \lim_{s\to\infty}(s+1)^{-1/2}\sum_{n=0}^{s}\exp(in\psi)\,|n\rangle\,.$$

5.5 Calculate the expectation value of the electric field operator $\hat{\vec{E}}(\vec{r}, t)$ of a single-mode field in the photon phase state $|\psi\rangle$. What would you conclude about the phase of the field?

5.6 Find the average value and variance of the Hermitian Pegg–Barnett phase operator

$$\hat{\psi} = \lim_{s\to\infty}\sum_{m=0}^{s}\psi_m|\psi_m\rangle\langle\psi_m| \tag{5.78}$$

in the photon number state $|n\rangle$.

5.7 For an arbitrary state $|\Phi\rangle$ calculate the probability distribution $|\langle\psi|\Phi\rangle|^2$ of the Pegg–Barnett phase state $|\psi\rangle$ as a function of the coefficients c_n of the decomposition of $|\Phi\rangle$ in the basis of photon number states $|n\rangle$.

5.8 Find the probability distribution $P(\psi) = |\langle\psi|\alpha\rangle|^2$ of the Pegg–Barnett phase state $|\psi\rangle$ in a coherent state α.

5.9 Using the result of Question 5.8, show that:

(a) The expectation value of the phase operator is

$$\langle\hat{\psi}\rangle = \lim_{s\to\infty}\langle\alpha|\hat{\psi}|\alpha\rangle = \theta,$$

where θ is the phase angle of the coherent state $\alpha = |\alpha|\exp(i\theta)$.

(b) Then calculate the fluctuations of the phase operator to show that

$$\langle(\Delta\hat{\psi})^2\rangle = \langle\hat{\psi}^2\rangle - \langle\hat{\psi}\rangle^2 = \frac{1}{3}\pi^2 - 4|\alpha|.$$

5.10 Show that in the limit of $s \to \infty$, the photon phase state

$$|\psi\rangle = (s+1)^{-1/2}\sum_{n=0}^{s}\exp(\mathrm{i}n\psi)\,|n\rangle$$

is an eigenstate of $\hat{V}^\dagger$.

Chapter 6

Squeezed States of Light

6.1 Introduction

In Chapter 4 we defined different representations of the electromagnetic (EM) field: the Fock, thermal and coherent state representations, and discussed in details properties of the EM field in these representations. We have seen that in the coherent state the fluctuations in the two quadratures of the EM field amplitudes are equal and minimize the uncertainty product given by the Heisenberg's uncertainty relation. In other words, the quantum fluctuations of the field in a coherent state are equal to the zero-point fluctuations and are randomly distributed in phase. These zero-point fluctuations represent the standard quantum limit to the reduction of noise in a signal. Even an ideal laser operating in a pure coherent state would still possess quantum noise due to zero-point fluctuations.

In this chapter, we consider special states of the EM field which have less fluctuations in one quadrature component than a coherent state at the expense of increased fluctuations in the other quadrature component. Such states are called *squeezed states*. The basic ideas underlying squeezed states of light involve quantum noise (or fluctuations) in the so-called quadrature components of

Quantum Optics for Beginners
Zbigniew Ficek and Mohamed Ridza Wahiddin

ISBN 978-981-4411-75-2 (Hardcover), 978-981-4411-76-9 (eBook)
www.panstanford.com

the electric field and the Heisenberg uncertainty principle. These concepts and the properties of squeezed states will be discussed in this chapter.

6.2 Definition of Squeezed States of Light

We introduce the concept of squeezed states of light using the single-mode representation of the EM field. This treatments will be later generalized to include multi-mode fields. Consider a single-mode electric field represented by the operator

$$\hat{E}(\vec{r}, t) = i\lambda\left[\hat{a}e^{-i\left(\omega t - \vec{k}\cdot\vec{r}\right)} - \hat{a}^{\dagger}e^{i\left(\omega t - \vec{k}\cdot\vec{r}\right)}\right], \tag{6.1}$$

where λ is a constant. In what follows the spatial dependence of the field will usually be suppressed for convenience.

We introduce two Hermitian operators

$$\hat{E}_1 = \frac{1}{2}\left(\hat{a} + \hat{a}^{\dagger}\right), \qquad \hat{E}_2 = \frac{1}{2i}\left(\hat{a} - \hat{a}^{\dagger}\right), \tag{6.2}$$

satisfying the commutation relation

$$\left[\hat{E}_1, \hat{E}_2\right] = \frac{i}{2}. \tag{6.3}$$

These two Hermitian operators are completely equivalent to the in-phase and out-off phase quadrature components, defined in Eq. (4.31). In terms of $\hat{E}_1$ and $\hat{E}_2$, the electric field operator (6.1) takes the form

$$\hat{E}(t) = 2\lambda\left[\hat{E}_1 \sin\left(\omega t - \vec{k}\cdot\vec{r}\right) - \hat{E}_2 \cos\left(\omega t - \vec{k}\cdot\vec{r}\right)\right] \tag{6.4}$$

in which we can identify $\hat{E}_1$ and $\hat{E}_2$ as the amplitudes of the two quadrature components of the field. In general, the electric field operator can be written as

$$\hat{E}(t) = 2\lambda\left[\hat{E}_{\phi} \sin\left(\omega t - \vec{k}\cdot\vec{r} + \phi\right) - \hat{E}_{\phi+\pi/2} \cos\left(\omega t - \vec{k}\cdot\vec{r} + \phi\right)\right], \tag{6.5}$$

where $\hat{E}_{\phi}$ and $\hat{E}_{\phi+\pi/2}$ are two quadrature components shifted in phase by $\pi/2$ and specified by the phase angle ϕ. In the following, we will consider the special case of $\phi = 0$, in which $E_0 = E_1$ and $E_{\pi/2} = E_2$. The results can be generalized to an arbitrary ϕ.

The non-commuting quadrature components satisfy the Heisenberg uncertainty relation

$$\sqrt{\left\langle\left(\Delta\hat{E}_1\right)^2\right\rangle\left\langle\left(\Delta\hat{E}_2\right)^2\right\rangle} \geq \frac{1}{4}, \tag{6.6}$$

where $\langle(\Delta\hat{E}_i)^2\rangle$ is the variance of the ith quadrature component of the field being in a state $|\Psi\rangle$ and the factor 1/4 determines the vacuum level of the fluctuations. The Heisenberg uncertainty relation predicts that it is never possible to be absolutely precise in measuring one of two non-commuting observables.

It follows from the Heisenberg uncertainty relation (6.6) that there is no restriction on the magnitude of each of the variances $\langle(\Delta\hat{E}_1)^2\rangle$ and $\langle(\Delta\hat{E}_2)^2\rangle$ as long as the inequality in (6.6) is satisfied. For example, we have shown in Chapter 4 that the variances of the field in a thermal or in a Fock state (random phase fluctuating field) are

$$\left\langle\left(\Delta\hat{E}_1\right)^2\right\rangle > \frac{1}{4} \quad \text{and} \quad \left\langle\left(\Delta\hat{E}_2\right)^2\right\rangle > \frac{1}{4}. \tag{6.7}$$

Thus, for the field in a thermal or in a Fock state the variances in both quadrature components are larger than that of the vacuum level.

For a field in the vacuum or coherent state, the variances are

$$\left\langle\left(\Delta\hat{E}_1\right)^2\right\rangle = \left\langle\left(\Delta\hat{E}_2\right)^2\right\rangle = \frac{1}{4}. \tag{6.8}$$

In this case, both variances are equal to the vacuum level of the fluctuations. Hence, the vacuum or coherent state of the field is a minimum uncertainty state with the fluctuations distributed symmetrically between the two quadratures, as illustrated in Fig. 6.1.

Now, we will define squeezed states of the EM field or a squeezed field, or simply squeezing. Namely, a squeezed field is characterized by fluctuations reduced below the vacuum level, that is, either

$$\left\langle\left(\Delta\hat{E}_1\right)^2\right\rangle < \frac{1}{4} \quad \text{or} \quad \left\langle\left(\Delta\hat{E}_2\right)^2\right\rangle < \frac{1}{4}, \tag{6.9}$$

such that the Heisenberg uncertainty relation is not violated.

In other words, squeezing is defined by the requirement that the variance of one of two non-commuting Hermitian operators must be less than half of the absolute value of their commutator. The variance of the other Hermitian operator is at the same

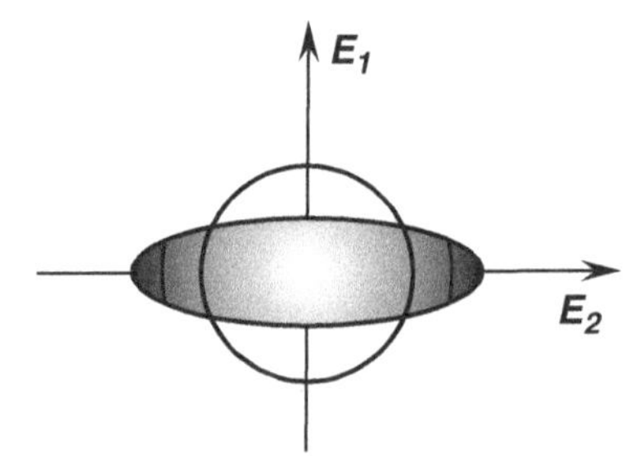

Figure 6.1 Graphical representation of the variances of a vacuum field and a squeezed vacuum field. The fluctuations of the vacuum field are isotropic which can be represented by a circle of radii 1/4. The squeezed vacuum field is characterized by reduced fluctuations in one of the quadrature components, which is represented by an ellipse with the length of the shorter axis reduced below 1/4 and the length of the longer axis respectively larger than 1/4.

time correspondingly larger than half of the absolute value of their commutator in order to preserve the Heisenberg uncertainty relation. This means that the reduction in the fluctuations of one of the quadrature components occurs at the expense of increased fluctuations in the other component. In the E_1, E_2 coordinates, squeezing is represented by an ellipse with the length of the shorter axis reduced below 1/4, as illustrated in Fig. 6.1.

The variances of the quadrature components can be expressed as

$$\left\langle\left(\Delta\hat{E}_i\right)^2\right\rangle = \left\langle : \left(\Delta\hat{E}_i\right)^2 : \right\rangle + \frac{1}{4}, \qquad i = 1, 2, \tag{6.10}$$

where the pair of colons (::) denote the normal ordering of the operators in which all the creation operators are placed to to the left of all the annihilation operators.

Proof.

$$\begin{aligned}\left\langle\left(\Delta\hat{E}_1\right)^2\right\rangle &= \frac{1}{4}\left\langle\left(\hat{a}+\hat{a}^\dagger\right)\left(\hat{a}+\hat{a}^\dagger\right)\right\rangle - \frac{1}{4}\left\langle\left(\hat{a}+\hat{a}^\dagger\right)\right\rangle^2 \\ &= \frac{1}{4}\left\langle\hat{a}\hat{a}+\hat{a}\hat{a}^\dagger+\hat{a}^\dagger\hat{a}+\hat{a}^\dagger\hat{a}^\dagger\right\rangle - \frac{1}{4}\left\langle\hat{a}+\hat{a}^\dagger\right\rangle^2 \\ &= \frac{1}{4}\left\langle 2\hat{a}^\dagger\hat{a}+\hat{a}\hat{a}+\hat{a}^\dagger\hat{a}^\dagger\right\rangle - \frac{1}{4}\left\langle\hat{a}+\hat{a}^\dagger\right\rangle^2 + \frac{1}{4} \\ &= \left\langle : \left(\Delta\hat{E}_1\right)^2 : \right\rangle + \frac{1}{4},\end{aligned} \tag{6.11}$$

as required. □

Similarly, we can show that

$$\left\langle \left(\Delta \hat{E}_2\right)^2 \right\rangle = \left\langle : \left(\Delta \hat{E}_2\right)^2 : \right\rangle + \frac{1}{4}. \tag{6.12}$$

Since the left-hand side of Eq. (6.10) has to be less than 1/4 for a squeezed field, another definition of a squeezed field is that the normally ordered variance of the quadrature component must be negative

$$\left\langle : \left(\Delta \hat{E}_i\right)^2 : \right\rangle < 0. \tag{6.13}$$

Consider the normally ordered variance $\langle : \left(\Delta \hat{E}_1\right)^2 : \rangle$, which can be written as

$$\begin{aligned}\left\langle : \left(\Delta \hat{E}_1\right)^2 : \right\rangle &= \frac{1}{4}\left\langle 2\hat{a}^\dagger \hat{a} + \hat{a}\hat{a} + \hat{a}^\dagger \hat{a}^\dagger \right\rangle - \frac{1}{4}\left\langle \hat{a} + \hat{a}^\dagger \right\rangle^2 \\ &= \frac{1}{2}\left\langle \Delta \hat{a}^\dagger \Delta \hat{a} \right\rangle + \frac{1}{4}\left(\left\langle \left(\Delta \hat{a}^\dagger\right)^2 \right\rangle + \langle (\Delta \hat{a})^2 \rangle\right) \\ &= \frac{1}{2}\left\langle \Delta \hat{a}^\dagger \Delta \hat{a} \right\rangle + \frac{1}{2}\left[|\langle \hat{a}\hat{a}\rangle| \cos\psi - |\langle \hat{a}\rangle|^2 \cos 2\theta\right],\end{aligned} \tag{6.14}$$

where $\Delta \hat{A} = \hat{A} - \langle \hat{A} \rangle$, and

$$\langle \hat{a}\hat{a}\rangle = |\langle \hat{a}\hat{a}\rangle|\, \mathrm{e}^{i\psi}, \quad \langle \hat{a}\rangle = |\langle \hat{a}\rangle|\, \mathrm{e}^{i\theta}. \tag{6.15}$$

It follows from Eq. (6.14) that the condition for squeezing is that the correlation functions $|\langle \hat{a}\hat{a}\rangle|$ and/or $|\langle \hat{a}\rangle|$ are non-zero. The variance might be negative if $0 > \cos\psi > -1$ and $1 > \cos(2\theta) > 0$, that is, when the second term in Eq. (6.14) is negative. The minimum value of $\langle : \left(\Delta \hat{E}_1\right)^2 : \rangle$, corresponding to maximum squeezing can be obtained for $\psi = \pi$ and $\theta = 0$. The correlation function $\langle \hat{a}\hat{a}\rangle$ can be produced by nonlinear two-photon processes whereas a non-zero amplitude $\langle \hat{a}\rangle$ can be produced by a coherent field. Thus, squeezing can be generated by two different processes, a nonlinear two-photon process or a coherent field.

Consider two simple examples illustrating the level of fluctuations in a single-mode EM field being in a photon number state and in a coherent state.

Example 6.1 (Field in a photon number state) *For the field in the photon number state $|n\rangle$, the correlation functions appearing in Eq. (6.14) are*

$$\left\langle \Delta \hat{a}^\dagger \Delta \hat{a} \right\rangle = n, \quad \langle \hat{a}\hat{a}\rangle = 0, \quad \langle \hat{a}\rangle = 0, \tag{6.16}$$

and then both the normally ordered variances are equal and positive

$$\left\langle : \left(\Delta \hat{E}_1\right)^2 : \right\rangle = \left\langle : \left(\Delta \hat{E}_2\right)^2 : \right\rangle = \frac{1}{2}n \geq 0. \qquad (6.17)$$

Hence, there is no squeezing in the field being in a Fock state. The fluctuations of the field amplitudes are isotropic and increase with the number of photons n. This property arises from the fact that a Fock state has complete uncertainty in phase.

Example 6.2 (Field in a coherent state) *As a second example, consider the field in the coherent state* $|\alpha\rangle$. *In this case, the correlation functions are*

$$\left\langle \Delta\hat{a}^{\dagger}\Delta\hat{a}\right\rangle = 0, \quad |\langle\hat{a}\hat{a}\rangle| = |\alpha|^2, \quad |\langle\hat{a}\rangle|^2 = |\alpha|^2. \qquad (6.18)$$

Hence, the normally ordered variances are

$$\left\langle : \left(\Delta \hat{E}_1\right)^2 : \right\rangle = \left\langle : \left(\Delta \hat{E}_2\right)^2 : \right\rangle = 0. \qquad (6.19)$$

Thus, for the field in the coherent state $|\alpha\rangle = \hat{D}(\alpha)|0\rangle$, *the fluctuations of the field amplitudes are isotropic and equal to the vacuum level of fluctuations. Note that the fluctuations are equal to the vacuum level independent of the amplitude* α *of the coherent field.*

6.3 Squeezed Coherent States

In connection with two-photon processes responsible for reduction of the field fluctuations, we may define squeezed states of the EM field in an alternative but equivalent way by introducing an unitary two-photon operator, called the *squeezed operator*

$$\hat{S}(s) = \exp\left\{\frac{1}{2}s^*\hat{a}^2 - \frac{1}{2}s\hat{a}^{\dagger 2}\right\}, \qquad (6.20)$$

where $s = r\exp(i\theta)$ and r is a real number. The parameter s determines the size of squeezing and depends on the type of the two-photon process.

The combined action of the squeezing operator $\hat{S}(s)$ and the displacement operator $\hat{D}(\alpha)$ on the vacuum state $|0\rangle$ generates a minimum uncertainty squeezed states, often called *squeezed coherent states*. There are, however, two equivalent but different definitions of the squeezed state. Yuen [27] defined the squeezed coherent states as

$$|\alpha, s\rangle = \hat{S}(s)\,\hat{D}(\alpha)\,|0\rangle\,, \tag{6.21}$$

whereas Caves [28] defined the squeezed states as

$$|\alpha, s\rangle = \hat{D}(\alpha)\,\hat{S}(s)\,|0\rangle\,. \tag{6.22}$$

Since the operators $\hat{D}(\alpha)$ and $\hat{S}(s)$ do not commute, the definitions (6.21) and (6.22) are not equal. However, the definitions are related as

$$\hat{S}(s)\,\hat{D}(\alpha) = \hat{S}(s)\,\hat{D}(\alpha)\,\hat{S}^{\dagger}(s)\,\hat{S}(s) = \hat{D}(\beta)\,\hat{S}(s)\,, \tag{6.23}$$

where

$$\beta = \alpha\cosh r + \alpha^{*}\mathrm{e}^{i\theta}\sinh r. \tag{6.24}$$

The Yuen and Caves definitions for coherent squeezed states can be used interchangeably, one or the other my be particularly convenient for a specific problem. The two definitions lead to the same results for squeezing, but produce different results for antibunching.

For $\alpha = 0$, the squeezed coherent states (6.21) and (6.22) reduce to the *squeezed vacuum state*

$$|0, s\rangle = \hat{S}(s)\,|0\rangle\,, \tag{6.25}$$

which has an interesting property, $\hat{b}\,|0, s\rangle = 0$, that it is an eigenstate with a zero eigenvalue of the annihilation operators $\hat{b}$, which can be obtained from the operators $\hat{a}$ and $\hat{a}^{\dagger}$ by the unitary transformation, called the Bogoliubov transformation

$$\begin{aligned}\hat{b} &= \hat{S}(s)\hat{a}\hat{S}^{\dagger}(s) = \hat{a}\cosh r + \hat{a}^{\dagger}\mathrm{e}^{i\theta}\sinh r,\\ \hat{b}^{\dagger} &= \hat{S}(s)\hat{a}^{\dagger}\hat{S}^{\dagger}(s) = \hat{a}^{\dagger}\cosh r + \hat{a}\mathrm{e}^{-i\theta}\sinh r.\end{aligned} \tag{6.26}$$

Proof. Applying the identity relation

$$\mathrm{e}^{\lambda\hat{A}}\,\hat{B}\mathrm{e}^{-\lambda\hat{A}} = \hat{B} + \frac{\lambda}{1!}[\hat{A}, \hat{B}] + \frac{\lambda^{2}}{2!}\left[\hat{A}, [\hat{A}, \hat{B}]\right] + \cdots \tag{6.27}$$

and introducing a notation $\hat{A} = (s^{*}\hat{a}^{2} - s\hat{a}^{\dagger 2})/2$, we can write the unitary transformation as

$$\hat{b} = \hat{S}(s)\hat{a}\hat{S}^{\dagger}(s) = \mathrm{e}^{\hat{A}}\hat{a}\mathrm{e}^{-\hat{A}} = \hat{a} + [\hat{A}, \hat{a}] + \frac{1}{2!}\left[\hat{A}, [\hat{A}, \hat{a}]\right] + \cdots \tag{6.28}$$

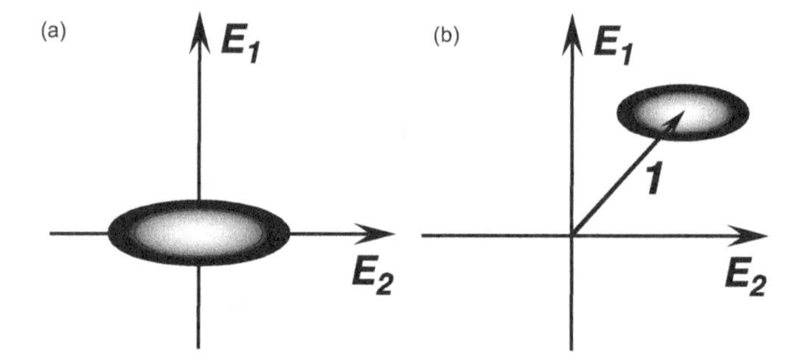

Figure 6.2 Example of (a) squeezed vacuum state and (b) squeezed coherent state with the variance of the quadrature component E_1 reduced (squeezed) below the vacuum level.

Using the well-known commutation relations

$$\left[\hat{a}, \left(\hat{a}^\dagger\right)^n\right] = n\left(\hat{a}^\dagger\right)^{n-1}, \qquad \left[\hat{a}^n, \hat{a}^\dagger\right] = n\hat{a}^{n-1}, \tag{6.29}$$

we get

$$[\hat{A}, \hat{a}] = \left[\frac{1}{2}s^*\hat{a}^2 - \frac{1}{2}s\hat{a}^{\dagger 2}, \hat{a}\right] = s\hat{a}^\dagger, \tag{6.30}$$

and

$$[\hat{A}, \hat{a}^\dagger] = \left[\frac{1}{2}s^*\hat{a}^2 - \frac{1}{2}s\hat{a}^{\dagger 2}, \hat{a}^\dagger\right] = s^*\hat{a}. \tag{6.31}$$

Upon substitution of the results (6.30) and (6.31) into Eq. (6.28), we obtain

$$\begin{aligned}\hat{b} &= \hat{a}\left(1 + \frac{r^2}{2!} + \frac{r^4}{4!} + \cdots\right) + \mathrm{e}^{i\theta}\hat{a}^\dagger\left(r + \frac{r^3}{3!} + \frac{r^5}{5!} + \cdots\right)\\ &= \hat{a}\cosh r + \hat{a}^\dagger \mathrm{e}^{i\theta}\sinh r,\end{aligned} \tag{6.32}$$

as required. □

Examples of squeezed vacuum and squeezed coherent states are shown in Fig. 6.2. The state is represented by a point located at the origin (vacuum state) or in a distance α from the origin (coherent state) surrounded by an ellipse representing the magnitude of the fluctuations.

We can use the operators $\hat{b}$ and $\hat{b}^\dagger$ to obtain the formula for the squeezed vacuum state (6.25) in the photon number representation. We expand the squeezed vacuum state in terms of the Fock states as

$$|0, s\rangle = \sum_n c_n|n\rangle. \tag{6.33}$$

Since

$$\hat{b}\,|0, s\rangle = \text{“0”}\,|0, s\rangle \tag{6.34}$$

and using Eq. (6.26), we obtain

$$\cosh(r)\sum_n c_n\sqrt{n}|n-1\rangle + \mathrm{e}^{i\theta}\sinh(r)\sum_n c_n\sqrt{n+1}|n+1\rangle$$
$$= \sum_n c_n\text{“0”}|n\rangle. \tag{6.35}$$

Applying the orthogonality property of the Fock states, we find that the coefficients c_n satisfy a recurrence relation

$$\cosh(r)c_{n+1}\sqrt{n+1} + \mathrm{e}^{i\theta}\sinh(r)c_{n-1}\sqrt{n} = \text{“0”}c_n. \tag{6.36}$$

We keep the zero term on the right-hand side of Eq. (6.36) to show that the recurrence relation for the coefficients c_n can be transformed to a recurrence relation identical to the familiar recurrence relation for the Hermite polynomials.

Substituting $\mu = \cosh(r)$, $\nu = \exp(i\theta)\sinh(r)$ and

$$c_n = \frac{1}{\sqrt{2^n n!}}\left(\frac{\nu}{\mu}\right)^{\frac{1}{2}n} a_n, \tag{6.37}$$

we obtain

$$\mu\sqrt{n+1}c_{n+1} = \sqrt{\frac{\mu\nu}{2}}\frac{1}{\sqrt{2^n n!}}\left(\frac{\nu}{\mu}\right)^{\frac{1}{2}n} a_{n+1},$$
$$\nu\sqrt{n}c_{n-1} = 2n\sqrt{\frac{\mu\nu}{2}}\frac{1}{\sqrt{2^n n!}}\left(\frac{\nu}{\mu}\right)^{\frac{1}{2}n} a_{n-1}. \tag{6.38}$$

Hence, the recurrence relation (6.36) takes a simple form

$$2na_{n-1} + a_{n+1} = 2za_n, \tag{6.39}$$

where $z = \text{“0”}/\sqrt{2\mu\nu}$.

The above recurrence relation is identical to that for the Hermite polynomials $H_n(z)$. Hence, $a_n \equiv H_n(0)$, and then

$$c_n = \frac{a_0}{\sqrt{2^n n!}}\left(\frac{\nu}{\mu}\right)^{\frac{1}{2}n} H_n(0), \tag{6.40}$$

where a_0 is the normalization constant.

Thus, the squeezed vacuum state can be written as

$$|0, s\rangle = \sum_n \frac{a_0}{\sqrt{2^n n!}}\left(\frac{\nu}{\mu}\right)^{\frac{1}{2}n} H_n(0)|n\rangle. \tag{6.41}$$

From the normalization of $|0, s\rangle$ and using the property of the Hermite polynomials

$$\sum_n \frac{1}{n!}\left(\frac{t}{2}\right)^n H_n^2(z) = \left(1-t^2\right)^{-\frac{1}{2}} \mathrm{e}^{\frac{2z^2t}{1-t^2}}, \tag{6.42}$$

we find that

$$1 = |a_0|^2 \sum_n \frac{1}{2^n n!}\left(\frac{\nu}{\mu}\right)^n H_n^2(0) = |a_0|^2\left(1-\frac{\nu^2}{\mu^2}\right)^{-\frac{1}{2}} = \mu|a_0|^2. \tag{6.43}$$

Hence, $a_0 = 1/\sqrt{\mu}$, and the squeezed vacuum state takes the form

$$|0, s\rangle = \sum_n \frac{1}{\sqrt{\mu}}\frac{1}{\sqrt{2^n n!}}\left(\frac{\nu}{\mu}\right)^{\frac{1}{2}n} H_n(0)|n\rangle. \tag{6.44}$$

However,

$$H_{2n}(0) = (-1)^n \frac{(2n)!}{n!} \quad \text{and} \quad H_{2n+1}(0) = 0, \tag{6.45}$$

and then we obtain

$$\begin{aligned} |0, s\rangle &= \sum_n \frac{1}{\sqrt{\mu}}\frac{1}{\sqrt{2^{2n}(2n)!}}\left(\frac{\nu}{\mu}\right)^n (-1)^n \frac{(2n)!}{n!}|2n\rangle \\ &= \sum_n \frac{\left(-\mathrm{e}^{i\theta}\tanh r\right)^n}{\sqrt{\cosh r}} \frac{(2n!)^{\frac{1}{2}}}{2^n n!}|2n\rangle. \end{aligned} \tag{6.46}$$

It follows from Eq. (6.46) that the squeezed vacuum state is a superposition of only *even* Fock states. This clearly shows the two-photon nature of the squeezed states.

Having the photon number representation of the squeezed vacuum state, we can find the coherent state representation of $|0, s\rangle$.

Inserting the completeness relation (4.87) into Eq. (6.44), we obtain

$$\begin{aligned} |0, s\rangle &= \frac{1}{\pi}\sum_n \frac{1}{\sqrt{\mu}}\frac{1}{\sqrt{2^n n!}}\left(\frac{\nu}{\mu}\right)^{\frac{1}{2}n} H_n(0)\int d^2\alpha \langle\alpha|n\rangle|\alpha\rangle \\ &= \frac{1}{\pi\sqrt{\mu}}\int d^2\alpha \sum_n \frac{\alpha^{*n}}{n!}\left(\frac{\nu}{2\mu}\right)^{\frac{1}{2}n} \mathrm{e}^{-\frac{1}{2}|\alpha|^2} H_n(0)|\alpha\rangle \\ &= \frac{1}{\pi\sqrt{\mu}}\int d^2\alpha \sum_n \frac{1}{n!}\left(\frac{\nu\alpha^{*2}}{2\mu}\right)^{\frac{1}{2}n} \mathrm{e}^{-\frac{1}{2}|\alpha|^2} H_n(0)|\alpha\rangle. \end{aligned} \tag{6.47}$$

Since

$$\sum_n H_n(0)\frac{t^n}{n!} = \mathrm{e}^{-t^2}, \tag{6.48}$$

we finally obtain

$$|0, s\rangle = \frac{1}{\pi\sqrt{\mu}}\int d^2\alpha \exp\left[-\frac{1}{2}|\alpha|^2 - \frac{\nu\alpha^{*2}}{2\mu}\right]|\alpha\rangle. \tag{6.49}$$

The dependence of the state on α^{*2} again confirms the two-photon nature of the squeezed vacuum states.

Using one of the definitions of the squeezed coherent states, Eq. (6.21) or Eq. (6.22), we find that expectation values of the field operators in the squeezed coherent state are given by

$$\begin{aligned}
\langle\hat{a}\rangle &= \alpha_s, \quad \langle\hat{a}^\dagger\rangle = \alpha_s^*, \\
\langle\hat{a}^\dagger\hat{a}\rangle &= |\alpha_s|^2 + \sinh^2(r), \\
\langle\hat{a}^2\rangle &= \alpha_s^2 - \mathrm{e}^{i\theta}\sinh(r)\cosh(r), \\
\langle\hat{a}^{\dagger 2}\rangle &= \alpha_s^{*2} - \mathrm{e}^{-i\theta}\sinh(r)\cosh(r),
\end{aligned} \tag{6.50}$$

where $\alpha_s = \alpha\cosh(r) - \alpha^*\mathrm{e}^{i\theta}\sinh(r)$.

Then the normally ordered variance (6.14) takes the form

$$\left\langle : \left(\Delta\hat{E}_1\right)^2 : \right\rangle = \frac{1}{2}\sinh^2(r) - \frac{1}{2}\cos(\theta)\sinh(r)\cosh(r). \tag{6.51}$$

Note that the variance (6.51) is independent of α despite the fact that the correlation functions (6.50) depend explicitly on α.

Simple manipulations with the sinh and cosh functions lead to

$$\begin{aligned}
\left\langle : \left(\Delta\hat{E}_1\right)^2 : \right\rangle &= \frac{1}{2}\sinh^2(r) - \frac{1}{2}\cos(\theta)\sinh(r)\cosh(r) \\
&= \frac{1}{2}\left\{\left(\frac{\mathrm{e}^r - \mathrm{e}^{-r}}{2}\right)^2 - \cos(\theta)\,\frac{\mathrm{e}^r - \mathrm{e}^{-r}}{2}\,\frac{\mathrm{e}^r + \mathrm{e}^{-r}}{2}\right\} \\
&= \frac{1}{8}\left\{\mathrm{e}^{2r} + \mathrm{e}^{-2r} - 2 - \left(\mathrm{e}^{2r} - \mathrm{e}^{-2r}\right)\cos(\theta)\right\} \\
&= \frac{1}{8}\left\{\mathrm{e}^{2r}\left[1 - \cos(\theta)\right] + \mathrm{e}^{-2r}\left[1 + \cos(\theta)\right] - 2\right\} \\
&= \frac{1}{4}\left\{\mathrm{e}^{2r}\sin^2\left(\frac{1}{2}\theta\right) + \mathrm{e}^{-2r}\cos^2\left(\frac{1}{2}\theta\right) - 1\right\}.
\end{aligned} \tag{6.52}$$

Hence, for $\theta = 0$, we find

$$\left\langle : \left(\Delta \hat{E}_1\right)^2 : \right\rangle = \frac{1}{4}\left(e^{-2r} - 1\right) < 0, \tag{6.53}$$

which shows that the fluctuations in the $\hat{E}_1$ component of the field are less than for the coherent field, thus the field is squeezed.

Similarly, we can show that

$$\left\langle : \left(\Delta \hat{E}_2\right)^2 : \right\rangle = \frac{1}{4}\left(e^{2r} - 1\right), \tag{6.54}$$

and then we can show that the squeezed coherent states satisfy the minimum uncertainty product. This fact justifies the use of the name *coherent* squeezed states.

Since the mean photon number in the squeezed vacuum state $\langle n\rangle = \langle \hat{a}^{\dagger}\hat{a}\rangle = \sinh^2 r$, the photon number probability distribution P_n for the squeezed vacuum state (6.46) is given by

$$\begin{aligned} P_{2n} &= \frac{1}{\cosh r}\frac{(2n)!}{(n!)^2 2^n}(\tanh r)^{2n} \\ &= \frac{1}{\sqrt{1+\langle n\rangle}}\frac{(2n)!}{(n!)^2 2^n}\left(\frac{\langle n\rangle}{1+\langle n\rangle}\right)^n, \\ P_{2n+1} &= 0. \end{aligned} \tag{6.55}$$

All of the probabilities corresponding to odd number of photons are zero.

Note that the probability distribution for even terms is the same as that for a thermal field except for a factor $(2n)!/(n!)^2 2^n$. This factor may give observable differences between the thermal and squeezed vacuum field distributions.

Moreover, non-zero values for even terms clearly show the two-photon nature of the squeezed vacuum field. Note that the probabilities sum to unity, as required, since the n-dependent factors in Eq. (6.55) are the terms of the binomial expansion of $(1 - \tanh^2 r)^{-1/2} = \cosh r$.

In Fig. 6.3, we plot the photon number distribution (6.55) and compare it with the photon number distribution of a thermal field. It is seen from Fig. 6.3 that the photon number distribution of the squeezed vacuum state exhibits unusual oscillations and its amplitude decays with n. The distribution peaks sharply at $n = 0$ and has a very long tail, similar to a thermal distribution. The

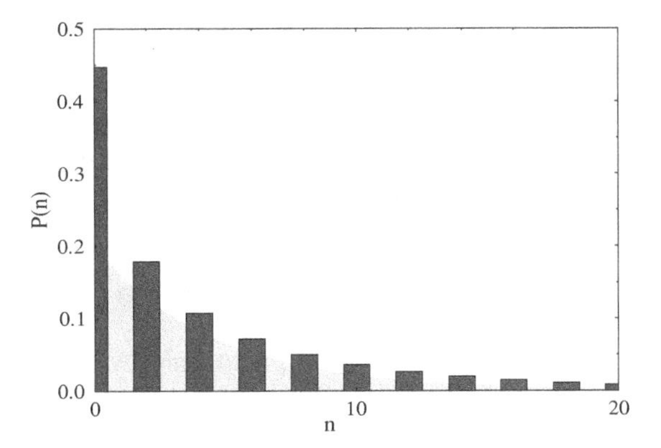

Figure 6.3 The photon number distribution of the squeezed vacuum state (plotted as bars) for $\langle n \rangle = 4$. The shaded area corresponds to the photon number distribution of a thermal field with the same mean photon number.

probabilities corresponding to odd number of photons are zero. This shows explicitly that squeezed photons are emitted in pairs and that these pairs are emitted at random.

Squeezed states are produced in two-photon nonlinear optical processes in which a classical laser field drives a nonlinear medium. These processes are distinguished by the simultaneous or nearly simultaneous production of a pair of photons in momentum conserving phase-matched modes. In the degenerate processes, where a single output mode plays the role of both signal and idler, squeezed states may be produced with reduced fluctuations in one of the output quadrature component. The situation is somewhat more complicated for non-degenerate processes, where the signal and idler are distinct. Here, the individual output modes display isotropic distribution of the fluctuations, similar to those usually associated with chaotic or thermal fields. However, the combined two-mode state can exhibit reduced (squeezed) fluctuations in modes formed by superposing the signal and idler modes.

We distinguish two different types of parametric down conversion processes in both the degenerate and non-degenerate cases. One is *parametric amplifier*, when the parametric process appears above the threshold, that is, the output modes have non-zero coherent amplitudes. The other type is a *parametric oscillator*, when the parametric process appear below the threshold, that is, the output modes are in a thermal state.

Example 6.3 (Squeezing in a degenerate parametric amplifier) *Consider a degenerate parametric amplifier (DPA) that can produce photon pairs in a single mode of frequency ω. The Hamiltonian of the DPA process can be written as*

$$\hat{H} = \hbar\omega\hat{a}^{\dagger}\hat{a} - \frac{1}{2}i\chi\hbar\left(\hat{a}^{2}e^{2i\omega t} - \hat{a}^{\dagger 2}e^{-2i\omega t}\right), \tag{6.56}$$

where the energy is provided by a pump field, which is treated classically.[a]

With the Hamiltonian (6.56), and working in the interaction picture, we find Heisenberg equations of motion for the field $\hat{a}$ and $\hat{a}^{\dagger}$ operators

$$\frac{d}{dt}\tilde{\hat{a}} = \frac{1}{i\hbar}\left[\tilde{\hat{a}}, \hat{H}\right] = \chi\tilde{\hat{a}}^{\dagger},$$
$$\frac{d}{dt}\tilde{\hat{a}}^{\dagger} = \frac{1}{i\hbar}\left[\tilde{\hat{a}}^{\dagger}, \hat{H}\right] = \chi\tilde{\hat{a}}, \tag{6.57}$$

where

$$\tilde{\hat{a}} = \hat{a}e^{i\omega t} \qquad \text{and} \qquad \tilde{\hat{a}}^{\dagger} = \hat{a}^{\dagger}e^{-i\omega t} \tag{6.58}$$

are slowly varying dynamical operators which are free from the rapid oscillations at the optical frequency ω.

The equations of motion (6.57) have simple solutions

$$\tilde{\hat{a}}(t) = \tilde{\hat{a}}(0)\cosh(\chi t) + \tilde{\hat{a}}^{\dagger}(0)\sinh(\chi t). \tag{6.59}$$

Using the equations of motion (6.57), we can write equations of motion for the Hermitian components $\hat{E}_1$ and $\hat{E}_2$ as

$$\frac{d}{dt}\hat{E}_1 = \chi\hat{E}_1, \qquad \text{and} \qquad \frac{d}{dt}\hat{E}_2 = -\chi\hat{E}_2. \tag{6.60}$$

Solutions of these equations are readily obtained, and are given by

$$\hat{E}_1(t) = \hat{E}_1(0)\,e^{\chi t},$$
$$\hat{E}_2(t) = \hat{E}_2(0)\,e^{-\chi t}. \tag{6.61}$$

Then, the variances of the field amplitudes are

$$\left\langle\left(\Delta\hat{E}_1(t)\right)^2\right\rangle = \left\langle\left(\Delta\hat{E}_1(0)\right)^2\right\rangle e^{2\chi t}, \tag{6.62}$$

---

[a] In the Hamiltonian, we have omitted the $\hbar\omega/2$ term since it is a constant and it does not affect the dynamics of the field mode.

and

$$\left\langle \left(\Delta \hat{E}_2(t)\right)^2 \right\rangle = \left\langle \left(\Delta \hat{E}_2\,(0)\right)^2 \right\rangle e^{-2\chi t}. \tag{6.63}$$

It is evident from the above equations that the noise reduction depends on the strength of the nonlinearity χ *and the interaction time. Taking that initially at* $t = 0$ *the field was in the vacuum (coherent) state,* $\langle (\Delta \hat{E}_1(0))^2 \rangle = \langle (\Delta \hat{E}_2(0))^2 \rangle = 1/4$*, the* $\hat{E}_2$ *quadrature component becomes squeezed for* $t > 0$*. Note the product of the variances satisfies the minimum uncertainty relation.*

6.4 Multi-Mode Squeezed States

Our analysis so far has been concerned with a single-mode squeezing, the reduction of fluctuations in single-mode fields. Consider now a non-degenerate parametric oscillator (NDPO) that can produce multi-mode squeezed fluctuations in the correlation between the signal and idler modes. Mathematically, one can create a multi-mode squeezed state using the unitary multi-mode squeezed operator

$$|\psi_k\rangle = D\,(\omega_k)\, S\,(s\,(\omega_k))\, |0_1, \ldots, 0_k, \ldots\rangle\,, \tag{6.64}$$

with

$$S\,(s\,(\omega_k)) = \sum_k \exp\left\{\frac{1}{2} s^*(\omega_k)\, \hat{a}\,(\omega_c + \omega_k)\, \hat{a}\,(\omega_c - \omega_k) - \text{H.c.}\right\}, \tag{6.65}$$

where $\hat{a}\,(\omega_c + \omega_k)$ and $\hat{a}\,(\omega_c - \omega_k)$ are the annihilation operators of the signal $(\omega_c + \omega_k)$ and idler $(\omega_c - \omega_k)$ modes, respectively, and ω_c is the frequency of the pumping field. In practice, multi-mode squeezed states can be generated in both NDPA and NDPO processes.

In the multi-mode squeezed state, apart from the single-mode correlation functions, there are non-zero correlation functions between two different modes k and k', $(k \neq k')$

$$\begin{aligned}
\langle \hat{a}_k \hat{a}_{k'} \rangle &= \alpha_k \alpha_{k'} - \mathrm{e}^{i\theta_k} \sinh(r_k) \cosh(r_{k'}), \\
\left\langle \hat{a}_k^\dagger \hat{a}_{k'}^\dagger \right\rangle &= \alpha_k^* \alpha_{k'}^* - \mathrm{e}^{-i\theta_k} \sinh(r_k) \cosh(r_{k'}).
\end{aligned} \tag{6.66}$$

However, for $k = k'$ the two-photon correlations are

$$\langle \hat{a}_k \hat{a}_k \rangle = \alpha_k^2, \qquad \left\langle \hat{a}_k^\dagger \hat{a}_k^\dagger \right\rangle = \alpha_k^{*2}, \tag{6.67}$$

and the average number of photons in the mode k is

$$\left\langle \hat{a}_k^\dagger \hat{a}_k \right\rangle = |\alpha_k|^2 + \sinh^2(r_k). \tag{6.68}$$

Let us assume that the field has no coherent amplitude ($\alpha_k = 0$). Then the only non-zero correlations are

$$\begin{aligned} \left\langle \hat{a}_k^\dagger \hat{a}_k \right\rangle &= \sinh^2(r_k) = N_k, \\ \langle \hat{a}_k \hat{a}_{k'} \rangle &= -\mathrm{e}^{i\theta_k} \sinh(r_k) \cosh(r_{k'}) = M_{kk'}, \\ \left\langle \hat{a}_k^\dagger \hat{a}_{k'}^\dagger \right\rangle &= -\mathrm{e}^{-i\theta_k} \sinh(r_k) \cosh(r_{k'}) = M_{kk'}^*. \end{aligned} \tag{6.69}$$

The parameters N_k and $M_{kk'}$ are not independent and satisfy an inequality

$$|M_{kk'}|^2 \leq N_k \left(N_{k'} + 1\right) = N_{k'} \left(N_k + 1\right). \tag{6.70}$$

Proof. Consider an operator $\hat{A} = z_k \hat{a}_k + z_{k'}^* \hat{a}_{k'}^\dagger$. Then from the definition of the mean value

$$\begin{aligned} \left\langle \hat{A}^\dagger \hat{A} \right\rangle = |z_k|^2 \left\langle \hat{a}_k^\dagger \hat{a}_k \right\rangle + z_k z_{k'} \langle \hat{a}_k \hat{a}_{k'} \rangle \\ + z_k^* z_{k'}^* \left\langle \hat{a}_k^\dagger \hat{a}_{k'}^\dagger \right\rangle + |z_{k'}|^2 \left\langle \hat{a}_{k'} \hat{a}_{k'}^\dagger \right\rangle \geq 0. \end{aligned} \tag{6.71}$$

The above equation can be written in a matrix form as

$$\begin{aligned} \left\langle \hat{A}^\dagger \hat{A} \right\rangle &= \left(z_k^* z_{k'}\right) \begin{pmatrix} \left\langle \hat{a}_k^\dagger \hat{a}_{k'} \right\rangle & \left\langle \hat{a}_k^\dagger \hat{a}_{k'}^\dagger \right\rangle \\ \langle \hat{a}_k \hat{a}_{k'} \rangle & \left\langle \hat{a}_{k'} \hat{a}_k^\dagger \right\rangle \end{pmatrix} \begin{pmatrix} z_k \\ z_{k'}^* \end{pmatrix} \\ &= \left(z_k^* z_{k'}\right) \begin{pmatrix} N_k & M_{kk'}^* \\ M_{kk'} & N_{k'} + 1 \end{pmatrix} \begin{pmatrix} z_k \\ z_{k'}^* \end{pmatrix}. \end{aligned} \tag{6.72}$$

Since $\left\langle \hat{A}^\dagger \hat{A} \right\rangle \geq 0$, the determinant of the 2×2 matrix is positive, so that

$$|M_{kk'}|^2 \leq N_k \left(N_{k'} + 1\right), \tag{6.73}$$

as required. □

We might ask whether it is enough to create two-photon correlations, $\langle \hat{a}_k \hat{a}_{k'} \rangle \neq 0$, in order to get squeezing between the modes.

To answer this question, consider a multi-mode field in a squeezed vacuum state

$$\hat{E}^{(+)} = \sum_k \lambda_k \hat{a}_k \, \mathrm{e}^{-i\omega_k t}, \tag{6.74}$$

where

$$\lambda_k = i\sqrt{\frac{\hbar\omega_k}{2\varepsilon_0 V}}\, \vec{e}_k \, \mathrm{e}^{i\vec{k}\cdot\vec{r}}. \tag{6.75}$$

We can define two Hermitian field operators, equivalent to the quadrature components

$$\hat{E}_1 = \frac{1}{2}\left(\hat{E}^{(+)} + \hat{E}^{(-)}\right), \qquad \hat{E}_2 = \frac{1}{2i}\left(\hat{E}^{(+)} - \hat{E}^{(-)}\right). \tag{6.76}$$

Then, in the squeezed vacuum state, the average amplitudes are

$$\left\langle \hat{E}_1 \right\rangle = 0, \qquad \left\langle \hat{E}_2 \right\rangle = 0, \tag{6.77}$$

and the normally ordered variance is

$$\left\langle : \left(\Delta \hat{E}_1\right)^2 : \right\rangle = \frac{1}{4}\{2\left\langle \hat{E}^{(-)}\hat{E}^{(+)}\right\rangle + \left\langle \hat{E}^{(-)}\hat{E}^{(-)}\right\rangle + \left\langle \hat{E}^{(+)}\hat{E}^{(+)}\right\rangle\}. \tag{6.78}$$

However, using the multi-mode description of the field, Eq. (6.74), and the correlation functions (6.69), we find that

$$\left\langle \hat{E}^{(-)}\hat{E}^{(+)}\right\rangle = \sum_k |\lambda_k|^2 N_k,$$
$$\left\langle \hat{E}^{(+)}\hat{E}^{(+)}\right\rangle = \sum_{kk'} \lambda_k \lambda_{k'} M_{kk'} \mathrm{e}^{-i(\omega_k+\omega_{k'})t}. \tag{6.79}$$

Assume that the modes are correlated with a central frequency ω_c such that $\omega_k + \omega_{k'} = 2\omega_c$. Then

$$\left\langle \hat{E}^{(+)}\hat{E}^{(+)}\right\rangle = \sum_k \lambda_k \lambda_{2k_c-k} M_{k,2k_c-k} \mathrm{e}^{-2i\omega_c t}. \tag{6.80}$$

For simplicity, we assume that $\lambda_k = \lambda_{2k_c-k}$, and $M_{k,2k_c-k} = M_k$. Then

$$\begin{aligned}\left\langle : \left(\Delta \hat{E}_1\right)^2 : \right\rangle &= \frac{1}{4}\sum_k |\lambda_k|^2 \left\{2N_k + M_k \mathrm{e}^{-2i\omega_c t} + M_k^* \mathrm{e}^{2i\omega_c t}\right\} \\ &= \frac{1}{4}\sum_k |\lambda_k|^2 \left\{2N_k + |M_k|\, \mathrm{e}^{i\psi} + |M_k|\, \mathrm{e}^{-i\psi}\right\} \\ &= \frac{1}{2}\sum_k |\lambda_k|^2 \left(N_k + |M_k| \cos\psi\right),\end{aligned} \tag{6.81}$$

where $\psi = \phi_s - 2\omega_c t$, and ϕ_s is the phase of the two-photon correlations.

Thus, for $\psi = \pi$, the normally ordered variance reduces to

$$\left\langle : \left(\Delta \hat{E}_1\right)^2 : \right\rangle = \frac{1}{2} \sum_k |\lambda_k|^2 \left(N_k - |M_k|\right). \qquad (6.82)$$

Hence, we get squeezing when the two-photon correlations over-weight the number of photons in the field modes, $|M_k| > N_k$, and there is no squeezing when $|M_k| \leq N_k$.

From the matrix representation of the mean value $\langle \hat{A}^\dagger \hat{A} \rangle$:

$$\langle \hat{A}^\dagger \hat{A} \rangle = \left(z_k^* z_{k'}\right) \begin{pmatrix} \left\langle \hat{a}_k^\dagger \hat{a}_{k'} \right\rangle & \left\langle \hat{a}_k^\dagger \hat{a}_{k'}^\dagger \right\rangle \\ \langle \hat{a}_k \hat{a}_{k'} \rangle & \left\langle \hat{a}_{k'} \hat{a}_k^\dagger \right\rangle \end{pmatrix} \begin{pmatrix} z_k \\ z_{k'}^* \end{pmatrix}, \qquad (6.83)$$

we see that with a classical field, for which $\langle \hat{a}_k^\dagger \hat{a}_{k'} \rangle = \langle a_k^* a_{k'} \rangle = N_k$ and $\langle \hat{a}_{k'} \hat{a}_k^\dagger \rangle = \langle a_{k'} a_k^* \rangle = N_k$, we obtain $|M_k|^2 \leq N_k^2$. However, for a quantum field, for which $\langle \hat{a}_k \hat{a}_{k'}^\dagger \rangle = N_k + 1$, we obtain $|M_k|^2 \leq N_k (N_k + 1)$. Therefore, squeezing which results from $|M_k| > N_k$ is a quantum effect.

In the literature, a field with $|M_k| \leq N_k$ is called *classically squeezed field*, and a field with $N_k < |M_k| \leq \sqrt{N_k (N_k + 1)}$ is called a *quantum squeezed field* [29]. The field with $|M_k| \leq N_k$ is called classically squeezed field since it exhibits anisotropic distribution of the noise, but the noise is not reduced below the vacuum level. In this sense, the classically squeezed field is always clearly distinguishable from its quantum mechanical counterpart.

In summary, we have seen that an anisotropic distribution of field fluctuations (noise) can be achieved with both classical and quantum fields, but reduction of the fluctuations below the vacuum (quantum) limit, $\langle : (\Delta \hat{E}_1)^2 : \rangle < 0$ or $\langle : (\Delta \hat{E}_2)^2 : \rangle < 0$, can be achieved *only* with a quantum field.

6.5 Squeezed States of Atomic Spin Variables

In the previous three sections, we dealt with squeezed states of the field boson annihilation and creation operators. The idea of squeezed states or squeezing can be extended to other quantum systems [30, 31]. Of particular interest is the description of squeezing in the atomic spin variables S_x, S_y and S_z. Since

$$[S_x, S_y] = iS_z, \tag{6.84}$$

the spin components satisfy the Heisenberg uncertainty relation

$$\sqrt{\langle(\Delta S_x)^2\rangle\langle(\Delta S_y)^2\rangle} \geq \frac{1}{2}|\langle S_z\rangle|. \tag{6.85}$$

This uncertainty can be thought of as due to the impossibility of simultaneous measurement of all three components of the atomic spin. Note that the uncertainty relation for the spin operators is fundamentally different from that for the boson operators. This is because for the spin operators the right-hand side of the uncertainty relation, that is equal to $|\langle S_z\rangle|/2$, depends on the state of the system. This means that the quantum level of the fluctuations of the spin components may vary during the evolution of the atomic system.

In analogy to squeezing in bosonic variables, squeezing in the spin atomic variables is defined as [32, 33]

$$\langle(\Delta S_i)^2\rangle < \frac{1}{2}|\langle S_z\rangle|, \qquad i = x, y. \tag{6.86}$$

Thus, squeezing in the atomic spin variables means reduction of quantum fluctuations in one of the components of the atomic dipole moment below the spin quantum limit $|\langle S_z\rangle|/2$, as illustrated in Fig. 6.4. In literature, squeezing in the context of the definition (6.86) is called the *natural* definition of squeezing. This has been introduced to distinguish between squeezing and spin squeezing, which is discussed in the next section.

As a measure of degree of natural squeezing in the atomic spin components, we can introduce a parameter

$$\xi_i = \frac{\langle(\Delta S_i)^2\rangle}{\frac{1}{2}|\langle S_z\rangle|}. \tag{6.87}$$

When $\xi_i < 1$, we say that the ith component of the atomic spin is squeezed, and $\xi_i = 0$ corresponds to maximum squeezing.

As an example illustrating the idea of squeezing in atomic variables, consider the fluctuations of the spin operators of a single two-level atom. The variance, for instance of the x component of the atomic spin can be written as

$$\begin{aligned}\langle(\Delta S_x)^2\rangle &= \frac{1}{4}\langle(S^+ + S^-)^2\rangle - \frac{1}{4}\langle S^+ + S^-\rangle^2 \\ &= \frac{1}{4} - \frac{1}{4}\langle S^+ + S^-\rangle^2.\end{aligned} \tag{6.88}$$

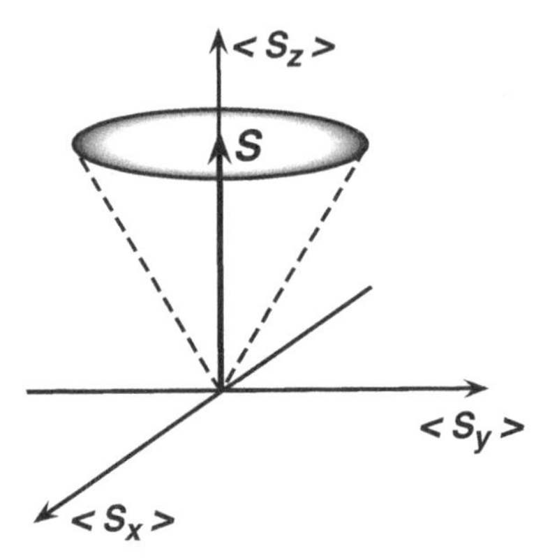

Figure 6.4 Graphical representation of squeezed fluctuations of an average atomic spin S oriented in the z-direction. The ellipse represents the fluctuations of the spin that are reduced (squeezed) in the x-direction with the corresponding increase of the fluctuations in the y-direction.

Since $\frac{1}{2}|\langle S_z\rangle| \leq \frac{1}{4}$, squeezing is possible only if the atom has a non-vanishing average dipole moment $\langle S^{\pm}\rangle$. If the atom is in the ground state $|1\rangle$ or in the excited state $|2\rangle$, the average dipole moment $\langle S^{\pm}\rangle = 0$, and there is no natural squeezing in the atomic spin components. In order to obtain a non-zero atomic dipole moment, $\langle S^{\pm}\rangle \neq 0$, we have to prepare the atom in a linear superposition of its ground and excited states. For example, if the atom is prepared in a linear superposition

$$|\Psi\rangle = \frac{1}{2}|1\rangle + \frac{\sqrt{3}}{2}|2\rangle, \tag{6.89}$$

the atomic dipole moment in this state is different from zero with $\langle S^{\pm}\rangle = \sqrt{3}/4$. Since in the state (6.89) the average inversion $\langle\Psi|S_z|\Psi\rangle = 1/4$, the atomic spin component prepared in the state (6.89) is squeezed to the degree of $\xi_x = 1/2$.

Note that the maximal squeezing that could be achieved in a two-level atom does not correspond to maximal coherence between the atomic levels. To illustrate this, let us consider an initial state

$$|\Psi\rangle = \frac{1}{\sqrt{2}}(|1\rangle + |2\rangle), \tag{6.90}$$

for which there is maximal initial coherence between the atomic levels, $\langle S^{\pm}\rangle = 1/2$. Since in the state (6.90) the average inversion $\langle\Psi|S_z|\Psi\rangle = 0$, squeezing of the dipole fluctuations requires

$\langle(\Delta S_i)^2\rangle < 0$, which is not possible to achieve, as the variance $\langle(\Delta S_i)^2\rangle$ is a positively defined quantity.

6.6 Spin Squeezing

In the preceding section, we have shown that a two-level atom can exhibit natural squeezing in the spin variables only if is prepared in a suitable linear superposition of its ground and excited states. The situation is different when we consider a multi-atom system.

As an example, consider two identical two-level atoms. In this case, the total dipole moment is $S^{\pm} = S_1^{\pm} + S_2^{\pm}$ and the total inversion is $S_z = S_{z1} + S_{z2}$. Hence, the variance $\langle(\Delta S_x)^2\rangle$ takes the form

$$\langle(\Delta S_x)^2\rangle = \frac{1}{2}\Big\{\langle\left(S_1^+ + S_2^+\right)\left(S_1^- + S_2^-\right)\rangle + \langle S_1^+ S_2^+\rangle + \langle S_1^- S_2^-\rangle - \frac{1}{2}\left[\langle S_1^+ + S_2^+\rangle + \langle S_1^- + S_2^-\rangle\right]^2\Big\}. \quad (6.91)$$

Equation (6.91) shows that the variance of the x-component of the atomic spin can be reduced below the spin quantum limit not only through the non-vanishing dipole moments $\langle S_i^{\pm}\rangle$, but also through the two-photon correlations $\langle S_1^+ S_2^+\rangle$ and $\langle S_1^- S_2^-\rangle$. This dependence suggests that there are two different processes that can lead to squeezing in multi-atom systems.

If the mean values of the spin components $\langle S_x\rangle$, $\langle S_y\rangle$ and $\langle S_z\rangle$ are different from zero, we can rotate the coordinate frame such that the mean value of one of the spin components, say $\langle S_{\vec{n}_k}\rangle$, will be different from zero, while the mean values of the other components $\langle S_{\vec{n}_i}\rangle$ and $\langle S_{\vec{n}_j}\rangle$ will be equal to zero. Here, $\vec{n}_i$, $\vec{n}_j$ and $\vec{n}_k$ are three mutually orthogonal unit vectors oriented along the rotated coordinate axis.

We can calculate the variances $\langle(\Delta S_{\vec{n}_i})^2\rangle$ and $\langle(\Delta S_{\vec{n}_j})^2\rangle$ of the spin components which are in the plane orthogonal to the mean spin direction $\vec{n}_k$. This is the main idea of spin squeezing introduced by Kitagawa and Ueda [34]. A system with the variance reduced below the standard quantum limit in one direction normal to the mean spin direction is called *spin squeezed*.

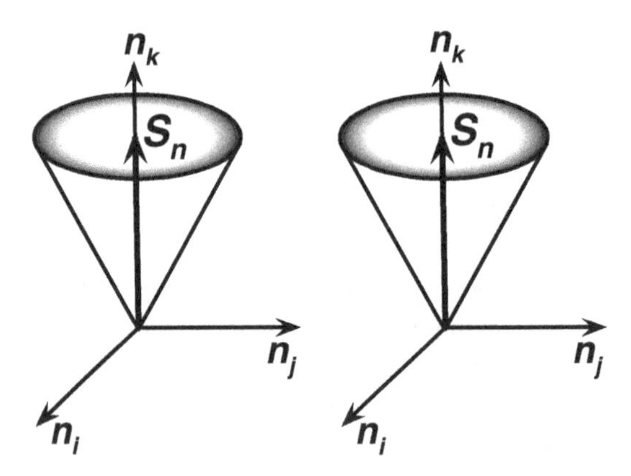

Figure 6.5 Two identical spins S_n oriented in the same n_k direction. The circle indicates the level of fluctuations of the rotating spins.

We can introduce a parameter $\xi_{\vec{n}_i}$ as a measure of degree of spin squeezing in the $\vec{n}_i$ direction

$$\xi_{\vec{n}_i} = \frac{\langle(\Delta S_{\vec{n}_i})^2\rangle}{(S/2)}, \tag{6.92}$$

where S is the maximal spin of the system ($S = 1$ for two two-level atoms).

A system with the variance reduced below the standard quantum limit in one direction normal to the mean spin direction is characterized by $\xi_{\vec{n}_i} < 1$, that is, spin squeezed in the direction $\vec{n}_i$. In Fig. 6.5, we show two identical spins oriented in one direction and that can have non-zero fluctuations around the mean spin direction $\vec{n}_k$. Since $\langle S_{\vec{n}_i}\rangle = \langle S_{\vec{n}_j}\rangle = 0$, single two-level atoms cannot be spin squeezed. However, an interaction between the atoms, which may create the two-photon correlations $\langle S_1^{\pm} S_2^{\pm}\rangle$, can lead to a spin squeezing.

There is another definition of spin squeezing, introduced by Wineland[a] *et al.* [35], called the spectroscopic spin squeezing, which involves an error in a measurement of a rotating spin oriented in the $\vec{n}_k$ direction

$$\frac{\langle(\Delta S_{\vec{n}_i})^2\rangle}{|\langle S_{\vec{n}_k}\rangle|^2}. \tag{6.93}$$

[a] David Wineland was granted the Nobel prize in 2012 for his experimental work that enabled measuring and manipulation of individual quantum systems.

It is easy to derive that the minimum value of the ratio (6.93) is $1/(2S)$. Therefore, we can introduce a parameter

$$\xi_{\vec{n}_i}^R = 2S\frac{\langle(\Delta S_{\vec{n}_i})^2\rangle}{|\langle S_{\vec{n}_k}\rangle|^2}, \tag{6.94}$$

which is a measure of degree of spin squeezing in the fluctuations relative to $|\langle S_{\vec{n}_k}\rangle|^2/(2S)$. The superscript R is used to signify the relative fluctuations. Since the mean value $|\langle S_{\vec{n}_k}\rangle| \leq S$, it follows that spin squeezing $\xi_{\vec{n}_i}^R < 1$ implies $\xi_{\vec{n}_i} < 1$, but not the vice versa.

Spin squeezing has been proposed as a measure of entanglement in multi-atom systems, which opens interesting applications in quantum information and quantum computation [36]. It has also been shown [37] that the parameter (6.92) is a better measure of entanglement than the spectroscopic spin squeezing parameter (6.94).

6.7 Squeezing Spectrum of the EM Field

In experiments the fluctuations are measured using the photon-counting technique where the variance $\langle(\Delta\hat{E}_\phi)^2\rangle$ is measured as a function of the phase angle ϕ. The condition $\langle(\Delta\hat{E}_\phi)^2\rangle < 1/4$, or equivalently $\langle: \ (\Delta\hat{E}_\phi)^2 \ :\rangle < 0$, refers to the squeezing of the total field. For multi-mode fields it does not exhaust all possible forms of squeezing. Therefore, another technique has been proposed of filtering the frequencies before detection and measure the fluctuations at the filtered frequencies. This form of squeezing is described in terms of the squeezing spectrum.

The squeezing spectrum of a stationary EM field is defined as the Fourier transform of the two-time normally ordered correlation function

$$S_\phi(\omega) = \lim_{t\to\infty}\int d\tau \left\langle : \hat{E}_\phi(t+\tau), \hat{E}_\phi(t) :\right\rangle e^{i\omega\tau}, \tag{6.95}$$

where

$$\left\langle\hat{E}_\phi(t+\tau), \hat{E}_\phi(t)\right\rangle = \left\langle\hat{E}_\phi(t+\tau)\hat{E}_\phi(t)\right\rangle - \left\langle\hat{E}_\phi(t+\tau)\right\rangle\left\langle\hat{E}_\phi(t)\right\rangle, \tag{6.96}$$

and ϕ is a phase angle that may be chosen at will.

A negative value of $S_\phi(\omega)$ is non-classical and indicates photocurrent noise at frequency ω below the quantum limit. In other words, it indicates squeezing at the frequency ω of the phase field component $\hat{E}_\phi$.

Integrating the squeezing spectrum $S_\phi(\omega)$ over all frequencies, we obtain the normally ordered variance of the total field

$$\int_{-\infty}^{+\infty} d\omega\, S_\phi(\omega) = \left\langle : \left(\Delta \hat{E}_\phi\right)^2 : \right\rangle . \quad (6.97)$$

It can happen that for a broadband field the spectrum $S_\phi(\omega)$ may dip below the shot noise at some frequencies ($S_\phi(\omega) < 0$) even though $\langle : (\Delta \hat{E}_\phi)^2 : \rangle$, which is the integral of $S_\phi(\omega)$ over all frequencies, is positive. We may refer to this as spectral component squeezing at selected frequencies. Also, when $\langle : (\Delta \hat{E}_\phi)^2 : \rangle < 0$, some selected modes may not exhibit squeezing or may exhibit more squeezing than the total field. Failure to recognize these frequency-dependent features led at the early stage of the research on squeezing to rather pessimistic predictions of attainable squeezing [38, 39].

6.8 Detection of Squeezed States of Light

We have already learnt that squeezing is a phase dependent phenomenon. How then one could identified that a given field is in a squeezed state? Direct photoelectric counting experiments are not sensitive to the phase-dependent nature such as squeezing in the incident field, but only to light intensity. Therefore, direct photon counting as a way of detecting squeezing is impractical. Hence, a phase sensitive measurement system is needed to observe squeezing. As we shall see, it is provided by phase-sensitive interference of a measured field with another optical field in a coherent state, followed by photoelectric detection of the resulting intensity fluctuations.

6.8.1 *Homodyne Detection Scheme*

Typical experimental schemes used to measure phase-dependent fields, which can be applied to detect squeezing, are homodyne and heterodyne detection techniques. An example of the experimental scheme is shown in Fig. 6.6. In these techniques the measured field

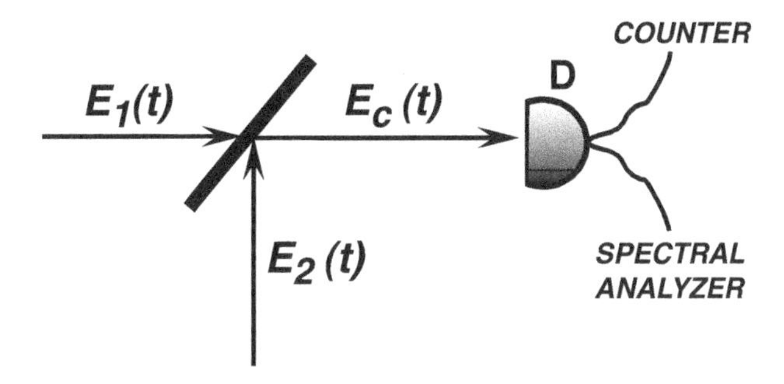

Figure 6.6 Schematic diagram of homodyne or heterodyne detection scheme. A field of a complex amplitude $E_1(t)$ is mixed with the highly coherent field of an amplitude $E_2(t)$. The mixing is accomplished by a beam splitter and the resulting field $E_c(t)$ is detected photoelectrically by a detector D and the resulting photocurrent is then analyzed by a photoelectron counter or a spectral analyzer. When both fields have the same frequency, we refer to the procedure as homodyning. Otherwise, when the frequencies are different, the procedure is referred to as heterodyning.

of an unknown amplitude $E_1(t)$ is mixed (beat) with the known strong coherent light of an amplitude $E_2(t)$ (local oscillator). The beats occurring in the superposed light of amplitude $E_c(t)$ are analysed by either photoelectric counting or photocurrent spectral measurements.

To illustrate phase dependence in homodyne or heterodyne detection scheme, consider two fields $\hat{E}_1(t)$ and $\hat{E}_2(t)$ of the same frequency (homodyne detection), combined on a lossless beam splitter of transmissivity β. The fields can be written in terms of the annihilation and creation operators as

$$\hat{E}_1(t) = \lambda\left(\hat{a}(t) - \hat{a}^\dagger(t)\right), \quad \hat{E}_2(t) = \lambda\left(\hat{b}(t) - \hat{b}^\dagger(t)\right), \quad (6.98)$$

and we assume that both fields have the same polarization.

The total field emerging from the beam splitter is

$$\hat{E}_c(t) = \lambda\left(\hat{c}(t) - \hat{c}^\dagger(t)\right), \quad (6.99)$$

where

$$\hat{c} = \sqrt{\beta}\,\hat{a} + i\sqrt{1-\beta}\,\hat{b}, \quad (6.100)$$

and the factor i indicates a $\pi/2$ phase shift between the reflected and transmitted fields. The detector D can be adjusted to respond to the intensity of the field, $I \sim \langle \hat{c}^\dagger(t)\,\hat{c}(t)\rangle$ or to the field fluctuations $\langle (\Delta \hat{c}^\dagger \hat{c})^2 \rangle$. The intensity at the detector is given by

$$I \sim \langle \hat{c}^\dagger \hat{c}\rangle = \beta \langle \hat{a}^\dagger \hat{a}\rangle + (1-\beta)\,\langle \hat{b}^\dagger \hat{b}\rangle \\ -i\sqrt{\beta\,(1-\beta)}\,\left(\langle \hat{a}\rangle\langle \hat{b}^\dagger\rangle - \langle \hat{a}^\dagger\rangle\langle \hat{b}\rangle\right), \quad (6.101)$$

where we have assumed that the fields $\hat{E}_1$ and $\hat{E}_2$ are not correlated.

Suppose, the field $\hat{E}_2$ is a coherent laser field (local oscillator) of a large amplitude α, whereas the signal (detected) field is weak. As a result, the terms proportional to $\langle \hat{b}^\dagger \hat{b}\rangle$ and $\langle \hat{b}^\dagger\rangle$, $\langle \hat{b}\rangle$ dominate over those without $\hat{b}^\dagger$ and $\hat{b}$. Hence, the terms independent of the amplitude of the local oscillator can be discarded. Thus, we can ignore the term $\beta\langle \hat{a}^\dagger \hat{a}\rangle$, and denoting $\langle \hat{b}\rangle = |\alpha|\exp(i\phi)$, $\langle \hat{b}^\dagger\rangle = |\alpha|\exp(-i\phi)$, the resultant light intensity at the detector is then

$$I \approx (1-\beta)\,|\alpha|^2 + 2\,|\alpha|\,\sqrt{\beta\,(1-\beta)}\,\langle \hat{E}_\phi\rangle, \quad (6.102)$$

where

$$\langle \hat{E}_\phi\rangle = \frac{1}{2i}\left(\langle \hat{a}\rangle \mathrm{e}^{-i\phi} - \langle \hat{a}^\dagger\rangle \mathrm{e}^{i\phi}\right), \quad (6.103)$$

and ϕ is the phase of the laser. The first term on the right-hand side of Eq. (6.102) is equal to the intensity of the reflected coherent beam. The second term is an interference term between the coherent and the signals beams. This term contains the phase-dependent quadrature amplitude of the signal beam.

Similarly, we can show that the variance of photoelectric counts can be expressed as

$$\langle (\Delta \hat{n}_c)^2\rangle \approx (1-\beta)^2\,|\alpha|^2 + |\alpha|^2\,\beta\,(1-\beta)\left\langle (\Delta \hat{E}_\phi)^2\right\rangle, \quad (6.104)$$

where $\hat{n}_c = \hat{c}^\dagger \hat{c}$ is the number of photons in the mode c. In the derivation of Eq. (6.104) we have retained terms of second order in $|\alpha|$. We see that the variance of the superposed field contains the reflected local oscillator noise, $(1-\beta)^2|\alpha|^2$, and the variance of the phase-dependent quadrature noise of the signal field, $\langle (\Delta \hat{E}_\phi)^2\rangle$. The detected fluctuations are determined by the fluctuations of the measured field $\langle (\Delta \hat{E}_\phi)^2\rangle$, the measured quadrature phase operator.

Figure 6.7 shows the variance $\langle (\Delta \hat{E}_\phi)^2\rangle$ as a function of ϕ for a squeezed input field. As the local oscillator phase ϕ is varied, the variance of photoelectric counts changes from being large to

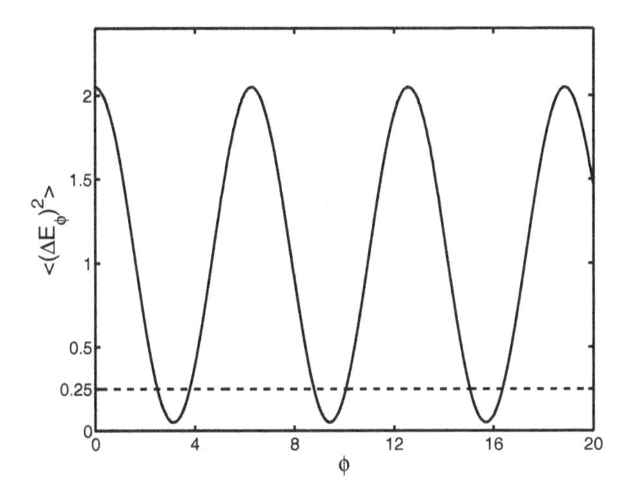

Figure 6.7 The variance $\langle(\Delta E_\phi)^2\rangle$ as a function of the laser phase ϕ for a squeezed input field. The dashed line indicates the vacuum level of the fluctuations.

very small values. It is seen that for certain values of the phase the variance is smaller than the vacuum level of the fluctuations.

In practice, the signal field is first blocked to determine the vacuum noise level. The signal field is then allowed to reach the beam spitter and the variance is determined with reference to the vacuum level. Note, however, that the intensity measurements in homodyne detection scheme are quite different from those in direct detection. In the homodyne detection intensity fluctuations directly measure the fluctuations in a quadrature of the input and the signal field and its variance depend upon the local oscillator phase angle, which is an external parameter.

In conclusion of this section, we point out that apart from the homodyne and heterodyne detection schemes, it is also possible to use a balanced homodyne detection [40, 41]. In a balance homodyne detection scheme, two output fields of the beamsplitter are detected by two identical photodetectors. Photocurrents from these photodetectors are subtracted electronically and the fluctuations of the difference current are analysed. In this scheme, the measured fluctuations are not affected by the noise of the local oscillator that is cancelled when the photocurrents are subtracted.

Exercises

6.1 Verify that the variances of the field quadrature operators are the same for the vacuum state when the field is in a coherent state $|\alpha\rangle$.

6.2 Consider the superposition state of two field modes

$$|\Psi\rangle = a|0\rangle + b|1\rangle,$$

where a and b are complex parameters satisfying the normalization condition, $|a|^2 + |b|^2 = 1$.

(a) Calculate the variances $\langle(\Delta\hat{X}_1)^2\rangle$ and $\langle(\Delta\hat{X}_2)^2\rangle$.

(b) Show that there exits values of the parameters a and b for which either of the quadrature variances is reduced below the vacuum level.

(c) For the cases where the quadrature variances are reduced below the vacuum level, verify if the uncertainty principle is not violated.

6.3 Calculate the variances $\langle(\Delta\hat{Y}_1)^2\rangle$ and $\langle(\Delta\hat{Y}_2)^2\rangle$ of the quadrature operators $\hat{Y}_1$ and $\hat{Y}_2$ defined as

$$\hat{Y}_1 = \frac{1}{2}\left(\hat{a}e^{-i\theta/2} + \hat{a}^\dagger e^{i\theta/2}\right),$$

$$\hat{Y}_2 = \frac{1}{2i}\left(\hat{a}e^{-i\theta/2} - \hat{a}^\dagger e^{i\theta/2}\right),$$

assuming that the field is in a squeezed coherent state $|\Psi\rangle = |\alpha, s\rangle$.

6.4 Consider two Hermitian operators corresponding to the real and imaginary parts of the square of the complex amplitude of the field

$$\hat{X}_1 = \frac{1}{2}\left(\hat{a}^2 + (\hat{a}^\dagger)^2\right), \quad \hat{X}_2 = \frac{1}{2i}\left(\hat{a}^2 - (\hat{a}^\dagger)^2\right).$$

Show that the squeezing condition for the above operators is

$$\left\langle(\Delta\hat{X}_i)^2\right\rangle < \left(\langle\hat{a}^\dagger\hat{a}\rangle + \frac{1}{2}\right), \qquad (i = 1, 2).$$

6.5 The Hamiltonian $\hat{H}$ of a degenerate parametric amplifier is given by

$$\hat{H} = \hbar\omega\hat{a}^\dagger\hat{a} - \frac{1}{2}i\hbar\kappa\left(\hat{a}^2 e^{2i\omega t} - \hat{a}^{\dagger 2} e^{-2i\omega t}\right).$$

(a) Calculate the intensity correlation function

$$g^{(2)}(0) = \langle \hat{a}^\dagger(t)\hat{a}^\dagger(t)\hat{a}(t)\hat{a}(t)\rangle / \langle \hat{a}^\dagger(t)\hat{a}(t)\rangle^2$$

assuming that initially ($t = 0$) the field was in a coherent state.

(b) Calculate the variances $\langle(\Delta\hat{Y}_1)^2\rangle$ and $\langle(\Delta\hat{Y}_2)^2\rangle$ defined in Exercise **6.3**, assuming that initially the field was in a coherent state.

6.6 Show that a field in a squeezed vacuum state $|0, s\rangle$ always exhibits super-Poissonian statistics.

6.7 Show that the degree of squeezing $\xi_x = 1/2$ obtained for the superposition state (6.89) is the optimum value of squeezing that can occur in a two-level atom.

6.8 Calculate the variance of photons in the squeezed vacuum state, and compare it with the variance of photons in a thermal state of the same mean photon number.

6.9 Consider a single-mode field in the superposition state

$$|\Psi\rangle = \frac{1}{\sqrt{2}}\left(|n\rangle + |n+2\rangle\right),$$

where $|n\rangle$ is a Fock state with n photons. For what value of n is

(a) the quadrature E_2 squeezed?

(b) the second-order correlation function $g^{(2)} < 1$?

6.10 Express the operators $\hat{n}^2$ and $\hat{n}^3$ in normal and in antinormal orders, where $\hat{n} = \hat{a}^\dagger\hat{a}$.

6.11 Consider an operator

$$\hat{b} = \mu\hat{a} + \nu\hat{a}^\dagger,$$

where $\hat{a}$, $\hat{a}^\dagger$ are the annihilation and creation operators of the field, and $|\mu|^2 - |\nu|^2 = 1$.

(a) Show that $\left[\hat{b}, \hat{b}^\dagger\right] = 1$.

(b) Calculate mean values $\langle\hat{a}\rangle$, $\langle\hat{a}^\dagger\rangle$, $\langle\hat{a}^2\rangle$ and $\langle\hat{a}^\dagger\rangle$ in the state $|\beta\rangle$, which is an eigenstate of $\hat{b}$ with an eigenvalue β.

6.12 Consider properties of squeezed vacuum states.

(a) Show that squeezed vacuum states, similar to the coherent states, are not mutually orthogonal.

(b) Find the overlap between the coherent state $|\alpha\rangle$ and the squeezed vacuum state $|0, s\rangle$. Are the coherent and squeezed states orthogonal?

Chapter 7

Phase Space Representations of the Density Operator

7.1 Introduction

In Chapter 4, we introduced the concept of the photon number representation of the electromagnetic (EM) field that is based on the number states, the eigenstates of the Hamiltonian of the EM field. However, this is not the only possible representation, in fact not always the most convenient. In this chapter, we introduce the concept of phase space representations of a quantum system which is in a state described by the density operator $\hat{\rho}$. The density operator of a given system encodes classical as well as non-classical (quantum) properties of the system. How to distinguish these two kinds of properties is of basic importance in quantum optics. Therefore, we shall address this issue by using representations of the density operator that are based on the parametric space of complex eigenvalues of the annihilation operator $\hat{a}$ in a coherent state $|\alpha\rangle$. We first determine what we mean by the density operator of a quantum system and discuss its basic properties. Next, we introduce different representations of the density operator in terms of coherent state projectors $|\alpha\rangle\langle\alpha|$. The representations show that

Quantum Optics for Beginners
Zbigniew Ficek and Mohamed Ridza Wahiddin

ISBN 978-981-4411-75-2 (Hardcover), 978-981-4411-76-9 (eBook)
www.panstanford.com

a state of the EM field or of an arbitrary quantum system may be regarded as a mixture of coherent states. Along the way, we will learn how to find the P, Q and Wigner representations of the density operator, what are their properties, how to interpret the properties and how to calculate relations between the different representations. Most importantly, certain field states exhibit non-classical features and these non-classical features can be manifested in the phase space representations. Accordingly, we will learn how the representations are very convenient tools to describe quantum states of simple systems.

7.2 Density Operator

Before studying the representations of the density operator, we first determine what we mean by the density operator and review its basic properties.

In statistical physics, an expectation or average or mean value of an arbitrary quantity A is obtained by weighting each measured value A_i by the associated probability P_i and summing over all the measurements. Thus

$$\langle A \rangle = \sum_i P_i A_i, \qquad i = 1, 2, \ldots, N, \tag{7.1}$$

where P_i is a probability of measuring the value A_i.

In quantum physics it is slightly different. A quantum system can be in a pure or mixed state. If the state of the system is determined by a single state vector $|\psi\rangle$, we say that the system is in the *pure state* $|\psi\rangle$. If the state of the system cannot be precisely specified, only the probabilities that the system is in a range of possible states can be determined, we say that the system is in a *mixed state*.

7.2.1 *Density Operator of a Pure State*

Consider an operator $\hat{A}$ acting on a pure state $|\psi\rangle$. Then, the quantum-mechanical expectation value of the operator $\hat{A}$ of a system in the pure state $|\psi\rangle$ is defined as

$$\begin{aligned}\langle\hat{A}\rangle &= \langle\psi|\hat{A}|\psi\rangle = \sum_n \langle\psi|\hat{A}|n\rangle\langle n|\psi\rangle \\ &= \sum_n \langle n|\psi\rangle\langle\psi|\hat{A}|n\rangle = \sum_n \langle n|\hat{\rho}\hat{A}|n\rangle = \mathrm{Tr}\left(\hat{\rho}\hat{A}\right), \qquad (7.2)\end{aligned}$$

where

$$\hat{\rho} = |\psi\rangle\langle\psi| \qquad (7.3)$$

is the density operator of the pure state. Thus, the density operator is essential in calculating expectation values of operators.

7.2.2 *Density Operator of a Mixed State*

If we do not have enough information to specify the state vector of a given system, but know only the probabilities P_i that the system is in a state $|\psi_i\rangle$, we then can introduce, in analogy to the statistical definition of the expectation value, the density operator of a mixed state as

$$\hat{\rho} = \sum_i P_i|\psi_i\rangle\langle\psi_i|, \qquad (7.4)$$

where P_i is the probability that a given system is in the state $|\psi_i\rangle$. This equation expresses ρ as an incoherent superposition of pure state density operators, $|\psi_i\rangle\langle\psi_i|$. If one of the $P_i = 1$ and all the others are zero, the density operator corresponds to a pure state. Note that the states $|\psi_i\rangle$ are assumed to be normalized but do not need to be orthogonal. The density operator provides a complete description of the statistical properties of the system spanned by the basis states $|\psi_i\rangle$.

7.2.3 *The Basic Properties of the Density Operator*

Density operator of an arbitrary physical system must satisfy the following properties:

Property 1. Hermitian: $\hat{\rho}^\dagger = \hat{\rho}$.

Proof.

$$\hat{\rho}^\dagger = \sum_i P_i\left(|\psi_i\rangle\langle\psi_i|\right)^\dagger = \sum_i P_i|\psi_i\rangle\langle\psi_i| = \hat{\rho},$$

as required. □

Property 2. Normalized: $\mathrm{Tr}\,(\hat{\rho}) = 1$.

Proof.

$$\begin{aligned}\mathrm{Tr}\,(\hat{\rho}) &= \sum_n \langle \psi_n | \,\hat{\rho}\, | \psi_n \rangle \\ &= \sum_n \sum_i P_i \langle \psi_n | \psi_i \rangle \langle \psi_i | \psi_n \rangle = \sum_n \sum_i P_i |\langle \psi_n | \psi_i \rangle|^2 \\ &= \sum_i P_i = 1.\end{aligned}$$

as required. □

Note that we have used the fact that the states $|\psi_i\rangle$ are normalized but have not assumed that the states are orthogonal.

Property 3. Positive operator: $\hat{\rho} \geq 0$.

Proof. If $\hat{\rho}$ is a positive operator, then for any state $|\phi\rangle$, the expectation value $\langle \phi | \hat{\rho} | \phi \rangle \geq 0$ is a positive number

$$\langle \phi | \hat{\rho} | \phi \rangle = \sum_i P_i \langle \phi | \psi_i \rangle \langle \psi_i | \phi \rangle = \sum_i P_i \, |\langle \phi \, | \psi_i \rangle|^2 \geq 0,$$

as required. □

Property 4. For a pure state: $\hat{\rho}^2 = \hat{\rho}$.

Proof.

$$\hat{\rho}^2 = \sum_{i,j} P_i P_j \, |\psi_i\rangle \langle \psi_i | \, \psi_j \rangle \langle \psi_j | = \sum_i P_i^2 \, |\psi_i\rangle \langle \psi_i |,$$

as required. □

Property 5. $\mathrm{Tr}\left(\hat{\rho}^2\right) \leq 1$.

Proof.

$$\mathrm{Tr}\left(\hat{\rho}^2\right) = \sum_n \sum_i P_i^2 \langle \psi_n | \psi_i \rangle \langle \psi_i | \psi_n \rangle = \sum_i P_i^2 \leq 1,$$

as required. □

The properties 1–3 can be alternatively expressed in terms of the eigenvalues λ_i of the density operators. Namely, the property 1 means that all the eigenvalues of $\hat{\rho}$ are real numbers, $\lambda_i \in \mathcal{R}$. The property 2 means that the sum of all the eigenvalues is equal to 1, $\sum_i \lambda_i = 1$, and the property 3 means that the eigenvalues are positive numbers, $\lambda_i \geq 0$. The property 5 is very often used to check whether a given system is in a pure or a mixed state. For a pure state $\text{Tr}\left(\hat{\rho}^2\right) = 1$, whereas $\text{Tr}\left(\hat{\rho}^2\right) < 1$ for a mixed state.

7.3 Number State Representation

The density operator may be represented in terms of arbitrary states of a given system. For example, the completeness relation for the photon number states

$$\sum_n |n\rangle\langle n| = 1, \tag{7.5}$$

can be used to represent the density operator in terms of the photon number states. This is done by multiplying the density operator by the unity given by Eq. (7.5) both on the right and the left, and then we obtain the density operator in terms of the projection operators $|n\rangle\langle m|$ as

$$\begin{aligned}\hat{\rho} &= \left(\sum_n |n\rangle\langle n|\right)\hat{\rho}\left(\sum_m |m\rangle\langle m|\right)\\ &= \sum_{n,m} \langle n|\,\hat{\rho}\,|m\rangle|n\rangle\langle m| = \sum_{n,m} a_{nm}\,|n\rangle\langle m|\,.\end{aligned} \tag{7.6}$$

The diagonal terms are probabilities $P_n = a_{nn} = \langle n\,|\hat{\rho}|\,n\rangle$, that a_{nn} is the probability of having n photons in the field being in the state $|n\rangle$. The off-diagonal elements $a_{nm}(n \neq m)$ are coherencies between two different photon number states.

The number states representation holds not only for the field states, but also for the system states [42, 43]. For example, $\hat{\rho}$ can be the density operator of a system (e.g., a two-level atom) and then $|n\rangle$, $|m\rangle$ represent the energy states of the system. In this case the diagonal terms are populations of the atomic states, while off-diagonal terms are atomic coherences equal to the mean value of the induced atomic dipole moments ($\rho_{nm} = \langle n|S^+|m\rangle$).

7.4 Coherent States P Representation

Since the coherent states are complete

$$\frac{1}{\pi}\int d^2\alpha\,|\alpha\rangle\langle\alpha| = 1, \tag{7.7}$$

we can use this relation to represent the density operator in terms of coherent states. If we formally multiply the density operator $\hat{\rho}$ by the unity given by Eq. (7.7) both on the right and the left, we obtain

$$\hat{\rho} = \frac{1}{\pi^2}\int\int \langle\alpha|\,\hat{\rho}\,|\beta\rangle\,|\alpha\rangle\langle\beta|\,d^2\alpha d^2\beta. \tag{7.8}$$

This is a complicated representation involving two integrals (in fact four integrals as $d^2\alpha = d\,(\mathrm{Re}\alpha)\,d\,(\mathrm{Im}\alpha)$).

We can now ask whether we can expand $\hat{\rho}$ in a simpler form

$$\hat{\rho} = \int d^2\alpha P\,(\alpha)\,|\alpha\rangle\langle\alpha|\,, \tag{7.9}$$

which is an analogue to $\hat{\rho} = \sum_i P_i\,|\psi_i\rangle\,\langle\psi_i|$ for discrete states, and whether we can call $P\,(\alpha)$ as a probability of finding the field in the coherent state $|\alpha\rangle$.

The above representation, called P representation or the Glauber–Sudarshan P representation [20, 44], is very useful for calculating expectation values of normally ordered field operators that appear, for example, in $\langle n\rangle$, $g^{(2)}$ or in $\langle :\,(\Delta E_\theta)^2\,:\rangle$.

For example, the average number of photons can be written as

$$\begin{aligned}\langle n\rangle &= \langle\hat{a}^\dagger\hat{a}\rangle = \mathrm{Tr}\left(\hat{\rho}\hat{a}^\dagger\hat{a}\right) = \mathrm{Tr}\left(\int d^2\alpha P\,(\alpha)\,|\alpha\rangle\langle\alpha|\,\hat{a}^\dagger\hat{a}\right)\\ &= \int d^2\alpha P\,(\alpha)\,\langle\alpha|\,\hat{a}^\dagger\hat{a}\,|\alpha\rangle = \int d^2\alpha P\,(\alpha)\,|\alpha|^2\,.\end{aligned} \tag{7.10}$$

Hence, the averages are calculated in the same way as that in classical statistics with $P\,(\alpha)$ playing the role of the probability distribution.

Since $\mathrm{Tr}\,(\hat{\rho}) = 1$, we obtain

$$1 = \mathrm{Tr}\,(\hat{\rho}) = \int d^2\alpha P\,(\alpha)\,\mathrm{Tr}\,(|\alpha\rangle\langle\alpha|) = \int d^2\alpha P\,(\alpha)\,. \tag{7.11}$$

Thus, $P\,(\alpha)$ is normalized as a classical probability distribution.

We know that the coherent states are the closest quantum states to the classical description of the field. Can we then treat $P\,(\alpha)$ as a quantum analogue of the classical probability distribution? In order to answer this question, we calculate an average

$$\langle\beta|\,\hat{\rho}\,|\beta\rangle = \int d^2\alpha\, P(\alpha)\,\langle\beta|\,\alpha\rangle\langle\alpha|\,\beta\rangle$$
$$= \int d^2\alpha\, P(\alpha)\,|\langle\beta|\,\alpha\rangle|^2 = \int d^2\alpha\, P(\alpha)\,\mathrm{e}^{-|\alpha-\beta|^2}. \quad (7.12)$$

Since $\exp(-\,|\alpha-\beta|^2)$ is not a δ-function, the diagonal elements

$$\langle\beta|\,\hat{\rho}\,|\beta\rangle \neq P(\beta), \quad (7.13)$$

that are not probabilities of finding the system in the coherent state $|\beta\rangle$.

Moreover, the average $\langle\beta|\,\hat{\rho}\,|\beta\rangle$ must be positive. However, the integral $\int d^2\alpha\, P(\alpha)\exp(-\,|\alpha-\beta|^2)$ does not require $P(\beta)$ to be positive.

The following two examples may help to clarify this observation.

Example 7.1 (Photon antibunching as a non-classical phenomenon) *Consider the normalized second-order correlation function*

$$g^{(2)} = 1 + \frac{\langle(\Delta\hat{n})^2\rangle - \langle\hat{n}\rangle}{\langle\hat{n}\rangle^2} = 1 + \frac{\langle:(\Delta\hat{n})^2:\rangle}{\langle\hat{n}\rangle^2}. \quad (7.14)$$

In the P representation, the normally ordered variance of the number of photons can be written as

$$\langle:(\Delta\hat{n})^2:\rangle = \langle\hat{a}^\dagger\hat{a}^\dagger\hat{a}\hat{a}\rangle - \langle\hat{a}^\dagger\hat{a}\rangle^2$$
$$= \int d^2\alpha\, P(\alpha)\,|\alpha|^4 - \left[\int d^2\alpha' P(\alpha')\,|\alpha'|^2\right]^2$$
$$= \int d^2\alpha\, P(\alpha)\left[|\alpha|^2 - \int d^2\alpha' P(\alpha')\,|\alpha'|^2\right]^2. \quad (7.15)$$

As the normally ordered variance must be negative for antibunching and the above equation involves the phase space integral of the product of the $P(\alpha)$ function with a function that is real and positive, it follows that antibunching is associated with a negative $P(\alpha)$ function. Thus, photon antibunching ($g^{(2)} < 1$) requires $P(\alpha) < 0$. We say that photon antibunching is non-classical by the criterion that $P(\alpha)$ for any classical field cannot be negative.

Example 7.2 (Squeezing as a non-classical phenomenon) *In the P representation, the normally ordered variance of the in-phase quadrature component can be written as*

$$\begin{aligned}\langle :\left(\Delta \hat{E}_1\right)^2 :\rangle &= \frac{1}{4}\langle 2\hat{a}^\dagger \hat{a} + \hat{a}\hat{a} + \hat{a}^\dagger \hat{a}^\dagger\rangle - \frac{1}{4}\langle \hat{a} + \hat{a}^\dagger\rangle^2 \\ &= \frac{1}{4}\left\{\int d^2\alpha P(\alpha)\left[2|\alpha|^2 + \alpha^2 + \alpha^{*2}\right]\right. \\ &\quad \left. - \left[\int d^2\alpha' P(\alpha')\left(\alpha' + \alpha'^*\right)\right]^2\right\} \\ &= \frac{1}{4}\int d^2\alpha P(\alpha)\left\{(\alpha + \alpha^*)\right. \\ &\quad \left. - \left[\int d^2\alpha' P(\alpha')\left(\alpha' + \alpha'^*\right)\right]\right\}^2 .\end{aligned} \tag{7.16}$$

Thus, similarly to photon antibunching, squeezing $\langle : (\Delta \hat{E}_1)^2 :\rangle < 0$ is associated with a negative $P(\alpha)$ function, $P(\alpha) < 0$, and therefore can also be regarded as non-classical.

Hence, unlike a classical probability, $P(\alpha)$ can take negative values and therefore $P(\alpha)$ is not a true probability distribution function. Sometimes $P(\alpha)$ is called as a quasi-distribution function or a quasi-probability function.

How do we find $P(\alpha)$?

Since we can write the density operator as

$$\hat{\rho} = \int d^2\alpha P(\alpha)|\alpha\rangle\langle\alpha|, \tag{7.17}$$

we then can find $P(\alpha)$ inverting the above equation. This is made possible using the relation

$$\begin{aligned}\mathrm{Tr}\left\{\hat{\rho} e^{iz^*\hat{a}^\dagger} e^{iz\hat{a}}\right\} &= \mathrm{Tr}\left\{\left[\int d^2\alpha P(\alpha)|\alpha\rangle\langle\alpha|\right] e^{iz^*\hat{a}^\dagger} e^{iz\hat{a}}\right\} \\ &= \int d^2\alpha P(\alpha) e^{iz^*\alpha^*} e^{iz\alpha}.\end{aligned} \tag{7.18}$$

The right-hand side of the above equation is just a two-dimensional Fourier transform of $P(\alpha)$. The inverse transform gives

$$P(\alpha) = \frac{1}{\pi^2}\int d^2z \mathrm{Tr}\left(\hat{\rho} e^{iz^*\hat{a}^\dagger} e^{iz\hat{a}}\right) e^{-iz^*\alpha^*} e^{-iz\alpha}. \tag{7.19}$$

However, the Tr under the integral is the characteristic function of the normally ordered $\hat{a}$, $\hat{a}^\dagger$ operators

$$\chi_{\mathrm{N}}(z) = \mathrm{Tr}\left(\hat{\rho}\mathrm{e}^{iz^*\hat{a}^\dagger}\mathrm{e}^{iz\hat{a}}\right). \quad (7.20)$$

Therefore

$$P(\alpha) = \frac{1}{\pi^2}\int d^2z\chi_{\mathrm{N}}(z)\,\mathrm{e}^{-iz^*\alpha^*}\mathrm{e}^{-iz\alpha}. \quad (7.21)$$

Thus, $P(\alpha)$ representation the is the Fourier transform of the normally ordered characteristic function.

The relation (7.20) represents a mapping from the operator $\hat{\rho}$, which is a function of the two operators $\hat{a}$ and $\hat{a}^\dagger$, to the scalar function $\chi_{\mathrm{N}}(z)$, which is a function of the complex variable z.

Example 7.3 ($P(\alpha)$ representation for a pure coherent state) *In this example, we illustrate how to find $P(\alpha)$ representation for a pure coherent state $|\alpha_0\rangle$, represented by the density operator $\hat{\rho} = |\alpha_0\rangle\langle\alpha_0|$.*

First, according to Eq. (7.21), we have to calculate the characteristic function

$$\chi_{\mathrm{N}}(z) = Tr\left\{|\alpha_0\rangle\langle\alpha_0|e^{iz^*\hat{a}^\dagger}e^{iz\hat{a}}\right\}$$
$$= \langle\alpha_0|e^{iz^*\hat{a}^\dagger}e^{iz\hat{a}}|\alpha_0\rangle = e^{iz^*\alpha_0^*+iz\alpha_0}. \quad (7.22)$$

Next, substituting the result for $\chi_{\mathrm{N}}(z)$ into Eq. (7.21), we obtain

$$P(\alpha) = \frac{1}{\pi^2}\int d^2ze^{-iz(\alpha-\alpha_0)}e^{-iz^*(\alpha^*-\alpha_0^*)}$$
$$= \delta^2(\alpha-\alpha_0), \quad (7.23)$$

that is, the P representation of a pure coherent state is a two-dimensional delta function.

7.5 Generalized *P* Representations

The generalized P representations are defined by the following expansion of the density operator using coherent state projectors

$$\hat{\rho} = \int d(\alpha,\beta)\,P(\alpha,\beta)\,\frac{|\alpha\rangle\langle\beta^*|}{\langle\beta^*|\alpha\rangle}, \quad (7.24)$$

where $d(\alpha,\beta)$ is the integration which can be chosen to define different representations. Note that this is non-diagonal expansion of $\hat{\rho}$.

A choice $d(\alpha, \beta) = \delta^2(\alpha^* - \beta)\, d^2\alpha d^2\beta$ gives the P representation. However, a choice

$$d(\alpha, \beta) = d^2\alpha d^2\beta \tag{7.25}$$

gives a positive P representation, introduced by Gardiner and Drummond [45]. Here the states denoted $|\alpha\rangle$ are the n-fold coherent states of n mode operators labelled $\hat{a}_1 \ldots \hat{a}_n$. That is, these states correspond to a superposition of different number states with the property $\hat{a}_k |\alpha\rangle = \alpha_k |\alpha\rangle$.

The initial distribution function $P(\alpha, \beta)$ can be chosen to be a positive function defined on the $4N$-dimensional phase space spanned by the complex coordinates α and β. This includes all diagonal and off-diagonal coherent states components of the density matrix. It is less obvious that a positive function exists in all cases, but it can be constructed, even for non-classical fields. A non-classical field necessarily corresponds to a superposition of coherent states. These are represented by the off-diagonal terms in the coherent expansion, in which $\alpha \neq \beta$.

In order to calculate an operator expectation value, there is a direct correspondence between the distribution and the normally ordered operator product

$$\langle \hat{a}_n^\dagger \ldots \hat{a}_m \rangle = \int\int d^2\alpha d^2\beta P(\alpha, \beta)\, \beta_n^* \ldots \alpha_m. \tag{7.26}$$

The important property here is the direct relation between the representation and the normally ordered moments, which are characteristic of photodetector measurements. If other types of moments are needed, then the operator commutation relations must be used to calculate them.

7.6 *Q* Representation

There are other orderings possible for the $\hat{a}^\dagger, \hat{a}$ operators. For example, we can use the antisymmetric ordering and define the antinormally ordered characteristic function

$$\chi_\mathrm{A}(z) = \mathrm{Tr}\left(\hat{\rho} e^{iz\hat{a}} e^{iz^*\hat{a}^\dagger}\right). \tag{7.27}$$

Its Fourier transform is called Q representation

$$Q(\alpha) = \frac{1}{\pi^2} \int d^2 z \chi_{\mathrm{A}}(z) \mathrm{e}^{-iz^*\alpha^*} \mathrm{e}^{-iz\alpha}. \tag{7.28}$$

The $Q(\alpha)$ representation, sometimes called the Husimi function, is a non-negative function and has a simple form

$$Q(\alpha) = \frac{1}{\pi} \langle \alpha | \hat{\rho} | \alpha \rangle \geq 0. \tag{7.29}$$

Thus, $\pi Q(\alpha)$ is strictly the probability of finding a system of the density operator $\hat{\rho}$ in the coherent state $|\alpha\rangle$.

Proof. From the definition of the Q function, and the completeness of the coherent states, we obtain

$$\begin{aligned} Q(\alpha) &= \frac{1}{\pi^2} \int d^2 z \chi_{\mathrm{A}}(z) \mathrm{e}^{-iz^*\alpha^*} \mathrm{e}^{-iz\alpha} \\ &= \frac{1}{\pi^2} \int d^2 z \mathrm{Tr} \left[\hat{\rho} \mathrm{e}^{iz\hat{a}} \frac{1}{\pi} \int d^2 \beta \, |\beta\rangle\langle\beta| \, \mathrm{e}^{iz^*\hat{a}^\dagger} \right] \mathrm{e}^{-iz^*\alpha^*} \mathrm{e}^{-iz\alpha}. \end{aligned} \tag{7.30}$$

Since

$$\begin{aligned} \mathrm{e}^{iz\hat{a}} |\beta\rangle &= \mathrm{e}^{iz\beta} |\beta\rangle, \\ \langle\beta| \mathrm{e}^{iz^*\hat{a}^\dagger} &= \langle\beta| \mathrm{e}^{iz^*\beta^*}, \end{aligned} \tag{7.31}$$

and

$$\frac{1}{\pi^2} \int d^2 z \mathrm{e}^{-iz(\alpha-\beta)} \mathrm{e}^{-iz^*(\alpha^*-\beta^*)} = \delta^2(\alpha - \beta), \tag{7.32}$$

we find

$$\begin{aligned} Q(\alpha) &= \frac{1}{\pi} \int d^2 \beta \langle\beta| \hat{\rho} |\beta\rangle \left[\frac{1}{\pi^2} \int d^2 z \mathrm{e}^{-iz(\alpha-\beta)} \mathrm{e}^{-iz^*(\alpha^*-\beta^*)} \right] \\ &= \frac{1}{\pi} \langle \alpha | \hat{\rho} | \alpha \rangle, \end{aligned} \tag{7.33}$$

as required. □

The Q representation has the advantage of existing for states where no P representation exists and unlike the P representation is always positive.

Example 7.4 (Q representation for a pure coherent state) *A pure coherent state is given by the density operator $\hat{\rho} = |\beta\rangle\langle\beta|$. Hence*

$$Q(\alpha) = \frac{1}{\pi}\langle\alpha\,|\hat{\rho}|\,\alpha\rangle = \frac{1}{\pi}\langle\alpha|\beta\rangle\langle\beta|\alpha\rangle = \frac{1}{\pi}|\langle\beta|\alpha\rangle|^2 = \frac{1}{\pi}e^{-|(\alpha-\beta)|^2}. \tag{7.34}$$

Thus, the $Q(\alpha)$ representation of the pure coherent state $|\beta\rangle$ is a Gaussian function centred at β.

Example 7.5 (Q representation for a Fock state) *For a Fock state $\hat{\rho} = |n\rangle\langle n|$ and then the $Q(\alpha)$ representation is*

$$Q(\alpha) = \frac{1}{\pi}\,|\langle n|\alpha\rangle|^2 = \frac{|\alpha|^{2n}}{\pi n!}e^{-|\alpha|^2}. \tag{7.35}$$

Thus, the $Q(\alpha)$ representation of the photon number state is a Poission function.

7.7 Wigner Representation

For symmetric or Weyl order of the operators, the characteristic function is defined as

$$\chi_S(z) = \mathrm{Tr}\left\{\hat{\rho}e^{iz^*\hat{a}^\dagger + iz\hat{a}}\right\}, \tag{7.36}$$

and its Fourier transform is called the Wigner[a] representation [46]

$$W(\alpha) = \frac{1}{\pi^2}\int d^2z\chi_S(z)\,e^{-iz^*\alpha^*}e^{-iz\alpha}. \tag{7.37}$$

The symmetric order of the $\hat{a}^\dagger$, $\hat{a}$ operators is the average (permutation) of all possible orderings of the operators

$$\begin{aligned}\left(\hat{a}^\dagger\hat{a}\right)_S &= \frac{1}{2}\left(\hat{a}^\dagger\hat{a} + \hat{a}\hat{a}^\dagger\right),\\ \left(\hat{a}^{\dagger 2}\hat{a}\right)_S &= \frac{1}{3}\left(\hat{a}^{\dagger 2}\hat{a} + \hat{a}^\dagger\hat{a}\hat{a}^\dagger + \hat{a}\hat{a}^{\dagger 2}\right).\end{aligned} \tag{7.38}$$

In order to illustrate the general method of finding Wigner representations, we consider two examples.

[a]Eugene Wigner was given the Nobel prize in 1963 for his contributions to the theory of the atomic nucleus and the elementary particles, particularly through the discovery and application of fundamental symmetry principles.

Example 7.6 (Wigner representation of the field in a pure coherent state) *As a first example, we calculate the Wigner function of the field in a pure coherent state given by the density operator* $\hat{\rho} = |\alpha_0\rangle\langle\alpha_0|$.

First, we calculate the symmetrically ordered characteristic function

$$\begin{aligned}\chi_S(z) &= Tr\left\{\hat{\rho}e^{-z^*\hat{a}^\dagger}e^{z\hat{a}}\right\}e^{-\frac{1}{2}|z|^2} \\ &= Tr\left\{|\alpha_0\rangle\langle\alpha_0|e^{-z^*\hat{a}^\dagger}e^{z\hat{a}}\right\}e^{-\frac{1}{2}|z|^2} \\ &= e^{-\frac{1}{2}|z|^2}e^{-z^*\alpha_0^*}e^{z\alpha_0},\end{aligned} \tag{7.39}$$

where we redefined $z \equiv iz$.

Next, substituting this into Eq. (7.37), we obtain

$$\begin{aligned}W(\alpha) &= \frac{1}{\pi^2}\int d^2z e^{z^*(\alpha^*-\alpha_0^*)}e^{-z(\alpha-\alpha_0)}e^{-\frac{1}{2}|z|^2} \\ &= \frac{1}{\pi^2}\int d^2z e^{z^*(\alpha^*-\alpha_0^*)-z(\alpha-\alpha_0)-\frac{1}{2}|z|^2}.\end{aligned} \tag{7.40}$$

The integral can be evaluated using the identity

$$\frac{1}{\pi}\int d^2z e^{-\gamma|z|^2+\mu z+\nu z^*} = \frac{1}{\gamma}e^{\frac{\mu\nu}{\gamma}}, \tag{7.41}$$

which holds for $Re(\gamma) > 0$ *and arbitrary* μ, ν.

This gives the Wigner function of the form

$$W(\alpha) = \frac{2}{\pi}e^{-2|(\alpha-\alpha_0)|^2}. \tag{7.42}$$

Since α is a complex number, it is convenient to express the Wigner function in terms of two (real) quadrature components $2x = \alpha + \alpha^*$, and $2y = -i(\alpha - \alpha^*)$. In the (x, y) representation, the Wigner function (7.42) takes the form

$$W(x, y) = \frac{2}{\pi}\mathrm{e}^{-2\left[(x-x_0)^2+(y-y_0)^2\right]}. \tag{7.43}$$

We can use the general form of a Gaussian function and write the Wigner function as

$$W(x, y) = \frac{1}{2\pi\sigma_x\sigma_y}\exp\left[-\frac{1}{2}\frac{(x-x_0)^2}{\sigma_x^2} - \frac{1}{2}\frac{(y-y_0)^2}{\sigma_y^2}\right], \tag{7.44}$$

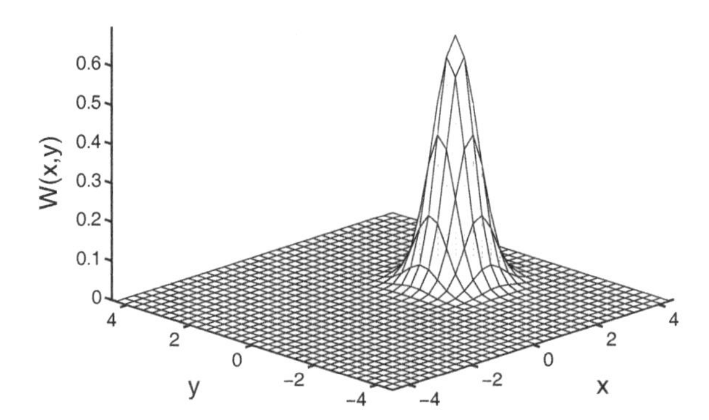

Figure 7.1 The Wigner function $W(x, y)$ for the pure coherent state with $x_0 = 2$ and $y_0 = 0$.

with $\sigma_x^2 = \sigma_y^2 = 1/4$, where $\sigma_x^2 = \langle(\Delta E_1)^2\rangle$ and $\sigma_y^2 = \langle(\Delta E_2)^2\rangle$ are the variances of the electric field amplitudes.

The Wigner function of the pure coherent state is a Gaussian centred at α_0, or equivalently at (x_0, y_0), see Fig. 7.1, with the fluctuations symmetrically distributed around (x_0, y_0). This simple example shows that the Wigner function is usually less singular than the corresponding P function. For the pure coherent state considered in this example, the P function is a two-dimensional delta function (singular function), whereas the Wigner function is a Gaussian.

Since the squeezed vacuum state is a special case of the squeezed coherent state, the Wigner function of the squeezed vacuum state is also a Gaussian, but with either $\sigma_x^2 < 1/4$ or $\sigma_y^2 < 1/4$. In Fig. 7.2, we plot the Wigner function of of the squeezed vacuum state. We assume that the fluctuations in the y-component are reduced in the expense of an increase in the fluctuations in the x-component, so that $\sigma_x^2 > 1/4$ and $\sigma_y^2 < 1/4$. Nevertheless, the product of the variances takes the minimum value, that is, $\sigma_x \sigma_y = 1/4$. The Wigner function of the squeezed vacuum state is always positive, and can be considered as a classical probability distribution.

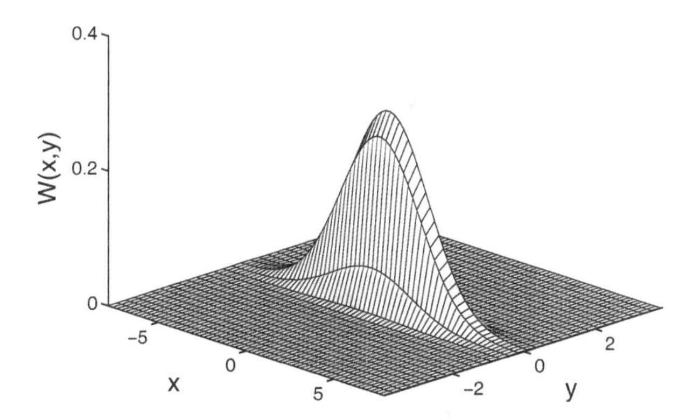

Figure 7.2 The Wigner function of the squeezed vacuum state with the fluctuations reduced in the y component with $\sigma_x^2 = 5$ and $\sigma_y^2 = 1/20$.

Example 7.7 (Wigner representation of the field in a Fock state.) *As a second example, we calculate the Wigner function of the field in the Fock state $|n\rangle$.*

In the pure Fock state $\hat{\rho} = |n\rangle\langle n|$, and then the symmetrically ordered characteristic function takes the form

$$\begin{aligned}\chi_S(z) &= Tr\left\{\hat{\rho}e^{-z^*\hat{a}^\dagger}e^{z\hat{a}}\right\}e^{-\frac{1}{2}|z|^2}\\ &= e^{-\frac{1}{2}|z|^2}\left(1 - n|z|^2 + \frac{1}{4}n(n-1)|z|^4 - \cdots\right).\end{aligned} \tag{7.45}$$

Substituting this into Eq. (7.37) and using Eq. (7.41), we obtain

$$W_{|n\rangle}(\alpha) = \frac{2}{\pi}(-1)^n e^{-2|\alpha|^2}\mathcal{L}_n(4|\alpha|^2), \tag{7.46}$$

where $\mathcal{L}_n(4|\alpha|^2)$ is the Laguerre polynomial of order n. This Wigner function can be negative for $n > 0$. It is shown in Fig. 7.3, where we plot $W_{|n\rangle}(\alpha)$ for the Fock state with $n = 0, 1$ and $n = 2$ photons. The negativities of the Wigner function $W_{|n\rangle}(\alpha)$ for $n > 0$ indicate the non-classicality of these quasi-probabilities.

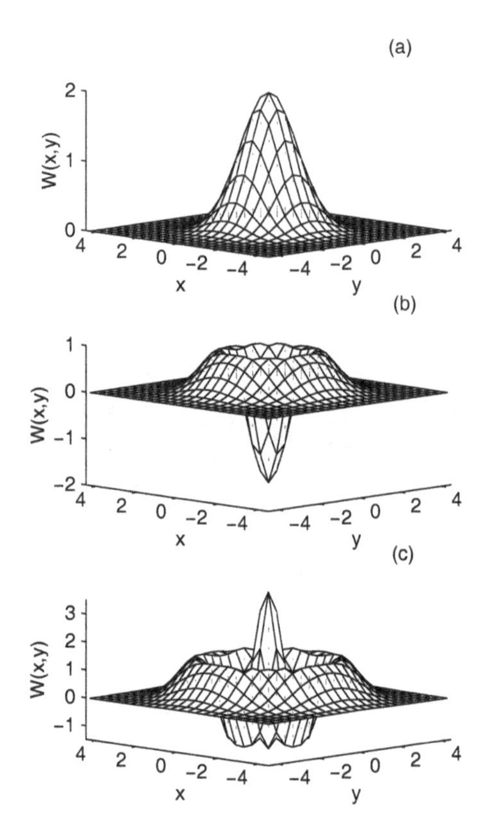

Figure 7.3 The Wigner function of the first three Fock states, that is (a) the vacuum state $|0\rangle$, (b) the Fock state $|1\rangle$ with one photon, (c) the Fock state $|2\rangle$ with two photons.

7.8 Relations between the Wigner, Q and P Representations

We now turn to a brief discussion of useful relations between the P, Q and Wigner representations.

A useful relation between the Wigner or Q representation and the P representation is obtained via the characteristic functions by replacing the density operator by

$$\hat{\rho} = \int d^2\beta P(\beta) |\beta\rangle\langle\beta| . \tag{7.47}$$

This leads to the following form of the Wigner function

$$W(\alpha) = \frac{2}{\pi}\int d^2\beta P(\beta)\, \mathrm{e}^{-2|\beta-\alpha|^2}, \tag{7.48}$$

which is a Gaussian convolution of the P functions.

We now further introduce the relation between the Q and the P representations, that is

$$Q(\alpha) = \frac{1}{\pi}\int d^2\beta P(\beta)\, \mathrm{e}^{-|\beta-\alpha|^2}. \tag{7.49}$$

Note that the Q function like the Wigner function is a Gaussian convolution of the P function. The width of the Gaussian is larger than for the Wigner function. It is $1/\sqrt{2}$ times the width of the Wigner function, which can be accounted for the rather more well-behaved properties.

The above equations show how to find $W(\alpha)$ and $Q(\alpha)$, if we know $P(\alpha)$. These equations also allow us to find $P(\alpha)$ if we know $W(\alpha)$ or $Q(\alpha)$. For example, if we know $Q(\alpha)$, we can write

$$Q(\beta) = \frac{1}{\pi}\int d^2\alpha P(\alpha)\, \mathrm{e}^{-|\alpha|^2}\mathrm{e}^{-|\beta|^2}\mathrm{e}^{\alpha\beta^*+\alpha^*\beta}. \tag{7.50}$$

Then

$$Q(\beta)\, \mathrm{e}^{|\beta|^2} = \frac{1}{\pi}\int d^2\alpha \left[P(\alpha)\, \mathrm{e}^{-|\alpha|^2}\right] \mathrm{e}^{\alpha\beta^*+\alpha^*\beta}. \tag{7.51}$$

Hence, the function $Q(\beta)\exp(|\beta|^2)$ is the Fourier transform of the function $P(\alpha)\exp(-|\alpha|^2)$. On taking the Fourier inverse of Eq. (7.51), we obtain

$$P(\alpha) = \frac{\mathrm{e}^{|\alpha|^2}}{\pi}\int d^2\beta \left[Q(\beta)\, \mathrm{e}^{|\beta|^2}\right] \mathrm{e}^{-\alpha\beta^*-\alpha^*\beta}. \tag{7.52}$$

We can summarize that the P, Q and Wigner representations are related to each other, and we can combine these relations by introducing a single mathematical formula linking the representations.

This can be done as follows. We can define an operator

$$\hat{T}(\alpha, u) = \frac{1}{\pi}\int d^2\beta \hat{D}(\beta, u)\, \mathrm{e}^{\alpha\beta^*-\alpha^*\beta}, \tag{7.53}$$

where

$$\hat{D}(\beta, u) = \hat{D}(\beta)\, \mathrm{e}^{u|\beta|^2/2} \tag{7.54}$$

is a generalized displacement operator.

A complex quasiprobability distribution can be defined as the expectation value of the $\hat{T}(\alpha, u)$ operator

$$U(\alpha, u) = \frac{1}{\pi}\mathrm{Tr}\left[\hat{\rho}\hat{T}(\alpha, u)\right], \tag{7.55}$$

where the complex number u determines the ordering of the field operators and hence, the type of the quasi-distribution function.

For $u = 0$, which determines the symmetric ordering, we get the Wigner function, while for $u = +1$ (normal ordering) and $u = -1$ (antinormal ordering), we obtain the P and Q functions, respectively.

7.9 Distribution Functions in Terms of Quadrature Components

We have introduced three representations of the density operator of the EM field and discussed relations between them. This would be a good time to give a second look at the representations. The phase space is the complex plane of eigenvalues α (Reα, Imα). Since α is the eigenvalue of a non-Hermitian operator, it is more convenient to represent the distribution functions in terms of eigenvalues of Hermitian operators.

We can express the annihilation operator $\hat{a}$ or the eigenvalue α in terms of the position and the momentum operators (eigenvalues) as

$$\hat{a} = \sqrt{\frac{m\omega}{2\hbar}}\hat{q} + i\sqrt{\frac{1}{2m\hbar\omega}}\hat{p},$$
$$\hat{a}^{\dagger} = \sqrt{\frac{m\omega}{2\hbar}}\hat{q} - i\sqrt{\frac{1}{2m\hbar\omega}}\hat{p}, \tag{7.56}$$

or in terms of quadrature operators

$$\hat{a} = \hat{x} + i\hat{y},$$
$$\hat{a}^{\dagger} = \hat{x} - i\hat{y}, \tag{7.57}$$

and present the distribution functions in the (p, q) or (x, y) coordinates.

For example, the Wigner function of the field in a pure coherent state (7.42) in terms of the quadratures ($\alpha = x + iy$) is given by

$$W(\alpha) = \frac{2}{\pi}\mathrm{e}^{-2\left[(x-x_0)^2+(y-y_0)^2\right]}, \tag{7.58}$$

which is a Gaussian centred at (x_0, y_0). The cross-section is a circle indicating that the the fluctuations in both quadratures of the coherent field are equal.

7.10 Summary

We have seen that each of the representations has advantages and disadvantages connected with its use. The P representation describes a quantum state in terms of the probability that the system is in a given coherent state. This distribution function is highly singular or negative for quantum states without classical analogues. This apparent disadvantage is often used as the signature of genuine quantum mechanical effects. The use of the generalized P distribution allows the calculations of normally ordered products without the attendant non-analyticity. The Wigner function may become negative for some quantum states, but it has the considerable advantage for squeezed states that its contours map out the variances of the field quadratures. The Q representation is a positive define distribution function, but its simple relation to antinormally ordered characteristic function makes it difficult to interpret in terms of conventional photon counting or squeezing measurements. The reader interested in a further study of the role of the representations in quantum optics problems is referred especially to the book by Schleich [47].

Exercises

7.1 Show that $\langle \hat{A} \rangle = \mathrm{Tr}(\rho \hat{A})$ for a general mixed state described by the density operator ρ.

7.2 Starting from the time-dependent Schrödinger equation derive the equation of motion for the density operator ρ, the Liouville–von Neumann equation

$$i\hbar \frac{\partial \rho}{\partial t} = [\hat{H}, \rho],$$

and show that the equation holds both for the pure and mixed states.

7.3 Starting from the Liouville–von Neumann equation show that:

(a) $\partial \mathrm{Tr}(\rho)/\partial t = 0$.

(b) $\mathrm{Tr}(\rho^2)$ is a constant of motion. In other words, there is no change in the purity of the state during the evolution of the system.

7.4 The entropy of a quantum system determined by the density operator ρ is defined by

$$S = -k_\mathrm{B} \mathrm{Tr}\left(\rho \ln \rho\right),$$

where S describes the degree of order in the system and k_B is the Boltzmann constant. Calculate the entropy of a pure state $\rho = |\psi\rangle\langle\psi|$ to show that for a pure state, $S = 0$.

7.5 Show that in the Glauber–Sudarshan P representation

$$\langle : (\Delta \hat{n})^2 : \rangle = \int d^2\alpha \; P\left(\alpha\right) \left[|\alpha|^2 - \int d^2\alpha' \; P\left(\alpha'\right) \left|\alpha'\right|^2 \right]^2,$$

and

$$\langle : \left(\Delta \hat{E}_1\right)^2 : \rangle = \frac{1}{4} \int d^2\alpha \; P\left(\alpha\right) \left\{ \left(\alpha + \alpha^*\right) - \left[\int d^2\alpha' P\left(\alpha'\right)\left(\alpha' + \alpha'^*\right) \right] \right\}^2.$$

7.6 Show that for arbitrary μ, ν and $\mathrm{Re}\,\gamma > 0$

$$\frac{1}{\pi} \int d^2 z \, \mathrm{e}^{-\gamma |z|^2 + \mu z + \nu z^*} = \frac{1}{\gamma} \mathrm{e}^{\frac{\mu\nu}{\gamma}}.$$

7.7 Write in the symmetric order the $\hat{a}^2 \hat{a}^{\dagger 2}$ product of the annihilation and creation operators.

7.8 Prove that a product of m annihilation operators and n creation operators can be symmetrically ordered in $(n + m)!/(n!m!)$.

7.9 Show that the Wigner function $W(\alpha)$ can be written as a Gaussian convolution of the P functions

$$W\left(\alpha\right) = \frac{2}{\pi} \int d^2\beta \; P(\beta) \mathrm{e}^{-2|\beta - \alpha|^2}.$$

7.10 Show that if in the Wigner representation the density operator is of the form

$$\rho = \frac{\hbar}{\Delta p \Delta q} \mathrm{e}^{-(q - q_0)^2/2(\Delta q)^2} \mathrm{e}^{-(p - p_0)^2/2(\Delta p)^2},$$

then the condition

$$\mathrm{Tr}(\rho^2) = \frac{1}{2\pi\hbar}\int_{-\infty}^{\infty}\int_{-\infty}^{\infty} dp\,dq\,\rho^2 \leq 1$$

is equivalent to the uncertainty relation

$$\Delta p \Delta q \geq \frac{\hbar}{2}.$$

7.11 Derive Eq. (7.46) for the Fock state with one photon ($n = 1$) to show that $W_{|1\rangle}(\alpha) < 0$ for $|\alpha|^2 < \frac{1}{4}$.

7.12 The density operator of a thermal field in the photon number representation can be written as

$$\hat{\rho} = \sum_{n=0}^{\infty} \frac{N^n}{(1+N)^{n+1}} |n\rangle\langle n|,$$

where N is the average number of photons in the field.

(a) Find the $Q(\alpha)$ representation of the thermal field.

(b) Using the relation between $Q(\beta)$ and the diagonal $P(\alpha)$ representation, show that $P(\alpha)$ for the thermal field is a Gaussian.

(c) Using the Gaussian form of the $P(\alpha)$ representation, find the variances $\sigma_x^2 = \langle(\Delta\hat{E}_1)^2\rangle$ and $\sigma_y^2 = \langle(\Delta\hat{E}_2)^2\rangle$ of a single mode thermal field amplitudes.

(d) Verify the result of the point (c) by calculating the variance from the definition of σ_x^2 using the correlation properties of the thermal field operators

$$\langle\hat{a}\rangle = \langle\hat{a}^\dagger\rangle = \langle\hat{a}^2\rangle = \langle(\hat{a}^\dagger)^2\rangle = 0,$$
$$\langle\hat{a}^\dagger\hat{a}\rangle = N,$$
$$\langle\hat{a}\hat{a}^\dagger\rangle = 1 + N.$$

Does the result for σ_x^2 agree with that found in the point (c)? Explain your answer.

(e) Show that the variance found in the point (d) agrees with that predicted from the Wigner distribution.

Chapter 8

Single-Mode Interaction

8.1 Introduction

In a quantum physics scenario, atoms whose quanta are fermions interact with force fields whose quanta are bosons. From this perspective, the interaction of a fermion with a boson is one of the fundamental systems in quantum physics. The interaction of a single mode of the electromagnetic (EM) field with a single two-level atom in the absence of any dissipation process, such as spontaneous emission, and any input or output from the cavity is the simplest example of interaction between fermions and bosons. This elementary model is an example of reversible system in which an initial energy is continuously and periodically exchanged between the atom and the cavity field. Mathematically, only two states are involved making the Hamiltonian of the system simple to diagonalize analytically.

An understanding and exploring of the properties of such a system is of great use since:

(1) It is a perennial problem in quantum optics.
(2) It has exact analytical solutions for arbitrary coupling constants.
(3) It exhibits certain periodic *collapse* and *revival* phenomena due to the quantum nature of the field.
(4) It has recently become possible to realize it experimentally.

Quantum Optics for Beginners
Zbigniew Ficek and Mohamed Ridza Wahiddin

ISBN 978-981-4411-75-2 (Hardcover), 978-981-4411-76-9 (eBook)
www.panstanford.com

In this chapter, we will examine this elementary model as a prelude to more complicated systems involving atoms interacting with multi-mode fields. We shall find that the state of the combined system is a pure entangled state between the atom and the field and study in details the time evolution of the atomic population for different initial states of the single-mode EM field.

8.2 The Jaynes–Cummings Model

The model of the interaction between a single two-level atom and a quantized single-mode field in the rotating wave approximation (RWA) was introduced independently by Jaynes and Cummings [48] and by Paul [49] in 1963, and is called the Jaynes–Cummings model. In this model the atom is coupled to a single-mode EM field of frequency ω_c being on resonance with the atomic transition, that is, $\omega = \omega_0$. The atom is represented by two energy states: the ground state $|1\rangle$ and the excited state $|2\rangle$, and is described by the spin operators S^+, S^- and S_z. The field mode is represented by a Fock state $|n\rangle$, and is described by the annihilation and creation operators $\hat{a}$ and $\hat{a}^\dagger$.

8.2.1 *The Jaynes–Cummings Hamiltonian*

Consider a two-level atom described by the Hamiltonian Eq. (2.15) interacting in the electric dipole approximation with a single-mode electric field. For the interaction of the atom with a single mode of the EM field, $\omega_k = \omega_c$ and under the rotating-wave approximation, the Hamiltonian of the system simplifies to

$$\hat{H} = \hbar\omega_0 S_z + \hbar\omega_c\left(\hat{a}^\dagger\hat{a} + \frac{1}{2}\right) - \frac{1}{2}i\hbar g\left(S^+\hat{a} - S^-\hat{a}^\dagger\right), \quad (8.1)$$

where

$$g = (\vec{\mu}\cdot\vec{e})\sqrt{\frac{2\omega_0}{\hbar\varepsilon_0 V}} \quad (8.2)$$

is the atom–field coupling constant, called one-photon Rabi frequency.

The Hamiltonian (8.1) is called the Jaynes–Cummings Hamiltonian. It describes the simplest model of the atom–field interaction, a single two-level atom interacting with a single-mode field.

8.2.2 *State Vector of the System*

In the absence of the interaction between the atom and the field, ($g = 0$), the Hilbert space of the system can be spanned by product states $|n, 1\rangle = |n\rangle \otimes |1\rangle$ and $|n-1, 2\rangle = |n-1\rangle \otimes |2\rangle$, where n is the number of photons in the field mode, and $|1\rangle$, $|2\rangle$ are the atomic states. The states group into manifolds, which we can label as $\mathcal{E}(n)$, where n is the number of excitations of the states in $\mathcal{E}(n)$. It can be easily verified that the manifold $\mathcal{E}(0)$ is a singlet $|0, 1\rangle$, whereas the manifolds corresponding to $n > 0$ are doublets. When $\omega_c = \omega_0$, the inter-doublet states are degenerate in energy whereas the states are non-degenerate when $\omega_c \neq \omega_0$, as it is illustrated in Fig. 8.1.

We now use these states as the basis states for the state vector of the total (interacting) system to find the time evolution of the state.

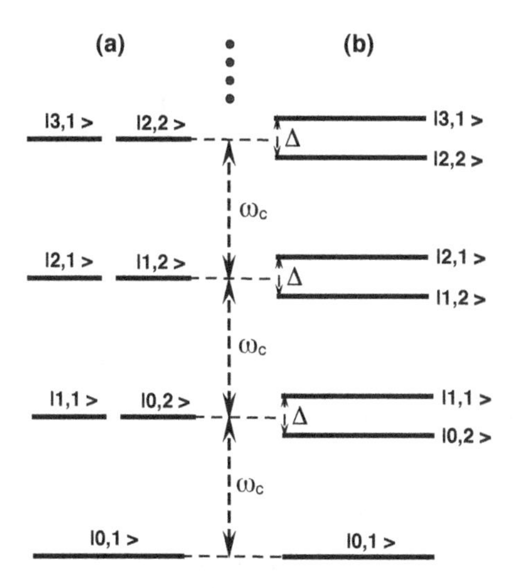

Figure 8.1 Energy levels of the non-interacting atom and the cavity mode for (a) degenerate frequencies $\Delta = \omega_c - \omega_0 = 0$, and (b) non-degenerate frequencies $\Delta \neq 0$.

Consider a state vector of the system at time t:

$$|\psi(t)\rangle = C_{1n}(t)|n, 1\rangle + C_{2n}(t)|n-1, 2\rangle, \tag{8.3}$$

where $C_{1n}(t)$ and $C_{2n}(t)$ are probability amplitudes of the states $|n, 1\rangle$ and $|n-1, 2\rangle$, respectively. The time evolution of the probability amplitudes is found from the time-dependent Schrödinger equation

$$i\hbar\frac{d}{dt}|\psi(t)\rangle = \hat{H}|\psi(t)\rangle, \tag{8.4}$$

which leads to the following differential equation for the probability amplitudes

$$i\hbar\left[\dot{C}_{1n}(t)|n, 1\rangle + \dot{C}_{2n}(t)|n-1, 2\rangle\right] = \hat{H}|\psi(t)\rangle. \tag{8.5}$$

To proceed further, we need the explicit form of the Hamiltonian $\hat{H}$. According to Eq. (8.1), the Hamiltonian is composed of three terms. Thus, we have three terms on the right-hand side of Eq. (8.5). Let us consider separately these three terms, which in the case of the resonant coupling ($\omega_c = \omega_0$) can be written as

$$\hbar\omega_0\left(\hat{a}^\dagger\hat{a} + \frac{1}{2}\right)|\psi(t)\rangle = \hbar\omega_0\left\{\left(n + \frac{1}{2}\right)C_{1n}(t)|n, 1\rangle + \left(n - \frac{1}{2}\right)C_{2n}(t)|n-1, 2\rangle\right\},$$

$$\hbar\omega_0 S^z|\psi(t)\rangle = \hbar\omega_0\left[-\frac{1}{2}C_{1n}(t)|n, 1\rangle + \frac{1}{2}C_{2n}(t)|n-1, 2\rangle\right],$$

$$-\frac{1}{2}ig\hbar(S^+\hat{a} - \hat{a}^\dagger S^-)|\psi(t)\rangle = -\frac{1}{2}i\hbar g\left\{C_{1n}(t)\sqrt{n}|n-1, 2\rangle - C_{2n}(t)\sqrt{n}|n, 1\rangle\right\}. \tag{8.6}$$

Projecting Eq. (8.5) onto $\langle 1, n|$ on the left results in a differential equation for the amplitude $C_{1n}(t)$:

$$i\hbar\dot{C}_{1n}(t) = \hbar n\omega_0 C_{1n}(t) + \frac{1}{2}i\hbar g\sqrt{n}C_{2n}(t). \tag{8.7}$$

Projecting Eq. (8.5) onto $\langle 2, n-1|$ on the left results in a differential equation for the amplitude $C_{2n}(t)$:

$$i\hbar\dot{C}_{2n}(t) = \hbar n\omega_0 C_{2n}(t) - \frac{1}{2}i\hbar g\sqrt{n}C_{1n}(t). \tag{8.8}$$

Thus, we have obtained two coupled differential equations that can be written in the form

$$\begin{cases} \dot{C}_{1n}(t) = -in\omega_0 C_{1n}(t) + \frac{1}{2}g\sqrt{n}C_{2n}(t), \\ \dot{C}_{2n}(t) = -in\omega_0 C_{2n}(t) - \frac{1}{2}g\sqrt{n}C_{1n}(t). \end{cases} \tag{8.9}$$

The set of the equations (8.9) can be solved for arbitrary initial conditions using, for example, the Laplace transform method. The solution for the amplitude $C_{2n}(t)$ is given by

$$\begin{aligned} C_{2n}(t) = \frac{1}{2}\mathrm{e}^{-in\omega_0 t}\Big\{&[iC_{2n}(0) - C_{1n}(0)]\,\mathrm{e}^{\frac{1}{2}i\Omega t} \\ &+ [iC_{2n}(0) + C_{1n}(0)]\,\mathrm{e}^{-\frac{1}{2}i\Omega t}\Big\}, \end{aligned} \tag{8.10}$$

where $\Omega = g\sqrt{n}$ is the Rabi frequency and

$$C_{in}(0) = \langle i, n|\psi(0)\rangle, \qquad i = 1, 2, \tag{8.11}$$

is the probability amplitude that the system was initially in the state $|i, n\rangle$.

Note that, in general, the state vector (8.3) is a superposition state, which cannot be written as a product of the atomic and field states. It is therefore an example of an entangled state between the atom and the field mode. Only for particular times at which $C_{1n}(t) = 0$ or $C_{2n}(t) = 0$, the state is in the form of a product states that at these particular times the atom and the field are independent of each other (disentangled).

8.2.3 *Population of the Atomic Excited State*

The probability amplitudes $C_{in}(t)$ continuously evolve in time indicating a continuous exchange of an excitation between the atom and the cavity mode. Let us look closely at the time evolution of the population of the atomic excited state.

Consider first the initial state of the system

$$|\psi(0)\rangle = |1, n_0\rangle, \tag{8.12}$$

in which the atom is in its ground state and n_0 photons are present in the cavity mode. In this case, the initial values of the probability amplitudes are

$$C_{2n}(0) = 0, \qquad C_{1n}(0) = \langle 1, n|1, n_0\rangle = \delta_{n,n_0}, \tag{8.13}$$

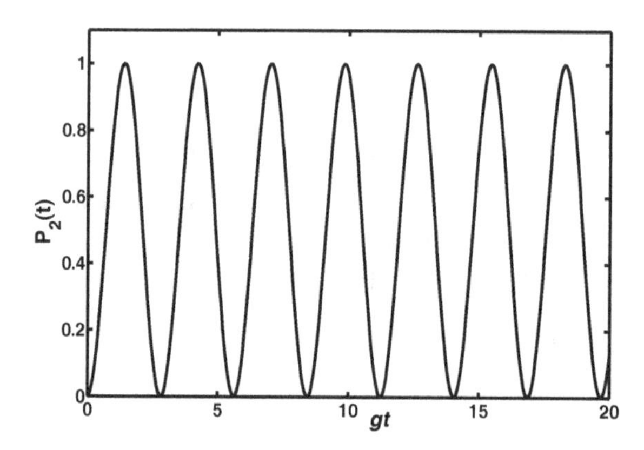

Figure 8.2 Time evolution of the population $P_2(t)$ for the atom initially in the ground state $|1\rangle$ and the definite (fixed) number of photons present in the cavity mode, $n_0 = 5$.

so that the time evolution of the amplitude $C_{2n}(t)$ takes the form

$$|C_{2n}(t)|^2 = \sin^2\left(\frac{1}{2}\Omega t\right)\delta_{n,n_0}. \tag{8.14}$$

Hence, the population of the state $|2\rangle$ is given by

$$P_2(t) = \sum_n |C_{2n}(t)|^2 = \sin^2\left(\frac{1}{2}\Omega t\right). \tag{8.15}$$

The time evolution of the population $P_2(t)$ for a definite number of photons present in the cavity mode is shown in Fig. 8.2. The population oscillates with the Rabi frequency and at certain times $P_2(t) = 1$, indicating that all the population is in the upper state (the total population inversion). Physically, Eq. (8.15) says that the atom initially in its ground state can absorb a photon from the initially excited field mode, and that the atom and the field then sinusoidally and continuously exchange this photon of energy.

8.3 Collapses and Revivals of the Atomic Evolution

The results derived in the previous section can be generalized to the case where the initial state of the field is not a state of definite photon number, but rather some superposition of such number states. For example, assume that the initial state of the system was $|\psi(0)\rangle = |1, \alpha\rangle$, that the atom was in the ground state $|1\rangle$, while the field was in the coherent state $|\alpha\rangle$. In this case, the initial values of the probability amplitudes are

$$C_{2n}(0) = 0, \qquad C_{1n}(0) = \langle n|\alpha\rangle, \tag{8.16}$$

which gives

$$|C_{1n}(0)|^2 = \frac{|\alpha|^{2n}}{n!} \mathrm{e}^{-|\alpha|^2}, \tag{8.17}$$

where α is a complex number. Then, the population P_2 takes the form

$$P_2(t) = \sum_n \frac{\langle n\rangle^n}{n!} \mathrm{e}^{-\langle n\rangle} \sin^2\left(\frac{1}{2}\sqrt{n}gt\right), \tag{8.18}$$

where $\langle n\rangle = |\alpha|^2$ is the mean number of photons in the cavity field.

No exact analytic expression for the sum in Eq. (8.18) exists, but one can notice that due to the Poisson distribution of the photon number n, there will be a spread in the Rabi frequency over different n. As a result the Rabi oscillations will dephase and next will collapse after some time t.

Figure 8.3 shows the time evolution of the population $P_2(t)$ for the mean photon number $\langle n\rangle = 20$. The oscillations collapse after

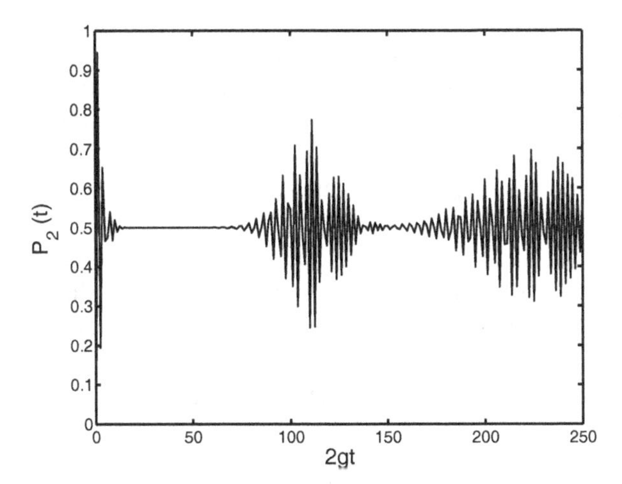

Figure 8.3 Time evolution of the population $P_2(t)$ for an initial coherent state of the field with the mean photon number $\langle n\rangle = 20$.

few Rabi periods, remain constant over a finite interval of time and then the oscillations reappear again. This revival then collapses and again after a finite time a new, but longer, revival begins. This process repeats but disappears after a long time, not shown in the figure.

This phenomenon is called *collapses* and *revivals* of the atomic evolution [50], and can be explained as follows:

Consider the population $P_2(t)$, given in Eq. (8.18). Usually, no analytical expression for the sum (8.18) exists, but the general behaviour of $P_2(t)$ can be described by noting that Eq. (8.18) can be rewritten as the product of a rapidly oscillating term and a slowly varying envelope function, that is

$$\begin{aligned} P_2(t) &= \frac{1}{2}\left[1 - \sum_n \frac{\langle n\rangle^n}{n!} e^{-\langle n\rangle} \cos\left(g\sqrt{n}t\right)\right] \\ &= \frac{1}{2}\left[1 - \sum_n \frac{\langle n\rangle^n}{n!} e^{-\langle n\rangle} \mathrm{Re}\left(e^{ig\sqrt{n}t}\right)\right] \\ &= \frac{1}{2}\left\{1 - \mathrm{Re}\left[e^{i\sqrt{\langle n\rangle}gt} D(t)\right]\right\}, \end{aligned} \tag{8.19}$$

where

$$D(t) = \sum_n \frac{\langle n\rangle^n}{n!} e^{-\langle n\rangle} e^{ig\left(\sqrt{n}-\sqrt{\langle n\rangle}\right)t} \tag{8.20}$$

is the slowly oscillating envelope function.

Note that $D(t)$ oscillates with two opposite phases, one (negative) for $n < \langle n\rangle$ and the other (positive) for $n > \langle n\rangle$. Interferences between the different components (with different phases) in the summation cause $D(t)$ to first approach zero (collapse) and then increase again (revival).

A collapse appears when the phases of the two components, $n < \langle n\rangle$ and $n > \langle n\rangle$, differ by π (are opposite). To show this explicitly, consider two phases which differ by the standard deviation Δn. These phases have the most significant effect on the modulation of $D(t)$. Thus, for two phases to get different by π:

$$\left(\sqrt{\langle n\rangle + \frac{1}{2}\Delta n} - \sqrt{\langle n\rangle}\right) gt \approx \left(\sqrt{\langle n\rangle - \frac{1}{2}\Delta n} - \sqrt{\langle n\rangle}\right) gt + \pi, \tag{8.21}$$

which gives the time

$$t_c = \frac{2\pi}{g}\frac{\sqrt{\langle n\rangle}}{\Delta n}, \tag{8.22}$$

that for a Poisson distribution reduces to

$$t_c = \frac{2\pi}{g}. \tag{8.23}$$

Revival time t_R is determined by how long it takes two phases differing by 1 to rephase, that is

$$\left(\sqrt{\langle n\rangle + 1} - \sqrt{\langle n\rangle}\right) gt \approx 2\pi, \tag{8.24}$$

which gives the time

$$t_R \approx \frac{4\pi}{g}\sqrt{\langle n\rangle}. \tag{8.25}$$

The ratio of the revival to the collapse time is

$$\frac{t_R}{t_C} = 2\Delta n. \tag{8.26}$$

Since $\Delta n \gg 1$, the revival time is longer compared to the collapse time.

We have illustrated the collapse and revival phenomena for the initial coherent state of the field. It is not difficult to extend the analysis to other initial field states, for example, thermal or squeezed states, which also produce collapses and revivals. The collapse and revival times then depend on the details of the photon number probability distributions of the field states.

An obvious question arises whether the Jaynes–Cummings model could be realized in practice. Recently it has become possible to create cavities composed of a single mode of sufficiently high quality factor Q. The quality factor indicates the lifetime of a single photon in the cavity Q/ω, where ω is the photon frequency. In the laboratory of Haroche[a], a high-Q microwave cavity, $Q \sim 6 \times 10^9$, was fabricated [51]. In such cavity, light travels 39,000 km before leaving the cavity. To increase the coupling of the atom to the cavity mode, Rydberg atoms were used. The Rydberg atoms are highly excited atoms which have large sizes and therefore can have huge dipole

[a]Serge Haroche was granted the Nobel prize in 2012 for his experimental work that enabled measuring and manipulation of individual quantum systems.

moments, proportional to n^2, where n is the principal quantum number [52]. Because the dipole moments of Rydberg atoms are so large, the atoms can strongly couple to the cavity mode.

Another possibility where the interaction between a two-level atom and a single photon can be studied is a superconducting qubit placed in a superconducting strip line cavity. The coplanar strip line cavity can be viewed as a squashed coaxial cable. The length of the cavity corresponds to the microwave wavelength. In the first realization of this system, by the group of Schoelkopf [53], the quality factor of $Q \sim 7500$ was achieved.

Exercises

8.1 **(a)** Show that the operator

$$\hat{N} = S_z + \hat{a}^\dagger \hat{a}$$

commutes with the Jaynes–Cummings Hamiltonian (8.1).

(b) What is the physical meaning of the operator $\hat{N}$ and the fact that it commutes with $\hat{H}$?

(c) What would be the result of the commutator $[\hat{N}, \hat{H}]$ if the Hamiltonian included the counter-rotating terms?

8.2 Show that in the basis of product states $|1, n\rangle$ and $|2, n-1\rangle$, the Jaynes–Cummings Hamiltonian (8.1) can be written in a matrix form as

$$\hat{H} = \hbar\omega\left(n + \frac{1}{2}\right) I + \hbar\begin{pmatrix} \frac{1}{2}\Delta & g\sqrt{n} \\ g\sqrt{n} & -\frac{1}{2}\Delta \end{pmatrix},$$

where I is the 2×2 unit matrix and $\Delta = \omega_0 - \omega$ is the detuning of the atomic transition frequency ω_0 from the field mode frequency ω.

8.3 Diagonalize the Hamiltonian given in Exercise 8.2 to find energies and the corresponding energy states of the Jaynes–Cummings model.

8.4 Solve the time-dependent Schrödinger equation with the Jaynes–Cummings Hamiltonian (8.1) to find the evolution operator

$$\hat{U}(t) = \mathrm{e}^{-i\hat{H}t/\hbar}.$$

8.5 Show that the free part $\hat{H}_0 = \hbar\omega_0 S_z + \hbar\omega(\hat{a}^\dagger\hat{a} + 1/2)$ and the interaction part $H_\mathrm{I} = -\frac{1}{2}i\hbar g(S^+\hat{a} - S^-\hat{a}^\dagger)$ of the Jaynes–Cummings Hamiltonian commute.

8.6 Calculate the evolution of the population $P_2(t)$ for a two-level atom interacting with a single-mode EM field of frequency ω detuned from the atomic frequency ω_0; that is, $\omega \neq \omega_0$. Assume the initial state of the system $|\psi(0)\rangle = |1, n_0\rangle$.

8.7 Using computer programs, plot the time evolution of the population $P_2(t)$ for the atom initially in the ground state $|1\rangle$, interacting with a resonant cavity field initially in

(a) thermal (chaotic) state with $\langle n\rangle = 1$.
(b) squeezed coherent state $|\alpha, r\rangle = D(\alpha)S(r)|0\rangle$.
(c) superposition state $|n_0\rangle = (|0\rangle + |2\rangle)/\sqrt{2}$.
(d) Does the population $P_2(t)$ experience the collapses and revival phenomena?

8.8 Evaluate the variance Δn for the squeezed coherent state $|\alpha, r\rangle = D(\alpha)S(r)|0\rangle$ to show that depending on the phase of α the collapse time may be shorter or longer than for the initial coherent state.

8.9 Show that an increase in the collapse time is accompanied by a delay in the onset of the revival.

8.10 Show that for a highly excited coherent state, $\langle n\rangle \gg 1$, the infinite sum in Eq. (8.18) can be converted into integrals that can be evaluated analytically by standard saddle point methods.

Chapter 9

Open Quantum Systems

9.1 Introduction

We have now reached a point where it is possible to consider more advanced and currently intensively developing topics in quantum optics. These particular topics have been chosen because of their importance in the development of modern laser physics, quantum atom optics, quantum information and computation. We start by considering the dynamics of an atom coupled in a free space to a multi-mode electromagnetic (EM) vacuum field. In free space an atom is coupled to many modes of the EM field and then an initial excitation in the system is irreversibly emitted from the atom into the field modes, never again to be reabsorbed by the atom, in contrast to a close system such as the Jaynes–Cummings model where an excitation was periodically exchanged between the atom and the field. The evolution of an excited atom coupled to a multi-mode EM field is a particular example of an open quantum system.

We will consider a two-level atom coupled to a multi-mode EM field and work in the electric dipole approximation. First, we will formulate the Hamiltonian of the system in which we will describe different subsystems and the interactions between them. Next, we will derive a master equation of the system that describes

Quantum Optics for Beginners
Zbigniew Ficek and Mohamed Ridza Wahiddin

ISBN 978-981-4411-75-2 (Hardcover), 978-981-4411-76-9 (eBook)
www.panstanford.com

dynamics of the atom interacting with a multi-mode EM field. The term *master equation* refers to an equation of motion for the reduced density operator of the atom interacting with the multi-mode EM field treated as an external reservoir. In many problems, the reservoir plays the role of environment upon which we have neither control nor influence. This leads to the irreversible process of spontaneous emission that is intrinsically connected with the process of decoherence.

9.2 Hamiltonian of the Multi-Mode Interaction

We start from the Hamiltonian of a combined system that is composed of a two-level atom driven by a single-mode laser field and coupled to the multi-mode EM field. The combined system is treated as a closed system and its Hamiltonian can be written as

$$\hat{H} = \hat{H}_0 + \hat{H}_{\mathrm{L}} + \hat{H}_{\mathrm{F}} + \hat{V} + \hat{H}_{\mathrm{int}}, \tag{9.1}$$

where

$$\hat{H}_0 = \hbar\omega_0 S^z \tag{9.2}$$

is the Hamiltonian of the free atom,

$$\hat{H}_{\mathrm{L}} = \hbar\omega_{\mathrm{L}}\left(\hat{a}^{\dagger}\hat{a} + \frac{1}{2}\right) \tag{9.3}$$

is the Hamiltonian of a single-mode laser field driving the atom,

$$\hat{H}_{\mathrm{F}} = \hbar\sum_k \omega_k\left(\hat{a}_k^{\dagger}\hat{a}_k + \frac{1}{2}\right) \tag{9.4}$$

is the Hamiltonian of the multi-mode field (reservoir) surrounding the atom,

$$\hat{V} = -\frac{1}{2}i\hbar\Omega\left(S^{+}\mathrm{e}^{-i\omega_{\mathrm{L}}t} - S^{-}\mathrm{e}^{i\omega_{\mathrm{L}}t}\right) \tag{9.5}$$

is the interaction between the laser (classical) field and the atom, and Ω is the Rabi frequency which, for simplicity, we assume is a real number, and the final term in Eq. (9.1)

$$\hat{H}_{\mathrm{int}} = -\frac{1}{2}i\hbar\sum_k\left[g_k S^{+}\hat{a}_k(t) - g_k^{*}S^{-}\hat{a}_k^{\dagger}(t)\right] \tag{9.6}$$

is the interaction Hamiltonian between the atom and the multi-mode field. We treat the interactions between the atom and the fields in the electric dipole approximation.

9.3 Derivation of the Master Equation

The combined system is usually in a mixed state. Therefore, the dynamics of a two-level atom coupled to an external field are conveniently studied in terms of the density operator of the combined atom–field system.[a] The density operator ρ_{T} of the combined atom–field system obeys the equation of motion, the Liouville–von Neumann equation [54]

$$i\hbar\frac{\partial}{\partial t}\rho_{\mathrm{T}}(t) = \left[\hat{H}, \rho_{\mathrm{T}}(t)\right]. \tag{9.7}$$

We first consider the interaction between the atom and the multimode field and assume that there is no coherent driving field. It is useful to work in the interaction picture in which

$$\begin{aligned}\tilde{\rho}_{\mathrm{T}} &= \mathrm{e}^{i\hat{H}_0' t/\hbar}\rho_{\mathrm{T}}\mathrm{e}^{-i\hat{H}_0' t/\hbar},\\ \tilde{\hat{H}}_{\mathrm{int}} &= \mathrm{e}^{i\hat{H}_0' t/\hbar}\hat{H}_{\mathrm{int}}\mathrm{e}^{-i\hat{H}_0' t/\hbar},\end{aligned} \tag{9.8}$$

where

$$\hat{H}_0' = \hat{H}_0 + \hat{H}_{\mathrm{L}} + \hat{V} + \hat{H}_{\mathrm{F}}. \tag{9.9}$$

This simplifies the equation of motion for the density operator to

$$i\hbar\frac{\partial}{\partial t}\tilde{\rho}_{\mathrm{T}}(t) = \left[\hat{H}_{\mathrm{int}}(t), \tilde{\rho}_{\mathrm{T}}(t)\right], \tag{9.10}$$

which shows clearly that the evolution of the density operator is governed by the interaction Hamiltonian $\hat{H}_{\mathrm{int}}(t)$ alone.

Formally integrating Eq. (9.10) with respect to time, we obtain

$$\tilde{\rho}_{\mathrm{T}}(t) = \tilde{\rho}_{\mathrm{T}}(0) + \frac{1}{i\hbar}\int_0^t dt' \left[\hat{H}_{\mathrm{int}}(t'), \tilde{\rho}_{\mathrm{T}}(t')\right], \tag{9.11}$$

This is a Volterra-type integral equation which can be solved by successive substitution in the form of an absolutely and uniformly convergent series.

Substituting Eq. (9.11) into the right-hand side of Eq. (9.10), we obtain

$$\begin{aligned}i\hbar\frac{\partial}{\partial t}\tilde{\rho}_{\mathrm{T}}(t) &= \left[\hat{H}_{\mathrm{int}}(t), \tilde{\rho}_{\mathrm{T}}(0)\right]\\ &\quad + \frac{1}{i\hbar}\int_0^t dt' \left[\hat{H}_{\mathrm{int}}(t), \left[\hat{H}_{\mathrm{int}}(t'), \tilde{\rho}_{\mathrm{T}}(t')\right]\right].\end{aligned} \tag{9.12}$$

---

[a] For a generalization of the procedure to the case of N multi-level atoms see Z. Ficek and S. Swain, *Quantum Interference and Coherence: Theory and Experiments* (Springer, Berlin, New York, Heidelberg, 2005).

We can continue the same procedure to obtain an infinite series of integral terms, which can be regarded as an exact explicit solution for $\tilde{\rho}_{\mathrm{T}}(t)$. In practice, the series is terminated at the second order, which gives a sufficiently good approximation to the solution.

To proceed further, we assume that no correlation exists between the atom and the EM field at the initial time $t = 0$, that is

$$\tilde{\rho}_{\mathrm{T}}(0) = \rho(0) \otimes \rho_{\mathrm{F}}(0), \tag{9.13}$$

where $\rho(0)$ is the density operator of the atom, so called reduced density operator, and $\rho_{\mathrm{F}}(0)$ is the density operator of the multi-mode vacuum field. No approximation is involved in this decorrelation.

We now employ *weak coupling or Born Approximation* in which the atom–field interaction is assumed to be weak, and there is no back reaction of the atom on the reservoir. In this approximation, the state of the reservoir does not change in time, remains unchanged during the interaction and retains its initial value. Moreover, the Born approximation involves treating the effects of the reservoir correct to order g_k^2 in the coupling constant. In this case, the time-dependent density operator of the combined system can be written as

$$\tilde{\rho}_{\mathrm{T}}(t') = \rho(t') \otimes \rho_{\mathrm{F}}(0). \tag{9.14}$$

It is equivalent to say that the future state of the system–reservoir density operator $\tilde{\rho}_{\mathrm{T}}(t')$ is determined by the state of the system $\rho(t')$, and is not a function of the history of the reservoir.

Under this assumption and then by tracing over the field variables, we can limit the calculation to the dynamics of the density operator of the atom alone:

$$\mathrm{Tr}_{\mathrm{F}}\left\{\tilde{\rho}_{\mathrm{T}}(t')\right\} = \mathrm{Tr}_{\mathrm{F}}\left\{\rho(t')\,\rho_{\mathrm{F}}(0)\right\} = \rho(t')\,\mathrm{Tr}_{\mathrm{F}}\left\{\rho_{\mathrm{F}}(0)\right\} = \rho(t'). \tag{9.15}$$

Thus, after tracing over the field variables in Eq. (9.12), the master equation takes the form

$$i\hbar\frac{\partial}{\partial t}\rho(t) = \mathrm{Tr}_{\mathrm{F}}\left\{\left[\hat{H}_{\mathrm{int}}(t), \rho(0)\,\rho_{\mathrm{F}}(0)\right]\right\} + \frac{1}{i\hbar}\int_0^t dt'\,\mathrm{Tr}_{\mathrm{F}}\left\{\left[\hat{H}_{\mathrm{int}}(t)\left[\hat{H}_{\mathrm{int}}(t'), \rho(t')\rho_{\mathrm{F}}(0)\right]\right]\right\}. \tag{9.16}$$

Substituting the explicit form of the interaction Hamiltonian $H_{\text{int}}(t)$, we get

$$\begin{aligned}\frac{\partial}{\partial t}\rho(t) &= -\frac{1}{2}\sum_k \{g_k S^+(t)\,\rho(0)\,\langle \hat{a}_k(t)\rangle - g_k^* S^-(t)\,\rho(0)\,\langle \hat{a}^\dagger(t)\rangle - \text{H.c.}\} \\ &\quad +\frac{1}{4}\sum_{k,k'}\int_0^t dt'\, g_k g_{k'}^* \Big\{S^+(t)\,S^-(t')\,\rho(t')\,\langle \hat{a}_k(t)\,\hat{a}_{k'}^\dagger(t')\rangle \\ &\quad + \left[\rho(t')\,S^+(t)\,S^-(t') - S^-(t)\,\rho(t')\,S^+(t')\right]\langle \hat{a}_k(t)\,\hat{a}_{k'}^\dagger(t')\rangle \\ &\quad - S^-(t')\,\rho(t')\,S^+(t)\,\langle \hat{a}_k(t')\,\hat{a}_{k'}^\dagger(t)\rangle\Big\} + \cdots 12 \text{ terms}. \end{aligned} \tag{9.17}$$

The 12 terms explicitly not listed in Eq. (9.17) involve combinations of the atomic operators giving the two-photon correlation functions $\langle \hat{a}_{k'}^\dagger(t)\,\hat{a}_{k'}^\dagger(t')\rangle$, $\langle \hat{a}_k(t)\,\hat{a}_{k'}(t')\rangle$ and number of photons $\langle \hat{a}_k^\dagger(t)\,\hat{a}_{k'}(t')\rangle$.

It is seen from Eq. (9.17) that the evolution of the density operator depends on the correlation functions of the field operators. In the following, we assume the temperature of the reservoir to be absolute zero that all modes of the EM field are in the ordinary vacuum state for which the correlation functions are given by

$$\begin{aligned} \langle \hat{a}_k\rangle &= \langle \hat{a}_{k'}^\dagger\rangle = 0, \\ \langle \hat{a}_{k'}^\dagger(t)\,\hat{a}_{k'}^\dagger(t')\rangle &= \langle \hat{a}_k(t)\,\hat{a}_{k'}(t')\rangle = 0, \\ \langle \hat{a}_k^\dagger(t)\,\hat{a}_{k'}(t')\rangle &= 0, \\ \langle \hat{a}_k(t)\,\hat{a}_{k'}^\dagger(t')\rangle &= \mathrm{e}^{i\omega_k t}\mathrm{e}^{-i\omega_{k'}t'}\delta_{kk'}. \end{aligned} \tag{9.18}$$

This step reduces the number of contributions that govern the evolution of the density operator. Effectively, the 16 terms contributing to the master equation (9.17) reduce to only four contributions. Care must be taken when selecting terms that are different from zero in the ordinary vacuum field. In Eq. (9.18), there are terms of the form

$$S^+\rho(t')S^-\mathrm{Tr}_\mathrm{F}\left\{\hat{a}_k(t)\,\rho_\mathrm{F}(0)\hat{a}_{k'}^\dagger(t')\right\}, \tag{9.19}$$

and

$$S^-\rho(t')S^+\mathrm{Tr}_\mathrm{F}\left\{\hat{a}_k^\dagger(t)\,\rho_\mathrm{F}(0)\hat{a}_{k'}(t')\right\}. \tag{9.20}$$

One can think that in the ordinary vacuum, the term (9.19) will make a non-zero contribution to the master equation as $\langle \hat{a}_k(t)\,\hat{a}_k^\dagger(t')\rangle \neq 0$. However,

$$\begin{aligned}\mathrm{Tr}_\mathrm{F}\left\{\hat{a}_k(t)\,\rho_\mathrm{F}(0)\ddot{a}_{k'}^\dagger(t')\right\} &= \mathrm{Tr}_\mathrm{F}\left\{\rho_\mathrm{F}(0)\hat{a}_{k'}^\dagger(t')\,\hat{a}_k(t)\right\} \\ &= \langle \hat{a}_{k'}^\dagger(t')\,\hat{a}_k(t)\rangle = 0, \end{aligned} \tag{9.21}$$

and

$$\mathrm{Tr_F}\left\{\hat{a}_k^\dagger(t)\,\rho_\mathrm{F}(0)\hat{a}_{k'}(t')\right\} = \mathrm{Tr_F}\left\{\rho_\mathrm{F}(0)\hat{a}_{k'}(t')\,\hat{a}_k^\dagger(t)\right\}$$
$$= \langle\hat{a}_{k'}(t')\,\hat{a}_k^\dagger(t)\rangle \neq 0. \qquad (9.22)$$

Thus, the terms of the form (9.19) are zero for the ordinary vacuum field, and only terms of the form (9.20) will contribute to the master equation.

Hence, for the ordinary vacuum field the master equation (9.17) reduces to

$$\frac{\partial}{\partial t}\rho(t) = -\frac{1}{4}\sum_k\int_0^t dt'|g_k|^2\left\{S^+S^-\rho(t')\,\mathrm{e}^{i(\omega_0-\omega_k)(t-t')}\right.$$
$$+\rho(t')\,S^+S^-\mathrm{e}^{-i(\omega_0-\omega_k)(t-t')} - S^-\rho(t')\,S^+\mathrm{e}^{i(\omega_0-\omega_k)(t-t')}$$
$$\left.-S^-\rho(t')\,S^+\mathrm{e}^{-i(\omega_0-\omega_k)(t-t')}\right\}. \qquad (9.23)$$

We now change the time variable to $\tau = t - t'$, which leads to

$$\frac{\partial}{\partial t}\rho(t) = -\frac{1}{4}\sum_k|g_k|^2\left\{S^+S^-\int_0^t d\tau\,\rho(t-\tau)\,\mathrm{e}^{i(\omega_0-\omega_k)\tau}\right.$$
$$+\int_0^t d\tau\,\rho(t-\tau)\,\mathrm{e}^{-i(\omega_0-\omega_k)\tau}S^+S^-$$
$$-\int_0^t d\tau\,S^-\rho(t-\tau)\,S^+\mathrm{e}^{i(\omega_0-\omega_k)\tau}$$
$$\left.-\int_0^t d\tau\,S^-\rho(t-\tau)\,S^+\mathrm{e}^{-i(\omega_0-\omega_k)\tau}\right\}. \qquad (9.24)$$

This is an integro-differential equation for ρ which includes the non-Markovian evolution of the density operator ρ, that $\rho(t)$ depends on the past $\rho(t-\tau)$.

We can eliminate the integral over τ by making the *Markov approximation*. In this approximation, we assume that in time τ the density operator $\rho(t-\tau)$ changes slowly compared to the exponents. Then we can write that $\rho(t-\tau) \approx \rho(t)$ and formally perform the integration. In a typical atom, the density operator changes on the time scale corresponding to the spontaneous emission rate, $t_s \sim 10^{-8}$s, and if ω_0 is an optical frequency, the exponents oscillate on the time scale $t_0 \sim 10^{-15}$s, which is much shorter than the atomic time scale. Thus, the Markov approximation is justified for atoms interacting with the ordinary vacuum field, and we can replace $\rho(t-\tau)$ by $\rho(t)$.

In the limit of large t, the integral

$$\int_0^t d\tau\, \mathrm{e}^{\pm i(\omega_0-\omega_k)\tau} \tag{9.25}$$

can be approximated by the function

$$\lim_{t\to\infty}\int_0^t d\tau\, \mathrm{e}^{\pm i(\omega_0-\omega_k)\tau} = \pi\delta(\omega_k-\omega_0) \pm i\frac{\mathcal{P}}{\omega_0-\omega_k}, \tag{9.26}$$

where the real part is a delta function whose area is π and the imaginary part is the Cauchy's principal value $\mathcal{P}$ of the integral. If we now introduce two frequency parameters defined as

$$\frac{\pi}{4}\sum_k |g_k|^2\delta(\omega_k-\omega_0) = \frac{1}{2}\Gamma, \tag{9.27}$$

$$\frac{1}{4}\sum_k |g_k|^2\frac{\mathcal{P}}{\omega_0-\omega_k} = \Delta, \tag{9.28}$$

we obtain the master equation of the form

$$\begin{aligned}\frac{\partial}{\partial t}\rho(t) &= -i\Delta\left[S^+S^-, \rho(t)\right] \\ &-\frac{1}{2}\Gamma\left[S^+S^-\rho(t) + \rho(t)S^+S^- - 2S^-\rho(t)S^+\right].\end{aligned} \tag{9.29}$$

Remember that the density operator $\rho(t)$ is in the interaction picture. Going back to the Schrödinger picture, we finally get

$$\begin{aligned}\frac{\partial}{\partial t}\rho &= -\frac{1}{\hbar}i\left[\hat{H}_0', \rho\right] - i\Delta\left[S^+S^-, \rho\right] \\ &-\frac{1}{2}\Gamma\left(S^+S^-\rho + \rho S^+S^- - 2S^-\rho S^+\right).\end{aligned} \tag{9.30}$$

The terms Γ and Δ arise from the non-zero correlations $\langle \hat{a}_k\hat{a}_{k'}^\dagger\rangle$ and therefore could be attributed to quantum fluctuations of the vacuum field. They might be regarded as characteristic consequences of field quantization and would be absent if the field was treated classically.

9.4 Spontaneous Emission and Decoherence

Our first problem using the master equation is the phenomenon of spontaneous emission from an excited two-level atom. It is well known that the coupling of the excited atom to the vacuum field will result in a spontaneous transition of the electron to the ground state of the atom. We have already shown that the coupling of the atom to the vacuum field results in a shift of the atomic levels and in the appearance of an incoherent part in the master equation of the atom. In view of this, the analysis of spontaneous emission from the excited atom will allow us to find the physical interpretation of the parameters Γ and Δ.

If only the vacuum field is involved in the interaction with the atom, that is, there is no coherent driving field, the Hamiltonian $\hat{H}_0'$ then reduces to $\hat{H}_0$, so that we can write the master equation as

$$\begin{aligned}\frac{\partial}{\partial t}\rho &= -i\left(\omega_0+\Delta\right)\left[S^+S^-,\rho\right]\\ &\quad-\frac{1}{2}\Gamma\left(S^+S^-\rho+\rho S^+S^- - 2S^-\rho S^+\right). \qquad (9.31)\end{aligned}$$

It is clear that the parameter Δ combines with the atomic transition frequency ω_0 and, therefore, represents a shift of the atomic energy levels.

9.4.1 *The Lamb Shift*

The shift of the atomic transition frequency, that appears in Eq. (9.31), can be identified with the Lamb[a] shift. Its magnitude is calculated from Eq. (9.28). In the continuous-mode approximation, in which we convert the mode sum over transverse plane waves into an integral

$$\Delta = \frac{1}{4}\mathcal{P}\int d\omega_k\,|g_k|^2\frac{1}{\omega_0-\omega_k}. \qquad (9.32)$$

The integral is non-zero, and in fact is infinite for an unbounded set of modes. In first sight, the infinite shift appears to be totally unphysical. However, a close look into Eq. (9.32) shows that the infinity is a consequence of the infinitely high frequencies in the integration, that in fact are not observable in practice. Hence, an approximate cutoff of the frequency at $\omega_k = \omega_{\max} \sim c/r_0$ needs to be added at an atomic dimension r_0, in order for the

[a]Willis Lamb was granted the Nobel prize in 1955 for his discoveries concerning the fine structure of the hydrogen spectrum.

present approximation to be valid and to obtain an approximate analytical formula for the Lamb shift. Moreover, in order to obtain a complete calculation of the Lamb shift, it is necessary to extend the calculations to higher order terms in the Hamiltonian including electron mass re-normalization [55], and to include effects of the other atomic levels [56, 57]. If these are included the standard non-relativistic vacuum Lamb shift result is obtained. In fact, it is only in the case of a fully relativistic Hamiltonian that the Lamb shift can be made finite, and even then, only after quantum electrodynamic re-normalization, which involves the removal of infinities.

In 1947, Lamb and Rutherford used a microwave frequency method to examine the finite structure of the $n = 2$ energy level of atomic hydrogen. Earlier, high resolution optical studies of the H_α line have indicated a discrepancy between experiment and the Dirac relativistic theory of the hydrogen atom. The Dirac theory predicts that the $2^2S_{1/2}$ and $2^2P_{1/2}$ energy levels should be degenerated. The early experiments suggested that these levels were not in fact degenerated but separated by about 0.033 cm^{-1}. Lamb and Rutherford used an elegant combination of atomic beam and microwave techniques and showed that the $2^2S_{1/2}$ level is higher in energy than the $2^2P_{1/2}$ level by about 1000 MHz. The lifting of the degeneracy was explained theoretically by Bethe as arising from the interaction of the bound electron with the vacuum fluctuations. These calculations predicted 1000 MHz for the shift. In the interaction with the vacuum field only the $2^2S_{1/2}$ is affected because non-relativistic atomic wavefunctions vanish at the origin except for the s-states with $l = 0$, see Eq. (2.25):

$$|\psi_{n00}|^2 = \frac{1}{\pi n^3 a_o^3}, \tag{9.33}$$

where a_o is the Bohr radius.

9.4.2 *Spontaneous Emission Rate and Decoherence*

Consider now an evolution of the atomic dipole moment $\langle S^+ \rangle$. Since

$$\langle S^+ \rangle = \text{Tr}\left(\rho S^+\right) = \text{Tr}\left(\rho\,|2\rangle\langle 1|\right) = \rho_{12}, \tag{9.34}$$

where $\rho_{12} = \langle 1|\rho|2\rangle$ is the coherence between the states $|1\rangle$ and $|2\rangle$, we can apply the master equation (9.31) to calculate the time evolution of the atomic dipole moment. We can find the equation of motion for $\langle S^+ \rangle$, or equivalently for the coherence ρ_{12}, by projecting the master equation (9.31) onto $|2\rangle$ on the right and $\langle 1|$ on the left

$$\begin{aligned}\frac{\partial}{\partial t}\langle S^+ \rangle = \frac{\partial}{\partial t}\rho_{12} &= -i(\omega_0 + \Delta)\langle 1| \left[S^+S^-, \rho\right] |2\rangle \\ &\quad -\frac{1}{2}\Gamma \langle 1| \left(S^+S^-\rho + \rho S^+S^- - 2S^-\rho S^+\right)|2\rangle \\ &= i(\omega_0 + \Delta)\rho_{12} - \frac{1}{2}\Gamma\rho_{12}. \end{aligned} \tag{9.35}$$

This differential equation has a simple solution

$$\rho_{12}(t) = \rho_{12}(0)\,\mathrm{e}^{-\frac{1}{2}\Gamma t}\mathrm{e}^{i(\omega_0+\Delta)t}, \tag{9.36}$$

where $\rho_{12}(0)$ is the initial coherence between the atomic states.

This shows that the dynamics of the atomic coherence is strongly influenced by Γ and Δ. The initially non-zero dipole moment oscillates in time at a shifted frequency $\omega_0 + \Delta$, and its amplitude is damped exponentially with the rate $\Gamma/2$. Thus, the obvious effect of having Γ is seen to be the damping of the atomic coherence. Physically, the damping is due to spontaneous emission and its pure exponential form is the result of the Markov approximation made in the derivation of the master equation.

The off-diagonal density matrix element ρ_{12} determines the coherence between the two atomic levels. Therefore, we may say that spontaneous emission causes decay of the coherence or, in other words, spontaneous emission is a source of decoherence. Note that the coherence ρ_{12} is different from zero only if the atom is in a superposition state of the ground and the excited states. Equivalently, we can say that spontaneous emission causes decay of the superposition state.

The role of spontaneous emission as a damping process is more evident if we consider the time evolution of the population of the state $|2\rangle$. The equation of motion for the population $\rho_{22} = \langle 2|\rho|2\rangle$ is found by projecting the master equation (9.31) onto $|2\rangle$ on both the left and the right

$$\begin{aligned}\frac{\partial}{\partial t}\rho_{22} &= -\frac{1}{2}\Gamma\langle 2| \left(S^+S^-\rho + \rho S^+S^- - 2S^-\rho S^+\right)|2\rangle \\ &= -\frac{1}{2}\Gamma\langle 2| S^+S^-\rho |2\rangle - \frac{1}{2}\Gamma\langle 2| \rho S^+S^- |2\rangle + \Gamma\langle 2| S^-\rho S^+ |2\rangle \\ &= -\Gamma\rho_{22}. \end{aligned} \tag{9.37}$$

The solution of the above equation is in a simple exponential form

$$\rho_{22}(t) = \rho_{22}(0)\,\mathrm{e}^{-\Gamma t}, \tag{9.38}$$

where $\rho(0)$ is the initial population of the state $|2\rangle$. The initial population of the excited state decays exponentially in time with the rate Γ. Equation (9.38) gives a physical interpretation of the coefficient Γ as the damping rate of the atomic excitation.

9.4.3 *Einstein's A Coefficient*

In order to complete our derivation of the master equation for a two-level atom and obtain a clear meaning of the parameters involved, we prove that the parameter Γ is equal to the Einstein's A coefficient for spontaneous emission.

The parameter Γ has been defined as

$$\Gamma = \frac{\pi}{2}\sum_{k} |g_k|^2 \delta\left(\omega_k - \omega_0\right), \tag{9.39}$$

where $k \equiv (\vec{k}, s)$. Inserting the explicit form of g_k, Eq. (2.41), we get

$$\Gamma = \frac{\pi}{\hbar\varepsilon_0 V}\sum_{k} \omega_k \left|\vec{\mu}\cdot\vec{e}_k\right|^2 \delta\left(\omega_k - \omega_0\right). \tag{9.40}$$

In order to evaluate the sum over k, we assume that $\vec{\mu}$ has only x-component, $\vec{\mu} = \mu[1, 0, 0]$, where $\mu = |\vec{\mu}|$, and will consider the polarization vectors $\vec{e}_k$ in spherical coordinates. If we take the unit propagation vector as

$$\vec{k} = [\sin\theta\cos\phi, \sin\theta\sin\phi, \cos\theta], \tag{9.41}$$

then the unit orthogonal polarization vectors $\vec{e}_{k1}$ and $\vec{e}_{k2}$ can be chosen as

$$\begin{aligned}\vec{e}_{k1} &= [-\cos\theta\cos\phi, -\cos\theta\sin\phi, \sin\theta],\\ \vec{e}_{k2} &= [\sin\phi, -\cos\phi, 0].\end{aligned} \tag{9.42}$$

With this choice of the polarization vectors, the sum over s appearing under the sum over k, becomes

$$\sum_{s=1}^{2} \left|\vec{\mu}\cdot\vec{e}_k\right|^2 = \mu^2\left(\cos^2\theta\cos^2\phi + \sin^2\phi\right). \tag{9.43}$$

We may assume a continuous distribution of the field modes, which is the case when the modes are redistributed in free space. This allows us to make the formal replacement of the sum over k by an integral

$$\sum_k \to \frac{V}{(2\pi c)^3}\int_0^\infty d\omega_k\,\omega_k^2\int_0^\pi d\theta\,\sin\theta\int_0^{2\pi} d\phi. \qquad (9.44)$$

All the integrations in the above equation can be performed analytically. Since

$$\int_0^{2\pi} d\phi\,\sin^2\phi = \int_0^{2\pi} d\phi\,\cos^2\phi = \pi, \qquad (9.45)$$

and

$$\int_0^\pi d\theta\,\left(1+\cos^2\theta\right)\sin\theta = \frac{8}{3}, \qquad (9.46)$$

we obtain for the damping rate

$$\Gamma = \frac{1}{4\pi\varepsilon_0}\frac{4\mu^2\omega_0^3}{3\hbar c^3}, \qquad (9.47)$$

which is the Einstein's A coefficient for spontaneous emission. The damping rate is given in terms of the atomic parameters, which comes from a fully quantum treatment of the atom–field interaction.

9.5 The Bloch–Siegert Shift: An Example of Non-RWA Effects

In Chapter 2, we showed that the exact interaction Hamiltonian between a two-level atom and an EM field contains the energy non-conserving terms called the counter-rotating terms. These terms are usually ignored as being rapidly oscillating over the time scale $t \sim 1/\omega_0$, and an obvious question arises whether there are situations where these terms could generate physical observable phenomena. In this section, we discuss the effect of the counter-rotating terms on spontaneous emission from a two-level atom coupled to a vacuum field. The counter-rotating terms are included into the interaction by not making the RWA on the interaction Hamiltonian between the atom and the vacuum field. As we shall see, the counter-rotating terms can produce a small shift of the atomic levels, known in the literature as the Bloch–Siegert shift.

The exact interaction Hamiltonian that includes the counter-rotating terms is of the form

$$\hat{H}_{\text{int}} = -\frac{1}{2}i\hbar\sum_k g_k\left[S^+\hat{a}_k(t) - S^-\hat{a}_k^\dagger(t) + S^-\hat{a}_k(t) - S^+\hat{a}_k^\dagger(t)\right]. \tag{9.48}$$

In addressing the question of the role of the counter-rotating terms, we derive, with the procedure outlined in Section 9.3, the master equation for the reduced density operator of the atom. The derivation, the details of which are left for the reader as a tutorial exercise, shows that the counter-rotating terms, appearing in the Hamiltonian (9.48), lead to additional terms in the master equation which takes the form

$$\begin{aligned}\frac{\partial\rho}{\partial t} &= -i\left(\omega_0+\Delta\right)\left[S^+S^-,\rho\right] - i\Delta\left(S^+\rho S^+ - S^-\rho S^-\right)\\ &\quad -\frac{1}{2}\Gamma\left\{S^+S^-\rho + \rho S^+S^- - 2S^-\rho S^+\right.\\ &\quad \left.-2S^+\rho S^+ - 2S^-\rho S^-\right\}.\end{aligned} \tag{9.49}$$

There are no terms present like $S^+S^+\rho$ or $S^-S^-\rho$, since $S^+S^+ = S^-S^- \equiv 0$.

An important modification of the master equation is the appearance of additional terms of the form $S^+\rho S^+$ and $S^-\rho S^-$, which indicates a two-photon nature of the counter-rotating terms. In the following, we will ignore the effect of the additional terms on the small Lamb shift, and we will check how the two extra terms in the dissipative part of the master equation modify the spontaneous emission.

To identify the role of the counter-rotating terms, we consider the evolution of the atomic dipole moment (coherence). Using the master equation (9.49), we obtain two coupled differential equations for the off-diagonal density matrix elements

$$\begin{aligned}\dot{\rho}_{12} &= i\omega_0\rho_{12} - \frac{1}{2}\Gamma\rho_{12} + \Gamma\rho_{21},\\ \dot{\rho}_{21} &= -i\omega_0\rho_{21} - \frac{1}{2}\Gamma\rho_{21} + \Gamma\rho_{12}.\end{aligned} \tag{9.50}$$

Thus, the additional terms brought by the counter-rotating terms couple the coherencies ρ_{12} to its conjugate ρ_{21}. This is the modification of the dynamics of the atoms due to the presence of the counter-rotating terms.[a]

We can solve the set of the coupled differential equations using, for example, the Laplace transform method, which allows us to transform the differential equations into a set of two coupled algebraic equations. We can write the set of the transformed equations in a matrix form as

$$\begin{pmatrix} \left(z+\frac{\Gamma}{2}-i\omega_0\right) & -\Gamma \\ -\Gamma & \left(z+\frac{\Gamma}{2}+i\omega_0\right) \end{pmatrix} \begin{pmatrix} \rho_{12}(z) \\ \rho_{21}(z) \end{pmatrix} = \begin{pmatrix} \rho_{12}(0) \\ \rho_{21}(0) \end{pmatrix}. \tag{9.51}$$

According to the Laplace transform method, the time evolution of the atomic coherence is determined by the roots of the determinant of the 2×2 matrix. The determinant is of the form

$$\begin{aligned} D(z) &= \left(z+\frac{\Gamma}{2}\right)^2 + \omega_0^2 - \Gamma^2 = \left(z+\frac{\Gamma}{2}\right)^2 + \omega_0^2\left(1-\frac{\Gamma^2}{\omega_0^2}\right) \\ &= \left[z+\frac{\Gamma}{2}+i\left(\omega_0-\frac{\Gamma^2}{2\omega_0}\right)\right]\left[z+\frac{\Gamma}{2}-i\left(\omega_0-\frac{\Gamma^2}{2\omega_0}\right)\right]. \end{aligned} \tag{9.52}$$

The roots of the polynomial $D(z)$ determine the time evolution of the atomic coherence such that the real parts of the roots contribute damping rates, while the imaginary parts contribute frequencies of the oscillations. According to the expression (9.52), the counter-rotating terms contribute to the imaginary parts $\left(\omega_0 - \Gamma^2/2\omega_0\right)$, that is, they give rise to a shift of the atomic resonance by the amount of $\Gamma^2/2\omega_0$. We identify this shift with the spontaneous emission Bloch–Siegert shift. Since typically $\Gamma \ll \omega_0$, it is apparent that the shift is very small.

In summary of this section, we may state that the counter-rotating terms (the energy non-conserving terms) can have a physical effect on the atomic dynamics. For an atom interacting with a multi-mode reservoir, the terms cause a small shift of the atomic levels, known as the Bloch–Siegert shift.

[a]The coupling of the atomic coherence to its conjugate is formally similar to that appearing in the equations of motion for a two-level atom interacting with a squeezed vacuum.

Exercises

9.1 Show that the master equation (9.30) for the reduced density operator preserves the basic properties of a density operator (normalization, hermiticity, etc.).

9.2 The state of a system is described by the density operator $\rho(t)$, and its evolution is determined by the Liouville–von Neumann equation

$$i\hbar\frac{\partial}{\partial t}\rho(t) = \left[\hat{H}(t), \rho(t)\right],$$

where $\hat{H}(t)$ is the Hamiltonian of the system, which in general can be time dependent. Show that the transformed density operator $\tilde{\rho}(t) = U(t)\rho(t)U^{\dagger}(t)$ evolves according to the Liouville–von Neumann equation

$$i\hbar\frac{\partial}{\partial t}\tilde{\rho}(t) = \left[\tilde{\hat{H}}(t), \tilde{\rho}(t)\right],$$

where

$$\tilde{\hat{H}}(t) = U(t)\hat{H}(t)U^{\dagger}(t) + i\hbar\dot{U}(t)U^{\dagger}(t).$$

9.3 Consider spontaneous emission from a two-level atom initially prepared in the excited state $|2\rangle$, that is, $\rho_{22}(0) = 1$.

(a) Find the time evolution of the density matrix elements of the system.

(b) Verify the conservation of the trace during the evolution, that is, show that $\mathrm{Tr}\rho(t) = 1$ for all t.

(c) Calculate the time evolution of $\mathrm{Tr}\rho^2(t)$. At which time $\mathrm{Tr}\rho^2(t)$ is minimal? What is the state of the atom at that time?

9.4 Show that the spontaneous emission rate Γ of a two-level atom is equal to the Einstein's A coefficient independent of the polarization of the atomic dipole moment.

9.5 Consider a three-level atom in the $\vee$ configuration with two degenerated upper states $|1\rangle$, $|3\rangle$ and a single ground state $|2\rangle$.

The master equation of this system is given by

$$\begin{aligned}\frac{d\rho}{dt} &= -\frac{1}{2}\Gamma\left(S_1^+S_1^-\rho + \rho S_1^+S_1^- - 2S_1^-\rho S_1^+\right)\\ &\quad -\frac{1}{2}\Gamma_{12}\left(S_1^+S_2^-\rho + \rho S_1^+S_2^- - 2S_2^-\rho S_1^+\right)\\ &\quad -\frac{1}{2}\Gamma_{12}\left(S_2^+S_1^-\rho + \rho S_2^+S_1^- - 2S_1^-\rho S_2^+\right)\\ &\quad -\frac{1}{2}\Gamma\left(S_2^+S_2^-\rho + \rho S_2^+S_2^- - 2S_2^-\rho S_2^+\right),\end{aligned}$$

where $S_1^+ = |1\rangle\langle 2|(S_1^- = |2\rangle\langle 1|)$, $S_2^+ = |3\rangle\langle 2|(S_2^- = |2\rangle\langle 3|)$, are the dipole raising (lowering) operators of the atomic transitions, and the parameters Γ and Γ_{12} are the spontaneous emission damping rates, such that $\Gamma_{12} \le \Gamma$.

(a) Calculate equations of motion of the following density matrix elements ρ_{11}, ρ_{33}, ρ_{13} and ρ_{31}.

(b) Under what condition the parameter

$$\alpha = \rho_{11} + \rho_{33} - \rho_{13} - \rho_{31}$$

is a constant of motion.

(c) Using the condition that α is a constant of motion find the time evolution of the population ρ_{11}.

(d) Find the stationary ($t \to \infty$) population of the state $|1\rangle$ assuming that initially $\rho_{11}(0) = 1$ and $\rho_{33}(0) = \rho_{13}(0) = \rho_{31}(0) = 0$.

(e) What would be the stationary population of the state $|1\rangle$ if the atom was initially prepared in a superposition state $|\Psi\rangle = (|1\rangle + |3\rangle)/\sqrt{2}$.

Chapter 10

Heisenberg Equations of Motion

10.1 Introduction

In the master equation method, we have already illustrated a powerful technique for the calculation of the dynamics of an atomic system interacting with the vacuum field. Another technique for calculating the dynamics of an atomic system coupled to the electromagnetic (EM) field involves Heisenberg equations of motion for the system's operators. A difference between the master equation and the Heisenberg equations is that the later involves dynamics of the operators, which allows to analyse the evolution of an atomic system in terms of the field and atomic operators. This creates some problems with handling the Heisenberg equations as, in general, operators do not commute and then in the course of solution of the equations we may face the problem of ordering of the operators. It is usually resolved by putting the operators in the normal order. We have gained some experience with the Heisenberg equations of motion in Chapter 6, Example 6.3, where we studied squeezing generation in the nonlinear degenerate parametric amplifier (DPA) process. Here, we illustrate the technique on the standard model of a two-level atom interacting with a multi-mode field. We then generalize the technique to some specific models such as Lorenz–

Quantum Optics for Beginners
Zbigniew Ficek and Mohamed Ridza Wahiddin

ISBN 978-981-4411-75-2 (Hardcover), 978-981-4411-76-9 (eBook)
www.panstanford.com

Maxwell and Langevin equations, the derivation of which involves some approximations that can be applied only in some limited cases. We also present in detail the Floquet approach, which is usually applied to problems determined by differential equations with time-dependent coefficients.

10.2 Heisenberg Equations of Motion

The Heisenberg equation of motion for an arbitrary operator of a given system is found from the Hamiltonian of the system, and has the form

$$\frac{d}{dt}\hat{A} = \frac{i}{\hbar}\left[\hat{H}, \hat{A}\right], \tag{10.1}$$

where $\hat{A}$ is an arbitrary operator of the system.

To illustrate the Heisenberg equation technique, we take a two-level atom interacting with the EM field. The atom and the field are described by the standard atomic spin and the field annihilation and creation operators. The Hamiltonian of the system is of the form

$$\begin{aligned}\hat{H} &= \hbar\omega_0 S_z + \sum_k \hbar\omega_k \left(\hat{a}_k^\dagger \hat{a}_k + \frac{1}{2}\right) \\ &-\frac{1}{2}i\hbar\sum_k g_k \left(S^+\hat{a}_k - \hat{a}_k^\dagger S^-\right),\end{aligned} \tag{10.2}$$

where, without loss of generality, we have assumed that the coupling constant g_k is a real number.

The Hamiltonian (10.2) generates the following Heisenberg equations of motion for the atomic dipole moment S^-, the atomic inversion S_z and the annihilation operator of the k mode of the EM field:

$$\frac{d}{dt}S^- = -i\omega_0 S^- + \sum_k g_k \hat{a}_k S_z, \tag{10.3}$$

$$\frac{d}{dt}S_z = -\frac{1}{2}\sum_k g_k \left(S^+\hat{a}_k + \hat{a}_k^\dagger S^-\right), \tag{10.4}$$

$$\frac{d}{dt}\hat{a}_k = -i\omega_k \hat{a}_k + \frac{1}{2}g_k S^-, \tag{10.5}$$

and the equations of motion for the S^+ and $\hat{a}^\dagger$ operators are obtained by taking the Hermitian conjugate of the equation of motion for the S^- and $\hat{a}$ operators, respectively.

We see that the operator's equations are in a form of nonlinear equations of motion, as they contain product terms $\hat{a}_k S_z$ and $S^+\hat{a}_k$, $\hat{a}_k^\dagger S^-$. The average values of the operators, however, produce linear set of equations of motion. For example, the equation of motion for the average dipole moment $\langle S^-\rangle$ depends on the correlation function $\langle \hat{a}_k S_z\rangle$. Of course, an exact solution of these equations is rather impossible, but we can make some approximations. In the following, we illustrate the commonly used approximate methods of solving the set of the Heisenberg equations of motion, Eqs. (10.3)–(10.5).

10.3 Lorenz–Maxwell Equations

We first illustrate a method of solving the Heisenberg equations of motion for the average atomic and field operators. After averaging the Heisenberg equations of motion over an arbitrary atomic and field state, we obtain

$$
\begin{aligned}
\frac{d}{dt}\langle S^-\rangle &= -i\omega_0\langle S^-\rangle + g\langle \hat{a} S_z\rangle, \\
\frac{d}{dt}\langle S_z\rangle &= -\frac{1}{2}g\left(\langle S^+\hat{a}\rangle + \langle \hat{a}^\dagger S^-\rangle\right), \\
\frac{d}{dt}\langle \hat{a}\rangle &= -i\omega\langle \hat{a}\rangle + \frac{1}{2}g\langle S^-\rangle,
\end{aligned} \tag{10.6}
$$

where, for simplicity, we have assumed that the atom is coupled to a single mode of the EM field. This can happen, for example, if the atom is located inside a cavity that tailors the EM field modes to a single mode, called the cavity mode.

As we have said earlier, the Heisenberg equations of motion for average values of the field and the atomic variables form a set of c-number linear equations. However, the set of equations is in fact composed of an infinite number of equations. It is easy to derive how the set of equations (10.6) develops into a set of infinite number of equations. For example, the equation of motion for $\langle S^-\rangle$ depends on the second-order correlation function $\langle \hat{a} S_z\rangle$.

Thus, we have to find the equation of motion for the second-order correlation function to determine $\langle S^- \rangle$. By writing the equation of motion for the second-order correlation function $\langle \hat{a} S_z \rangle$, we will find that the equation depends on a third-order correlation function. This procedure continues up to infinity.

How to deal with this problem? One of the possible approaches is to truncate the set of equations by factoring out the field and atom variables

$$\langle \hat{a} S_z \rangle = \langle \hat{a} \rangle \langle S_z \rangle, \qquad \langle \hat{a} S^+ \rangle = \langle \hat{a} \rangle \langle S^+ \rangle, \quad \text{etc.} \tag{10.7}$$

This factorization is called the *semiclassical approximation* that the field and the atom evolutions are independent of each other without any quantum correlations between them. It is valid when the field amplitude is large, $|\langle a \rangle| \gg 1$.

The semiclassical approximation closes the equations. In the next step, we add phenomenologically damping rates to the right-hand sides of the equations of motion, as the atom and field can be treated as two independent classical damped oscillators, and obtain[a]

$$\begin{aligned}
\frac{d}{dt}\langle S^- \rangle &= -\left(i\omega_0 + \Gamma_{\mathrm{p}}\right)\langle S^- \rangle + g\langle \hat{a} \rangle \langle S_z \rangle, \\
\frac{d}{dt}\langle S_z \rangle &= -\Gamma_{\mathrm{d}} - \Gamma_{\mathrm{d}}\langle S_z \rangle - \frac{1}{2} g\left(\langle \hat{a} \rangle \langle S^+ \rangle + \langle \hat{a}^\dagger \rangle \langle S^- \rangle\right), \\
\frac{d}{dt}\langle \hat{a} \rangle &= -\left(i\omega + \kappa\right)\langle \hat{a} \rangle + \frac{1}{2} g \langle S^- \rangle,
\end{aligned} \tag{10.8}$$

where Γ_{p} is the damping rate of the atomic polarization, Γ_{d} is the damping rate of the atomic inversion and κ is the damping rate of the field mode. To remove the effect of the fast oscillations with the frequencies ω_0 and ω, we introduce a rotating frame through the relations

$$\langle \tilde{\hat{a}} \rangle = \langle \hat{a} \rangle e^{i\omega t}, \quad \langle \tilde{\hat{a}}^\dagger \rangle = \langle \hat{a}^\dagger \rangle e^{-i\omega t}, \quad \langle \tilde{S}^\pm \rangle = \langle S^\pm \rangle e^{\mp i\omega_0 t}. \tag{10.9}$$

After substituting Eq. (10.9) into Eq. (10.8) and introducing a notation $P(t) = \langle \tilde{S}^\pm \rangle$, $D(t) = \langle S_z \rangle$, and $E(t) = \langle \tilde{\hat{a}} \rangle = \langle \tilde{\hat{a}}^\dagger \rangle$, the

[a]Note that by inclusion of the damping rates, the Heisenberg equations of motion account, partially, for quantum fluctuations.

equations of motion (10.8) take the form

$$\begin{aligned}
\frac{d}{dt}P(t) &= -\Gamma_{\rm p}P(t) + gE(t)D(t),\\
\frac{d}{dt}D(t) &= -\Gamma_{\rm d} - \Gamma_{\rm d}D(t) - gP(t)E(t),\\
\frac{d}{dt}E(t) &= -\kappa E(t) + \frac{1}{2}gP(t),
\end{aligned} \tag{10.10}$$

where, we have assumed that the field frequency is resonant with the atomic transition frequency, $\omega = \omega_0$.

These equations are known in literature as the Lorenz–Maxwell equations. Despite the decorrelation approximation, these equations are still too complicated to be handled analytically, and are usually solved by a numerical integration. The decorrelation converted the infinite set of linear equation into a finite set of nonlinear equations.

An important property of the equations is that their nonlinear character can lead to chaotic instabilities (classical chaos) in the atomic and field dynamics. This is illustrated in Fig. 10.1, where

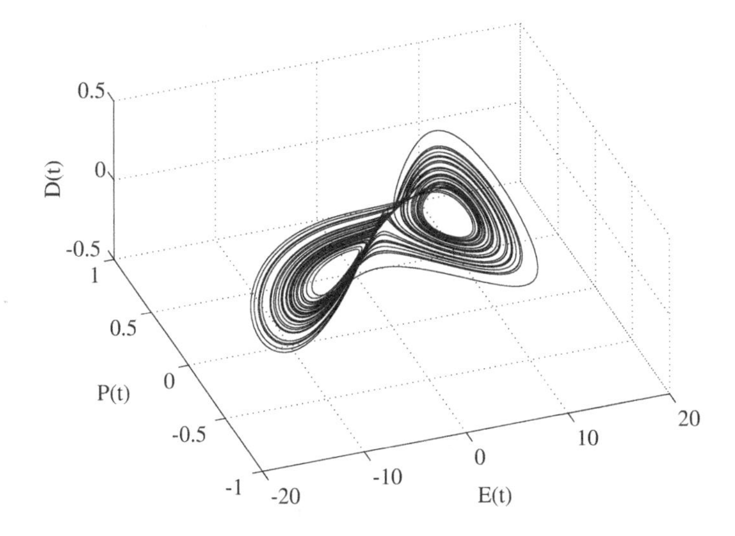

Figure 10.1 Three-dimensional plot of the time evolution of the polarization $P(t)$, inversion $D(t)$, and the field amplitude $E(t)$ for $g/\kappa = 46$, $\Gamma_{\rm p}/\kappa = 0.66$ and $\Gamma_{\rm d}/\kappa = 0.16$. (Figure courtesy Greg Kociuba).

we plot the time evolution of the atomic and field variables. The evolution forms attractors display the multi-stability of the atomic and field variables.

10.4 Langevin Equations

Here we illustrate an approximate technique of solving the Heisenberg equations of motion in which we *adiabatically* eliminate the field operators $\hat{a}_k$ and $\hat{a}_k^{\dagger}$ from the Heisenberg equations of motion, Eqs. (10.3)–(10.5), leaving the analyses of the dynamics to the atomic operators only. The adiabatic approximation is equivalent to assume that the field does not change during the evolution, which is a good approximation when the field is composed of a large number of modes.

The technique proceeds as follows. In the first step, we solve approximately the Heisenberg equation of motion for the field operator $\hat{a}_k$. Note from Eq. (10.5) that the equation of motion for $\hat{a}_k$ contains no operator products. Therefore, we can formally integrate the equation to obtain

$$\hat{a}_k(t) = \hat{a}_k^o(t) + \frac{1}{2}g_k \int_0^t dt' S^-\left(t'\right) \mathrm{e}^{-i\omega_k(t-t')}, \qquad (10.11)$$

where the first part

$$\hat{a}_k^o(t) = \hat{a}_k(0)\,\mathrm{e}^{-i\omega_k t} \qquad (10.12)$$

is the free evolution part of the field that is not disturbed by the atom. The second part of the solution (10.11) is a contribution from the atom. It is the source field or radiation reaction field of the atomic dipole and gives a field generated by the emitting atom.

In the next step, we substitute the solution (10.11) into the equations of motion for the atomic operators. Before doing this, we introduce few assumptions about the evolution of the field amplitude that will allow us to perform the integration in Eq. (10.11). First, we change the time variable under the integral to $\tau = t - t'$, giving

$$\hat{a}_k(t) = \hat{a}_k^o(t) + \frac{1}{2}g_k \int_0^t d\tau S^-(t-\tau)\,\mathrm{e}^{-i\omega_k \tau}. \qquad (10.13)$$

There is a atomic operator under the integral and its time evolution is unknown. Therefore, to remove the operator from the integral we make an approximation, which in fact is equivalent to the Markov approximation. In this approximation we assume that the evolution of the atomic operator is close to a harmonic evolution with the atomic frequency ω_0, so that we can write

$$S^-(t-\tau) \simeq S^-(t)\,e^{i\omega_0\tau}. \tag{10.14}$$

Then, we obtain the following equation for the time evolution of the field operator

$$\hat{a}_k(t) = \hat{a}_k^o(t) + \frac{1}{2}g_k S^-(t)\int_0^t d\tau\, e^{i(\omega_0-\omega_k)\tau}, \tag{10.15}$$

Thus, a major simplification has been achieved by the Markov approximation. The remaining integral can be handled analytically and, as before in the derivation of the master equation, the integral can be approximated, in the limit of $t \to \infty$, by the zeta function, Eq. (9.26).

We now substitute the solution for $\hat{a}_k(t)$ into the Heisenberg equations of motion for S^- and S^z, we find that the equation of motion for the atomic operators become

$$\dot{S}^-(t) = -i(\omega_0+\Delta)S^-(t) - \frac{1}{2}\Gamma S^-(t) + \sum_k g_k\,\hat{a}_k^o(t)\,S_z(t), \tag{10.16}$$

$$\begin{aligned}\dot{S}_z(t) &= -\frac{1}{2}\Gamma - \Gamma S_z(t)\\ &\quad -\frac{1}{2}\sum_k g_k\left[\hat{a}_k^o(t)\,S^+(t) + \hat{a}_k^{o\dagger}(t)\,S^-(t)\right].\end{aligned} \tag{10.17}$$

The last terms on the right-hand side of Eqs. (10.16) and (10.17) depend on the state of the free field. For the vacuum state, $\hat{a}_k^o(t)|0\rangle = 0$, and then the terms vanish. For a coherent field, the terms give the coherent field amplitude, which will lead to the Rabi frequency of the field.

The operator equations that we have just derived for $\dot{S}^-(t)$ and $\dot{S}_z(t)$ are known in the literature as the *quantum Langevin equations* for a two-level atom. The equations are usually written in the form

$$\begin{aligned}\dot{S}^-(t) &= -i(\omega_0+\Delta)S^-(t) - \frac{1}{2}\Gamma S^-(t) + S_z(t)F(t),\\ \dot{S}_z(t) &= -\frac{1}{2}\Gamma - \Gamma S_z(t) - \frac{1}{2}S^+(t)F(t) - \frac{1}{2}F^+(t)S^-(t),\end{aligned} \tag{10.18}$$

where

$$F(t) = \sum_k g_k \hat{a}_k^o(t) \tag{10.19}$$

is called the Langevin operator.

The Langevin equations are commonly used in the study of the effects of external fields on atoms where higher order statistics are important as thermal or squeezed fields. For these fields $\langle \hat{a}_k \rangle$ and $\langle \hat{a}_k^{\dagger} \rangle$ are zero and the higher order moments are important.

10.5 Optical Bloch Equations

Optical properties of a coherently driven two-level atom are often studied in terms of the optical Bloch equations. These equations are in fact equations of motion for the expectation values of the atomic spin operators obtained from the Heisenberg equations of motion averaged over the initial state of the atom and the field. The equations can be written as

$$\begin{aligned}
\frac{d}{dt}\langle S^{+}\rangle &= -\left(\frac{1}{2}\Gamma - i\omega_0\right)\langle S^{+}\rangle + \Omega(t)\langle S_z\rangle, \\
\frac{d}{dt}\langle S^{-}\rangle &= -\left(\frac{1}{2}\Gamma + i\omega_0\right)\langle S^{-}\rangle + \Omega^{*}(t)\langle S_z\rangle, \\
\frac{d}{dt}\langle S_z\rangle &= -\frac{1}{2}\Gamma - \Gamma\langle S_z\rangle - \frac{1}{2}\left(\Omega^{*}(t)\langle S^{+}\rangle + \Omega(t)\langle S^{-}\rangle\right),
\end{aligned} \tag{10.20}$$

where the interaction of the atom with the multi-mode vacuum field results in damping of the atomic dipole moment $\langle S^{\pm}\rangle$ with rate $\Gamma/2$ and the atomic population inversion $\langle S_z\rangle$ with rate Γ. The interaction of the atom with the driving laser field is determined by the time-dependent Rabi frequency $\Omega(t)$.

The optical Bloch equations (10.20) are coupled first-order differential equations with time-dependent coefficients. In order to solve the set of the Bloch equations, we have to know explicitly the time dependence of $\Omega(t)$. For a monochromatic laser field

$$\Omega(t) = \Omega\, e^{i(\omega_{\mathrm{L}} t + \phi_{\mathrm{L}})}, \tag{10.21}$$

where ω_{L} is the frequency of the laser field and ϕ_{L} is its phase. In this case, one can find a rotating frame in which the coefficients will be independent of time. To show this, we introduce new (rotated) variables for the expectation values of the atomic operators that are free from the rapid oscillations at optical frequencies

$$\langle \tilde{S}^{\pm}(t)\rangle = \langle S^{\pm}\rangle e^{\mp i(\omega_{\mathrm{L}} t + \phi_{\mathrm{L}})}. \tag{10.22}$$

In terms of the new variables, the optical Bloch equations become

$$\begin{aligned}
\frac{d}{dt}\langle \tilde{S}^{+}\rangle &= -\left(\frac{1}{2}\Gamma - i\delta_{\mathrm{L}}\right)\langle \tilde{S}^{+}\rangle + \Omega\langle S_z\rangle,\\
\frac{d}{dt}\langle \tilde{S}^{-}\rangle &= -\left(\frac{1}{2}\Gamma + i\delta_{\mathrm{L}}\right)\langle \tilde{S}^{-}\rangle + \Omega\langle S_z\rangle,\\
\frac{d}{dt}\langle S_z\rangle &= -\frac{1}{2}\Gamma - \Gamma\langle S_z\rangle - \frac{1}{2}\Omega\left(\langle \tilde{S}^{+}\rangle + \langle \tilde{S}^{-}\rangle\right),
\end{aligned} \tag{10.23}$$

where $\delta_{\mathrm{L}} = \omega_0 - \omega_{\mathrm{L}}$ is the detuning of the laser frequency from the atomic transition frequency ω_0. Equations (10.23) are first-order differential equations with time-independent coefficients. In principle, the equations can be solved analytically by direct integration, or by the Laplace transformation to an easily solvable algebraic equations.

We can rewrite Eqs. (10.23) in terms of components of the atomic spin vector by introducing a real vector $\vec{B} = (\langle S_x\rangle, \langle S_y\rangle, \langle S_z\rangle)$, called the Bloch vector. The components of the Bloch vector satisfy the following equations of motion

$$\begin{aligned}
\frac{d}{dt}\langle S_x\rangle &= -\frac{1}{2}\Gamma\langle S_x\rangle - \delta_{\mathrm{L}}\langle S_y\rangle + \Omega\langle S_z\rangle,\\
\frac{d}{dt}\langle S_y\rangle &= -\frac{1}{2}\Gamma\langle S_y\rangle + \delta_{\mathrm{L}}\langle S_x\rangle,\\
\frac{d}{dt}\langle S_z\rangle &= -\frac{1}{2}\Gamma - \Gamma\langle S_z\rangle - \Omega\langle S_x\rangle.
\end{aligned} \tag{10.24}$$

The components $\langle S_x\rangle$ and $\langle S_y\rangle$ are, respectively, the real and imaginary parts of the coherence between the atomic levels, and $\langle S_z\rangle$ is the population inversion. In terms of the components of the Bloch vector, and in the absence of damping ($\Gamma = 0$), Eq. (10.24) may be written as [3, 4]

$$\frac{d\vec{B}}{dt} = \vec{\Omega}_B \times \vec{B}, \tag{10.25}$$

where $\vec{\Omega}_B = (0, \Omega, \delta_{\mathrm{L}})$ is the pseudo-field vector of magnitude $|\vec{\Omega}_B| = (\Omega^2 + \delta_{\mathrm{L}}^2)^{1/2}$.

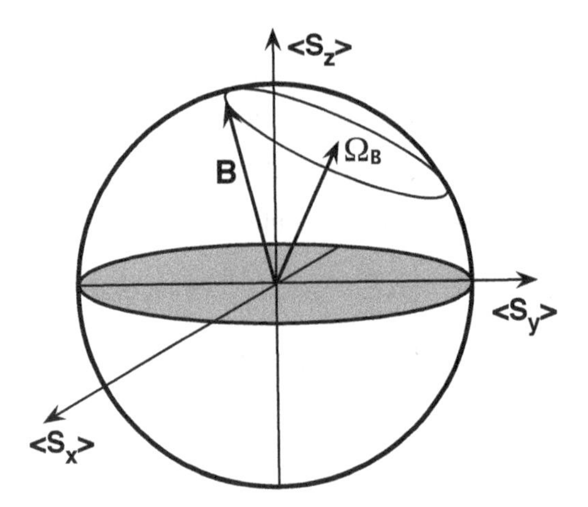

Figure 10.2 The Bloch sphere showing the Bloch vector $\vec{B}$ precessing about the pseudo-field vector $\vec{\Omega}_B$ for the case of an off-resonant driving with $\delta_{\rm L} \neq 0$.

The solutions of Eq. (10.25) describe the precession of the atomic spin vector around the $\vec{\Omega}_B$ axis. It is clear that the evolution of $\vec{B}$ depends on its orientation relative to $\vec{\Omega}_B$. In general, $\vec{B}$ precesses about $\vec{\Omega}_B$ in such a way that the relative angle between the two vectors remains constant, that is, $\vec{B}$ precesses in a cone about $\vec{\Omega}_B$, as illustrated in Fig. 10.2. For the resonant driving ($\delta_{\rm L} = 0$), the Bloch vector precesses in the $(\langle S_x\rangle, \langle S_z\rangle)$ plane. For $\delta_{\rm L} \gg 0$, the Bloch vector remains almost constant in time and pointing towards the south pole of the Bloch sphere.

If we prepare the system in a state such that $\vec{B}$ is initially parallel or antiparallel to $\vec{\Omega}_B$, the Bloch vector will stay in this position forever. This phenomenon is sometimes referred to as spin locking [60]. Thus, the simple representation of the system in terms of the Bloch vector provides a useful picture for understanding dynamics of a driven two-level atom.

10.6 Floquet Method

For non-stationary (time-dependent) fields the expectation values of the atomic operators can depend on time even in the stationary (steady-state) limit. In this section, we illustrate a method of solving the optical Bloch equations for time-dependent driving fields.

If the driving field is composed of more than one frequency component, the time dependence of the Rabi frequency $\Omega(t)$ is quite complicated and in general can involve many different parameters. In this case, it is not possible to find a rotating frame in which the coefficients of the Bloch equations would be time independent. This renders the problem difficult to solve, except for those cases in which the time dependence involves only few, two or three, frequencies or many frequencies separated by a constant detuning.

An example of such fields is a bichromatic field containing two components of different frequencies

$$\begin{aligned}\Omega(t) &= \Omega\left[e^{i(\omega_{L1}t+\phi_{L1})} + e^{i(\omega_{L2}t+\phi_{L2})}\right]\\ &= \Omega\left(1 + e^{i(\delta t+\delta\phi)}\right)e^{i(\omega_{L1}t+\phi_{L1})},\end{aligned} \tag{10.26}$$

where $\delta = \omega_{L2} - \omega_{L1}$ is the frequency difference between the frequency components, and $\delta\phi = \phi_{L2} - \phi_{L1}$ is the difference between initial phases of the fields.

Another example is an amplitude-modulated field [58]

$$\Omega(t) = \Omega\left[1 \pm a\cos(\delta t)\right], \tag{10.27}$$

where $a = 2\Omega_m/\Omega$ is the modulation amplitude, or a phase-modulated field

$$\Omega(t) = \Omega\left[1 \pm ia\cos(\delta t)\right], \tag{10.28}$$

where Ω_m is the Rabi frequency of the modulating (sideband) fields, and δ is the modulation frequency.

The optical Bloch equations with time-dependent (time-periodic) coefficients are solved by the Floquet method, in which the atomic dynamics are described in terms of Fourier harmonics of the expectation values of the atomic spin operators. In this approach, we make the Fourier decomposition of the expectation values

$$X_k(t) = \sum_{l=-\infty}^{\infty} X_k^{(l)}(t)e^{il\delta t}, \quad k = 1, 2, 3, \tag{10.29}$$

where $X_1 = \langle\tilde{S}^+\rangle$, $X_2 = \langle\tilde{S}^-\rangle$, $X_3 = \langle S_z\rangle$ and $X_k^{(l)}(t)$ are slowly varying harmonic amplitudes. The Fourier decomposition (10.29) shows that the atomic variables will respond at harmonics of the modulation frequency δ, and knowledge of $X_k^{(l)}(t)$ gives all the information about the system evolution.

Let us illustrate the Floquet method for the case of a two-level atom driven by a bichromatic field. We choose a frame rotating at frequency ω_{L1} and, for simplicity, we assume $\delta_{L1} = 0$. Hence, substituting Eq. (10.29) into Eq. (10.23) and comparing coefficients of the same powers in $l\delta$, we obtain the following set of infinite number of coupled first-order differential equations for the harmonic amplitudes

$$\begin{aligned}
\frac{\partial}{\partial t}X_1^{(l)} &= -\left(\frac{1}{2}\Gamma + il\delta\right)X_1^{(l)} + \Omega\left(X_3^{(l)} + X_3^{(l-1)}\right), \\
\frac{\partial}{\partial t}X_2^{(l)} &= -\left(\frac{1}{2}\Gamma + il\delta\right)X_2^{(l)} + \Omega\left(X_3^{(l)} + X_3^{(l+1)}\right), \\
\frac{\partial}{\partial t}X_3^{(l)} &= -\frac{1}{2}\Gamma\delta_{l,0} - \left(\Gamma + il\delta\right)X_3^{(l)} \\
&\quad -\frac{1}{2}\Omega\left(X_1^{(l)} + X_1^{(l+1)} + X_2^{(l)} + X_2^{(l-1)}\right), \qquad (10.30)
\end{aligned}$$

where $\delta = \omega_{L2} - \omega_{L1}$, and $\delta_{l,0}$ is the Kronecker delta function.

Thus, the Floquet method transforms the three Bloch equations with time-dependent coefficients into an infinite number of equations with time-independent coefficients. We can solve Eq. (10.30) by using the continued fraction technique, or we can write Eq. (10.30) in a matrix form and solve by matrix inversion. In both techniques, we have to use truncated basis rather than infinite basis of the harmonic amplitudes. The validity of the truncation is ensured by requiring that the solution does not change as the number of truncated harmonics increases or decreases by one.

Equation (10.30) can be written in the form of a recurrence relation

$$\begin{aligned}
\frac{d}{dt}\vec{X}^{(\ell)}(t) &= -\frac{\Gamma}{2}\delta_{\ell,0}\vec{I} \\
&\quad -A_\ell\vec{X}^{(\ell)}(t) - B_\ell\vec{X}^{(\ell-1)}(t) - D_\ell\vec{X}^{(\ell+1)}(t), \qquad (10.31)
\end{aligned}$$

where $\vec{I}$ is a column vector with the components $I_1 = I_2 = 0, I_3 = 1$, and A_ℓ, B_ℓ and D_ℓ are the matrices

$$A_\ell = \begin{pmatrix} \left(\frac{1}{2}\Gamma + i\ell\delta\right) & 0 & -\Omega \\ 0 & \left(\frac{1}{2}\Gamma + i\ell\delta\right) & -\Omega \\ \frac{1}{2}\Omega & \frac{1}{2}\Omega & \left(\Gamma + i\ell\delta\right) \end{pmatrix}, \tag{10.32}$$

$$B_\ell = \begin{pmatrix} 0 & 0 & -\Omega \\ 0 & 0 & 0 \\ 0 & \frac{1}{2}\Omega & 0 \end{pmatrix}, \quad D_\ell = \begin{pmatrix} 0 & 0 & 0 \\ 0 & 0 & -\Omega \\ \frac{1}{2}\Omega & 0 & 0 \end{pmatrix}. \tag{10.33}$$

One method of solving a recurrence relation is to use continued fractions. However, we choose instead to solve in terms of the eigenvalues and eigenvectors of the infinite-dimensional (Floquet) matrix, which we construct by arranging the amplitudes $X^{(\ell)}(t)$ in the order

$$\vec{Y}(t) = \begin{pmatrix} \vdots \\ X^{(1)}(t) \\ X^{(0)}(t) \\ X^{(-1)}(t) \\ \vdots \end{pmatrix}. \tag{10.34}$$

Equation (10.31) can then be written as the matrix differential equation

$$\frac{d}{dt}\vec{Y}(t) = \bar{K}\vec{Y}(t) + \vec{P}, \tag{10.35}$$

where $\bar{K}$ is an infinite-dimensional tridiagonal (Floquet) matrix composed of the 3×3 matrices A_ℓ, B_ℓ and D_ℓ, and $\vec{P}$ is an infinite-dimensional vector with the non-zero component $-\frac{1}{2}\Gamma\delta_{\ell,0}I$.

The matrix equation (10.35) is a simple differential equation with time-independent coefficients that can be solved by direct integration. For an arbitrary initial time t_0, the integration of Eq. (10.35) leads to the following formal solution for $\vec{Y}(t)$:

$$\vec{Y}(t) = \vec{Y}(t_0)\,\mathrm{e}^{\bar{K}t} - \left(1 - \mathrm{e}^{\bar{K}t}\right)\bar{K}^{-1}\vec{P}. \tag{10.36}$$

In order to proceed further, we have to truncate the dimension of the vector $\vec{Y}(t)$. The validity of the truncation is ensured by requiring that the solution (10.36) does not change as the dimension of $\vec{Y}(t)$ increases or decreases by one. Because the determinant of the finite-dimensional (truncated) matrix $\bar{K}$ is different from zero, there exists a complex invertible matrix $\bar{T}$ which diagonalises $\bar{K}$, and $\lambda = \bar{T}^{-1}\bar{K}\bar{T}$ is the diagonal matrix of complex eigenvalues. By introducing $\vec{L} = \bar{T}^{-1}\vec{Y}$ and $\vec{R} = \bar{T}^{-1}\vec{P}$, we can rewrite Eq. (10.36) as

$$\vec{L}(t) = \vec{L}(t_0)\,\mathrm{e}^{\lambda t} - \left(1 - \mathrm{e}^{\lambda t}\right)\lambda^{-1}\vec{R}, \tag{10.37}$$

or, in component form

$$L_i(t) = L_i(t_0)\,\mathrm{e}^{\lambda_i t} - \sum_{j=1}^{q} \left(\lambda^{-1}\right)_{ij}\left(1 - \mathrm{e}^{\lambda_j t}\right) R_j, \tag{10.38}$$

where q is the dimension of the truncated matrix. To obtain solutions for the components $X_i^{(\ell)}(t)$, we determine the eigenvalues λ_i and eigenvectors $L_i(t)$ by a numerical diagonalization of the matrix $\bar{K}$.

The steady state values of the harmonics $X_i^{(\ell)}(t)$ can be found from Eq. (10.38) by taking $t \to \infty$, or more directly by setting the left-hand side of Eq. (10.35) equal to zero. Thus

$$Y_i(\infty) = -\sum_{j=1}^{q} \left(\bar{K}^{-1}\right)_{ij} P_j. \tag{10.39}$$

The quantity $X_3^{(0)}(\infty)$ has an important interpretation in terms of directly measurable quantities, besides being the stationary energy expectation value of the atom in units of $\hbar\omega_0$.

The stationary intensity $I_s = \langle E^{(-)}(\vec{R},\infty)E^{(+)}(\vec{R},\infty)\rangle$ of the fluorescence field radiated by the atom and detected by a photodetector at a point $\vec{R}$ in the far field zone may be expressed with help of the commutation relations (2.18) in the form

$$I_s = \left[\frac{1}{2} + X_3^{(0)}(\infty)\right]. \tag{10.40}$$

It follows that the quantity $\frac{1}{2} + X_3^{(0)}(\infty)$ is a measure of the light intensity in the far-field zone.

We now present some numerical calculations that illustrate the behaviour of the stationary intensity of the fluorescence field.

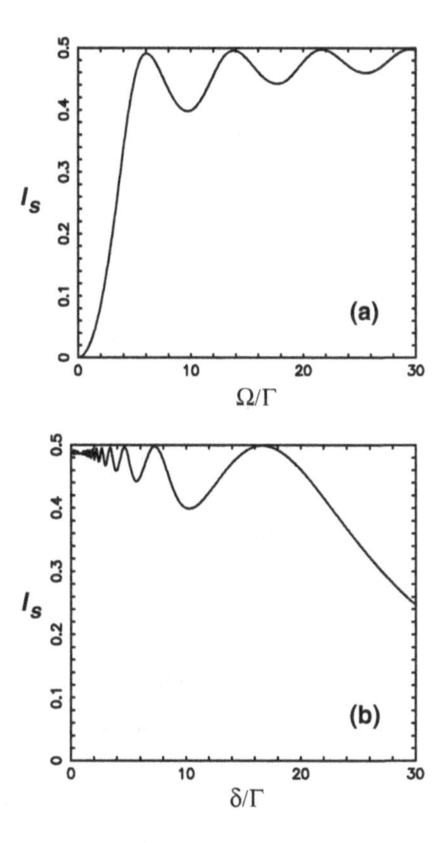

Figure 10.3 The stationary fluorescence intensity for a bichromatic driving field plotted in frame (a) as a function of Ω/Γ and constant $\delta = 5\Gamma$, and in frame (b) as a function of δ/Γ and constant $\Omega = 20\Gamma$.

Figure 10.3(a) shows the stationary intensity as a function of Ω for a fixed δ, whereas Fig. 10.3(b) shows the variation of the intensity with δ for a constant Ω. We see that the periodic modulation of the Rabi frequency introduces new features. The intensity no longer increases steadily to the saturation value when the Rabi frequency Ω increases, as in the case of the monochromatic driving with a constant Rabi frequency. The intensity exhibits oscillations at the harmonics of the modulation frequency δ. The intensity is far from the saturation, $I_s = 1/2$, when δ is large, even when Ω is much larger than saturating values for the monochromatic driving [59].

Exercises

10.1 Consider the Jaynes–Cummings Hamiltonian

$$\hat{H} = \hbar\omega_0 S_z + \hbar\omega\left(\hat{a}^{\dagger}\hat{a} + \frac{1}{2}\right) - \frac{1}{2}i\hbar g\left(S^{+}\hat{a} - S^{-}\hat{a}^{\dagger}\right).$$

(a) Find the Heisenberg equations of motion for the atomic and field operators.

(b) Show that in the case of exact resonance ($\omega = \omega_0$),

$$2\ddot{S}_z + g^2 S_z + 2g^2 S_z \hat{N} = 0,$$

where $\hat{N} = S_z + \hat{a}^{\dagger}\hat{a}$ is a constant of motion, $\hat{N}(t) = \hat{N}(0)$.

10.2 Equations of motion for average values of the atomic spin operators S^+, S^- and S_z of a two-level atom driven by a coherent laser field are

$$\frac{d}{dt}\langle S^{+}\rangle = -\left(\frac{1}{2}\Gamma - i\omega_0\right)\langle S^{+}\rangle + \Omega\langle S_z\rangle e^{i\omega_{\rm L}t},$$

$$\frac{d}{dt}\langle S^{-}\rangle = -\left(\frac{1}{2}\Gamma + i\omega_0\right)\langle S^{-}\rangle + \Omega\langle S_z\rangle e^{-i\omega_{\rm L}t},$$

$$\frac{d}{dt}\langle S_z\rangle = -\frac{1}{2}\Gamma - \Gamma\langle S_z\rangle - \frac{1}{2}\Omega\left(\langle S^{+}\rangle e^{-i\omega_{\rm L}t} + \langle S^{-}\rangle e^{i\omega_{\rm L}t}\right),$$

where $\omega_{\rm L}$ is the laser frequency, ω_0 is the atomic transition frequency and Ω is the Rabi frequency of the laser field.

(a) Transform the equations of motion to a rotating frame, where the equations of motion for the transformed average values of the atomic spin operators have no explicit time dependence.

(b) Find the time evolution of the average atomic dipole moment $\langle S^{+}(t)\rangle$ assuming that initially the atom was in its ground state.

(c) Show that there is a threshold value for the Rabi frequency Ω above which the dipole moment of the atom oscillates with three different frequencies: $\omega_{\rm L}$ and $\omega_{\rm L} \pm \Omega$. For simplicity, assume that $\omega_{\rm L} = \omega_0$.

10.3 Using the results of Exercise 10.2 for the time evolution of $\langle S^{\pm}(t)\rangle$ and $\langle S_z(t)\rangle$, show that in the absence of the spontaneous emission ($\Gamma = 0$), the magnitude of the Bloch vector is conserved, that is, $\langle S_x(t)\rangle^2 + \langle S_y(t)\rangle^2 + \langle S_z(t)\rangle^2 = 1$ for all times t.

10.4 Verify, using the results of Exercise 10.2 that in the steady state ($t \to \infty$) the expression for $\langle S_z(t) \rangle$ is of the form of a Lorentzian centred at $\delta_L = \omega_0 - \omega_L = 0$. The expression is known as the absorption spectrum of the atom or stationary line shape.

10.5 Show that the configurations with the Bloch vector $\vec{B}$ parallel or antiparallel to $\vec{\Omega}_B$ correspond to an excitation of the atom to particular superposition states of the atomic states $|1\rangle$ and $|2\rangle$.

10.6 Show that an arbitrary 2×2 matrix M can be written as

$$M = u_0 I + \vec{u} \cdot \vec{\sigma},$$

where $u_0 = \frac{1}{2}\text{Tr}(M)$, $\vec{u} = \frac{1}{2}\text{Tr}(M\vec{\sigma})$, I is the 2×2 unit matrix and $\vec{\sigma} = (\sigma_x, \sigma_y, \sigma_z)$ are the Pauli matrices.

10.7 Using the result of Exercise 10.6, show that the density matrix of a two-level atom can be written as

$$\rho = \frac{1}{2}\left(I + \vec{B} \cdot \vec{\sigma}\right),$$

where all information about the state of the atom is contained in the Bloch vector $\vec{B}$.

10.8 The most general state of a two-level atom is described by the density matrix

$$\rho = \begin{pmatrix} \rho_{11} & \rho_{12} \\ \rho_{21} & \rho_{22} \end{pmatrix},$$

where the diagonal matrix elements ρ_{11}, ρ_{22} are populations of the atomic states and the off-diagonal elements ρ_{12}, ρ_{21} are coherences between them.

(a) Diagonalize the density matrix to show that the density operator can be written in diagonal form

$$\rho = \rho_{\alpha\alpha}|\alpha\rangle\langle\alpha| + \rho_{\beta\beta}|\beta\rangle\langle\beta|,$$

where $|\alpha\rangle$, $|\beta\rangle$ are the eigenstates and $\rho_{\alpha\alpha}$, $\rho_{\beta\beta}$ are the corresponding eigenvalues (energies) of the density matrix ρ.

(b) The diagonal states $|\alpha\rangle$ and $|\beta\rangle$ are superposition states that can be written as

$$|\alpha\rangle = \cos\theta|1\rangle + \sin\theta\, e^{i\phi}|2\rangle,$$
$$|\beta\rangle = -\sin\theta|1\rangle + \cos\theta\, e^{i\phi}|2\rangle.$$

Find $\cos\theta$ in terms of the density matrix elements.

(c) Show that the state $|\alpha\rangle$ can be written in an alternative form as

$$|\alpha\rangle = (1 + |z|^2)^{-\frac{1}{2}} \exp(zS^+)|1\rangle,$$

if we choose $z = \tan\theta \exp(i\phi)$.

Chapter 11

Dressed-Atom Model

11.1 Introduction

In this chapter, we present a powerful method of solving the Bloch equations or the master equation of a driven system called the dressed-atom model. The method is valid in situations where the Rabi frequency of the applied driving field is much larger than the spontaneous emission rate of the atom, $\Omega \gg \Gamma$. Under such conditions, one can make the secular approximation that consists of dropping terms oscillating in time with frequencies 2Ω and higher. These terms, if kept in the master equation, would make corrections to the dynamics of the system of the order of Γ/Ω, and thus are negligible. Although limited in the range of parameters for which it can be used, the dressed-atom model provides a physical insight into the properties and dynamics of the system. Within this model, one can explicitly calculate energy states and transition rates between them in a relatively simple way. The knowledge of the energy states and transition rates is for most of the problems enough to fully understand the underlying physics. There are two mathematically different approaches to the dressed-atom model, but giving the same results, depending on whether we treat the driving field classically or quantum mechanically. These are the semiclassical and quantum

Quantum Optics for Beginners
Zbigniew Ficek and Mohamed Ridza Wahiddin

ISBN 978-981-4411-75-2 (Hardcover), 978-981-4411-76-9 (eBook)
www.panstanford.com

dress-atom models. In the following we explain in details these two dressed-atom models.

11.2 Semiclassical Dressed-Atom Model

First, we illustrate the concept of the semiclassical dressed-atom model in which the atom is treated as a quantum system composed of two energy states $|1\rangle$ and $|2\rangle$, but the applied driving field is treated classically, that is, the field is treated as a c-number not as an operator. We have encountered this situation in the optical Bloch equations (10.23), which describe the interaction of a two-level atom with a classical coherent laser field.

By introducing a complex Bloch vector $\vec{Y} = (\langle S^+\rangle, \langle S^-\rangle, \langle S_z\rangle)$, we can put the optical Bloch equations (10.23) into a matrix form

$$\frac{d\vec{Y}}{dt} = A\vec{Y} - \Gamma G\vec{Y} + \Gamma\vec{F}, \tag{11.1}$$

where A and G are 3×3 matrices of the form

$$A = \begin{pmatrix} i\delta_{\rm L} & 0 & \Omega \\ 0 & -i\delta_{\rm L} & \Omega \\ -\frac{1}{2}\Omega & -\frac{1}{2}\Omega & 0 \end{pmatrix}, \quad G = \begin{pmatrix} \frac{1}{2} & 0 & 0 \\ 0 & \frac{1}{2} & 0 \\ 0 & 0 & 1 \end{pmatrix}, \tag{11.2}$$

and $\vec{F}$ is a column vector with the components $F_1 = F_2 = 0$, $F_3 = -1/2$. The matrix G is composed of the damping rates Γ and for this reason can be called as a dissipative matrix. The matrix A is composed of the parameters $\delta_{\rm L}$ and Ω that are characteristic of the driving field. For this reason we can call the matrix as a driving matrix.

Note that the matrix A is not diagonal in the basis of the complex components $(\langle S^+\rangle, \langle S^-\rangle, \langle S_z\rangle)$. We may find *a new* basis in which the dynamics of the system would be determined by diagonal matrices. The new basis is called the semiclassical dressed-atom basis that determines dynamics between semiclassical dressed states of the system. The dressed states are found from noting that the Hamiltonian of the two-level atom driven by a classical field, which leads to the non-dissipative part of the Bloch equations

$$\hat{H} = \hbar\delta_{\rm L}S_z + \frac{1}{2}\hbar\Omega\left(S^- + S^+\right) \tag{11.3}$$

is not diagonal in the basis of the atomic states $|1\rangle$ and $|2\rangle$. Diagonalization of the Hamiltonian (11.3) results in the semiclassical dressed states

$$\begin{aligned} |\tilde{1}\rangle &= \cos\phi|1\rangle + \sin\phi|2\rangle, \\ |\tilde{2}\rangle &= \sin\phi|1\rangle - \cos\phi|2\rangle, \end{aligned} \tag{11.4}$$

where $\cos^2\phi = (1 + \delta_{\mathrm{L}}/\tilde{\Omega})/2$ with $\tilde{\Omega} = (4\Omega^2 + \delta_{\mathrm{L}}^2)^{1/2}$ and the angle ϕ defined such that $0 \le \phi \le \pi/2$. Let us introduce the raising, lowering and population difference operators in the dressed-atom basis

$$R_{21} = |\tilde{2}\rangle\langle\tilde{1}|, \quad R_{12} = |\tilde{1}\rangle\langle\tilde{2}|, \quad R_3 = |\tilde{2}\rangle\langle\tilde{2}| - |\tilde{1}\rangle\langle\tilde{1}|. \tag{11.5}$$

Using Eq. (11.4) it is easy to verify that dressed-atom operators satisfy the commutation relations

$$[R_{21}, R_{12}] = R_3, \quad [R_3, R_{21}] = 2R_{21}, \quad [R_3, R_{12}] = -2R_{21}. \tag{11.6}$$

The Hamiltonian and the master equation of a given system written in terms of the dressed-atom operators are usually simpler in form, easier to deal with mathematically and to interpret the physics involved.

11.2.1 *Dressing Transformation on the Interaction Hamiltonian*

Let us illustrate the *dressing* transformation on the interaction Hamiltonian of the driven two-level atom interacting with a multimode (reservoir) vacuum field, Eq. (9.6). First, we replace the atomic operators by the dressed-state operators

$$\begin{aligned} S^- &= -\frac{1}{2}\sin(2\phi)R_3 + \sin^2\phi R_{21} - \cos^2\phi R_{12}, \\ S^+ &= -\frac{1}{2}\sin(2\phi)R_3 + \sin^2\phi R_{12} - \cos^2\phi R_{21}, \\ S_z &= -\cos(2\phi)R_3 + \sin(2\phi)\left(R_{12} + R_{21}\right), \end{aligned} \tag{11.7}$$

where $R_{ij} = |\tilde{i}\rangle\langle\tilde{j}|$ are the dressed-atom dipole operators and $R_3 = R_{22} - R_{11}$. Then, we make the following unitary transformation

$$\tilde{H}_{\mathrm{int}} = \exp(i\tilde{H}_0 t)H_{\mathrm{int}}\exp(-i\tilde{H}_0 t), \tag{11.8}$$

with

$$\tilde{H}_0 = \Omega R_3 + \sum_{\lambda} \Delta_\lambda a_\lambda^\dagger a_\lambda, \tag{11.9}$$

where $\Delta_\lambda = \omega_\lambda - \omega_L$, and obtain the interaction Hamiltonian between the dressed atom and the vacuum field

$$\tilde{H}_{\text{int}} = i\hbar \sum_{\lambda} g_\lambda \left(sca_\lambda^\dagger R_3 e^{i\Delta_\lambda t} + c^2 a_\lambda^\dagger R_{12} e^{i(\Delta_\lambda - 2\Omega)t} \right.$$
$$\left. - s^2 a_\lambda^\dagger R_{21} e^{i(\Delta_\lambda + 2\Omega)t} - \text{H.c.} \right), \tag{11.10}$$

in which $s = \sin\phi$ and $c = \cos\phi$.

In the dressed-atom picture, the vacuum modes are tuned to the dressed-state transitions that occur at three characteristic frequencies, Δ_λ and $\Delta_\lambda \pm 2\Omega$. If one considers a broadband reservoir that is characterized by bandwidth much larger than Ω, then all of the reservoir modes couple with the same strengths to the dressed-atom transition frequencies. However, when the reservoir field has a finite bandwidth that is much smaller than Ω, the vacuum field modes then couple to the dressed-atom frequencies with unequal strengths. In this case, spontaneous emission can be dynamically suppressed by Rabi shifting the atomic transition away from the reservoir central frequency. In this way, one can control spontaneous emission from the driven atom by a suitable matching of the frequencies of the reservoir field to the dressed-atom frequencies [61].

11.2.2 *Master Equation in the Dressed-Atom Basis*

Let us now derive, using the Hamiltonian (11.10) given in the dressed-atom basis, the master equation for the reduced density operator of the system coupled to a vacuum field reservoir. We shall assume that the reservoir has a finite bandwidth that is broad enough for the Markov approximation to be valid but much smaller than the Rabi frequency of the driving field. On carrying out this procedure in the dressed-atom basis, it is found that in the dissipative part of the master equation certain terms are slowly varying in time while the others are oscillating with frequencies 2Ω and 4Ω. Since we are interested in the case where the Rabi frequency Ω is much larger than the atomic damping rate, $\Omega \gg \Gamma$, we can invoke the secular approximation that consists of dropping these rapidly oscillating terms. These terms, if kept in the master equation, would make corrections to the dynamics of the system of the order of Γ/Ω, and thus completely negligible. After discarding the rapidly oscillating terms in the dissipative part of the master equation, the time evolution of the reduced density operator takes the form

$$\begin{aligned}\frac{\partial \rho}{\partial t} &= \frac{1}{2}\Gamma_0\left(R_3\rho R_3 - \rho\right) + \frac{1}{2}\Gamma_-\left(R_{21}\rho R_{12} - R_{12}R_{21}\rho\right) \\ &\quad + \frac{1}{2}\Gamma_+\left(R_{12}\rho R_{21} - R_{21}R_{12}\rho\right) + \text{H.c.} \end{aligned} \tag{11.11}$$

The parameters

$$\begin{aligned}\Gamma_0 &= s^2c^2\Gamma|D(\omega_{\rm L})|^2, \\ \Gamma_- &= s^4\Gamma|D(\omega_{\rm L} - 2\Omega)|^2, \\ \Gamma_+ &= c^4\Gamma|D(\omega_{\rm L} + 2\Omega)|^2, \end{aligned} \tag{11.12}$$

determine the damping rates between the dressed states of the system. They also include the frequency-dependent density of the vacuum modes, which may arise from a finite bandwidth of the reservoir field. It is represented by the frequency-dependent function $D(\omega_\lambda)$, which is also known as the transfer function of the reservoir. The absolute value square of $D(\omega_\lambda)$ can be identified as the Airy function of a frequency-dependent radiation reservoir. The coefficient Γ_0 corresponds to spontaneous emission occurring at two transitions of the dressed atom. One from the dressed state $|\tilde{1}\rangle$ to the state $|\tilde{1}\rangle$ of the manifold below, and the other from the dressed state $|\tilde{2}\rangle$ to the state $|\tilde{2}\rangle$ of the manifold below. These transitions occur at frequency $\omega_{\rm L}$. The coefficient Γ_+ corresponds to spontaneous emission from the upper dressed state to the lower dressed state of the manifold below and occurs at frequency $\omega_{\rm L}+2\Omega$, whereas the coefficient Γ_- corresponds to spontaneous emission from the lower dressed state to the upper dressed state of the manifold below and occurs at frequency $\omega_{\rm L} - 2\Omega$.

An important feature of the master equation (11.11), derived in the limit of $\Omega \gg \Gamma$, is that spontaneous transitions occur at three well-separated frequencies, $\omega_{\rm L}$, $\omega_{\rm L} + 2\Omega$ and $\omega_{\rm L} - 2\Omega$. Thus, each transition can be considered as a single two-level system, which makes the master equation simple to solve.

11.3 Quantum Dressed-Atom Model

We now illustrate the fully quantum-mechanical dressed-atom model of a driven two-level atom first introduced by Cohen–Tannoudji[a] and Reynaud [62, 63]. As we shall see, the quantum description in which both the atom and field are treated as quantum systems is more elegant than the semiclassical model that gives a better insight into the processes involved in the dynamics of the system. In the quantum description we clearly see the meaning of dressing. The atoms are *dressed* in photons of the applied field to form an effective single-cascade multi-level quantum system.

The fully quantum mechanical Hamiltonian of the system is

$$\hat{H} = \hat{H}_0 + \hat{V}, \tag{11.13}$$

where

$$\hat{H}_0 = \hbar\omega_0 S_z + \hbar\omega_L \left(\hat{a}^\dagger\hat{a} + \frac{1}{2}\right) \tag{11.14}$$

is the non-interacting atom-plus-field Hamiltonian and $\hat{V}$ is the interaction (in the rotating wave approximation (RWA)) between the atom and the laser field

$$\hat{V} = \frac{1}{2}\hbar g \left(\hat{a}^\dagger S^- + S^+\hat{a}\right), \tag{11.15}$$

where the coefficient g describes the strength of the coupling between the atom and the field.

11.4 Atom–Field Entangled States

The basis states for the quantum description of the system are the eigenstates of the non-interacting atom-plus-field Hamiltonian $\hat{H}_0$, which are the product states of the atomic and the field states

$$|\psi_0\rangle = |i\rangle \otimes |n\rangle \equiv |i, n\rangle, \tag{11.16}$$

where $|i\rangle$ $(i = 1, 2)$ is an atomic state, and $|n\rangle$ is the photon number state of the field. We will call the product states (11.16) the *undressed* states of the system.

[a]Claude Cohen–Tannoudji was granted the Nobel prize in 1997 for development of methods to cool and trap atoms with laser light.

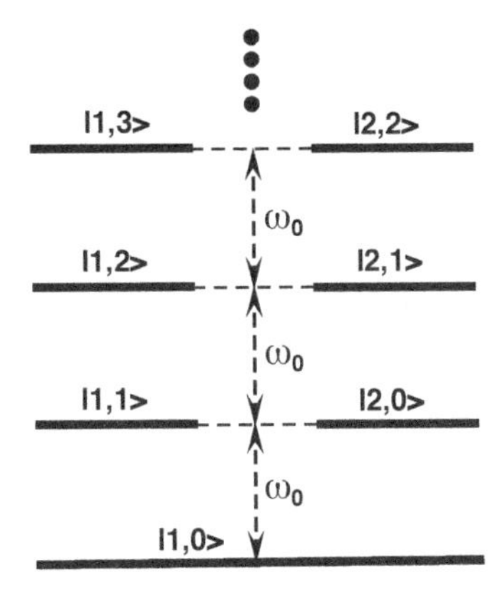

Figure 11.1 Energy level diagram of undressed states of the system composed of a two-level atom and a resonant ($\Delta = 0$) quantum field. The ground state is a singlet state, whereas the excited ($n > 0$) states form doublets of degenerate states separated in frequency by ω_0.

In the energy level diagram, the undressed states group into manifolds. The lowest manifold is composed of a single state $|1, 0\rangle$, and the higher energy manifolds are composed of degenerate (when $\Delta = \omega_L - \omega_0 = 0$) or non-degenerate (when $\Delta \neq 0$) doublets. Figure 11.1 shows the energy levels of the undressed states of the system for the resonant case of $\Delta = 0$. We see that in the basis of the undressed states the energy levels form a ladder of degenerate doublets separated by ω_0.

The next step in the dressed-atom procedure is to include the interaction $\hat{V}$ and find the matrix representation of the total Hamiltonian of the system in the basis of the undressed states. This interaction appears as a perturber to the undressed states and we will analyse the effect of the perturber on the undressed states of the system. We show the detailed procedure for manifold n composed two undressed states $|1, n\rangle$ and $|2, n - 1\rangle$. In the state $|1, n\rangle$, the atom is in the ground state $|1\rangle$, and there are n photons present in the laser mode. In the state $|2, n - 1\rangle$, the atom is in the excited state $|2\rangle$, with $n - 1$ photons present in the laser mode.

We find the matrix representation of the Hamiltonian $\hat{H}$ in the basis of the undressed states $|1, n\rangle$ and $|2, n - 1\rangle$. It is easy to show that in the basis of the undressed states, the matrix elements of $\hat{H}$

are

$$\langle 1, n|\hat{H}|1, n\rangle = \hbar n\omega_{\rm L} + \frac{1}{2}\hbar\Delta,$$
$$\langle 2, n-1|\hat{H}|2, n-1\rangle = \hbar n\omega_{\rm L} - \frac{1}{2}\hbar\Delta,$$
$$\langle 1, n|\hat{H}|2, n-1\rangle = \langle 2, n-1|\hat{H}|1, n\rangle = \frac{1}{2}\hbar g\sqrt{n}. \quad (11.17)$$

Note that the diagonal elements are determined solely by the free Hamiltonian $\hat{H}_0$, whereas the off-diagonal elements are determined solely by the interaction $\hat{V}$.

11.4.1 *Resonant Field, $\Delta = 0$*

With the matrix elements (11.17) and at $\Delta = 0$, the Hamiltonian of the system written in the basis of the undressed states is given by a 2×2 matrix

$$\hat{H} = \hbar \begin{pmatrix} n\omega_0 & \frac{1}{2}g\sqrt{n} \\ \frac{1}{2}g\sqrt{n} & n\omega_0 \end{pmatrix}. \quad (11.18)$$

The eigenvalues (energies) and eigenstates of $\hat{H}$ are found by the diagonalization of the matrix (11.18). From the diagonalization, we find that the matrix has two non-degenerate eigenvalues

$$E_\pm = \hbar n\omega_0 \pm \frac{1}{2}\hbar g\sqrt{n}, \quad (11.19)$$

which indicates that the interaction $\hat{V}$ lifts the degeneracy and leads to new non-degenerate states $|\psi_n\rangle$, called the dressed states of the system, that satisfy the eigenvalue equation

$$\hat{H}\,|\psi_n\rangle = E_\pm\,|\psi_n\rangle. \quad (11.20)$$

Note that the splitting between the states $|\psi_n\rangle$ depends on n and increases with n. However, for large n the splitting is almost constant and we can replace n by $\langle n\rangle$.

In order to find the explicit form of the eigenvectors (dressed states) $|\psi_n\rangle$, consider a linear superposition

$$|\psi_n\rangle = a\,|1, n\rangle + b\,|2, n-1\rangle. \quad (11.21)$$

For the eigenvalue E_+, the eigenvalue equation (11.20) written in a matrix form

$$\begin{pmatrix} \hbar n\omega_0 & \frac{1}{2}\hbar g\sqrt{n} \\ \frac{1}{2}\hbar g\sqrt{n} & \hbar n\omega_0 \end{pmatrix} \begin{pmatrix} a \\ b \end{pmatrix} = E_+ \begin{pmatrix} a \\ b \end{pmatrix}, \quad (11.22)$$

yields the relation

$$\hbar n\omega_0 a + \frac{1}{2}\hbar g\sqrt{n}b = \hbar n\omega_0 b + \frac{1}{2}\hbar g\sqrt{n}a, \tag{11.23}$$

from which we find that $a = b$. Hence, we write

$$|\psi_{n+}\rangle = a\left(|1, n\rangle + |2, n-1\rangle\right), \tag{11.24}$$

where the remaining constant a is found from the normalization, which gives $a = 1/\sqrt{2}$.

Thus, the eigenstate corresponding to the eigenvalue E_+ is of the form

$$|\psi_{n+}\rangle = \frac{1}{\sqrt{2}}\left(|1, n\rangle + |2, n-1\rangle\right). \tag{11.25}$$

Similarly, we find that the eigenstate corresponding to the eigenvalue E_- is of the form

$$|\psi_{n-}\rangle = \frac{1}{\sqrt{2}}\left(|1, n\rangle - |2, n-1\rangle\right). \tag{11.26}$$

The eigenstates $|\psi_{n+}\rangle$ and $|\psi_{n-}\rangle$ are called the *quantum dressed states* of the system. In other words, the laser field *dresses* the atom in photons, and forms along with it a single, entangled quantum system. Physically, this reflects the fact that photons are exchanged between the atom and the driving field via absorption and stimulated emission processes many times between successive spontaneous emissions by the atom into the vacuum modes.

The dressed states of the system are shown in Fig. 11.2. In the dressed state representation, the atom and the driving field evolve as a single system, where the states cannot be written as a product of the atomic and the field states. Since the dressed states (11.25) and (11.26) are given in a form of linear superpositions of two product states with equal amplitudes, we call them *maximally entangled* states of the system. Note that the dressed states result from the presence of off-diagonal terms (coherencies) in the matrix representation of the Hamiltonian. Thus, the entanglement results from the presence of the coherence between the atom and the field.

11.4.2 *Vacuum Rabi Splitting and AC Stark Effect*

The splitting of the first pair ($n = 1$) of the states is called the *vacuum Rabi splitting*. Using the master equation of the driven and

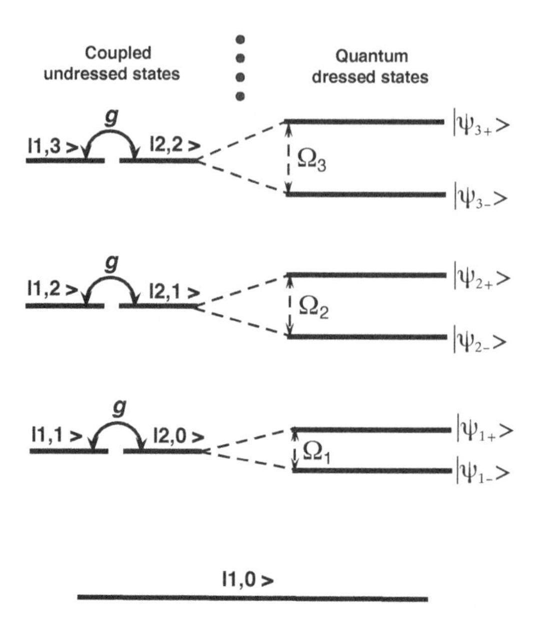

Figure 11.2 Energy level diagram of the coupled undressed states and the resulting quantum dressed states of the system. The coupling between the atom and the field lifts the degeneracy between the undressed states resulting in two non-degenerate dressed states separated in frequency by the Rabi frequency $\Omega_n = \sqrt{n}g$.

spontaneously damped atom, we can find spontaneous dynamics of the dressed system.

Before going into detailed calculations, we may notice interesting effects just by looking into the energy structure of the dressed states. For example, spontaneous transitions from the first pair of the dressed states to the ground state, $n = 1 \rightarrow n = 0$, show the vacuum Rabi doublet, as illustrated in Fig. 11.3, whereas the transitions between two neighbouring doublets with $n \gg 1$ show the Mollow triplet, as illustrated in Fig. 11.4. The constant splitting in the Mollow triplet is equal to the Rabi frequency $\Omega = g\sqrt{\langle n\rangle}$. The splitting of the dressed states by the Rabi frequency Ω is sometimes called the ac Stark effect.

It is interesting to note that the dressing process reduces the spontaneous emission rate (decoherence). To show this more explicitly, recall that the damping rate of an atom is proportional to

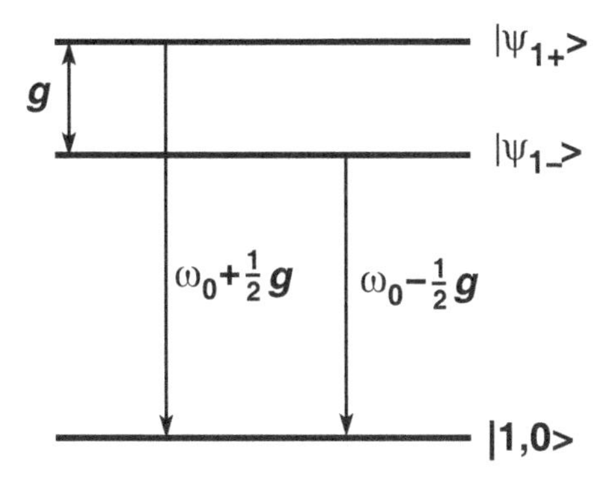

Figure 11.3 Spontaneous transitions from $n = 1$ doublet to the ground state $|1, 0\rangle$ that give rise to the vacuum Rabi doublet.

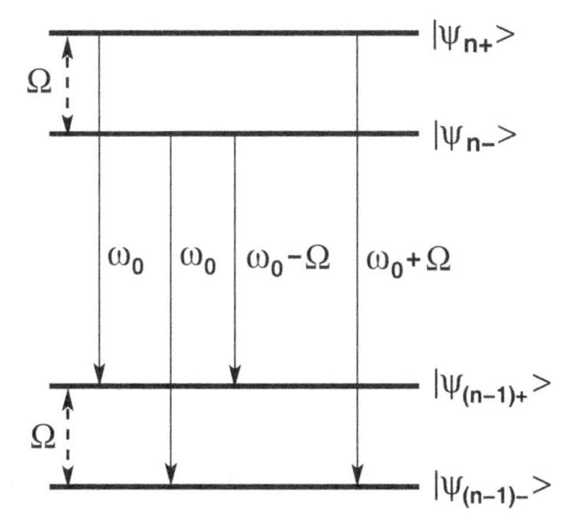

Figure 11.4 Spontaneous transitions from the manifold $n\,(n \gg 1)$ to the manifold below, $n-1$, that give rise to the Mollow triplet. For a large n, the Rabi frequencies do not vary much with n, so that $\Omega_n \approx \Omega_{n-1} \equiv \Omega$.

the dipole moment between the two atomic levels

$$\Gamma_a \sim |\mu_{12}|^2, \tag{11.27}$$

where $\mu_{12} = \langle 1|\mu|2\rangle$.

In analogy, the damping rate between the dressed state $|\psi_{1+}\rangle$ and the ground state $|1, 0\rangle$ is proportional to

$$\Gamma_+ \sim |\mu_{+0}|^2, \tag{11.28}$$

where

$$\mu_{+0} = \langle \psi_+|\mu|1, 0\rangle = \frac{1}{\sqrt{2}}\left(\langle 1, 1| + \langle 2, 0|\right)\mu|1, 0\rangle$$
$$= \frac{1}{\sqrt{2}}\langle 1|\mu|2\rangle = \frac{1}{\sqrt{2}}\mu_{12}. \tag{11.29}$$

Thus, the damping rate is given by

$$\Gamma_+ = \frac{1}{2}\Gamma_a. \tag{11.30}$$

Similarly, we can show that the damping rate between the dressed state $|\psi_{1-}\rangle$ and the ground state $|1, 0\rangle$ is given by

$$\Gamma_- = \frac{1}{2}\Gamma_a. \tag{11.31}$$

Hence, the damping rates between the $n = 1$ dressed states and the ground state is a half of that between the bare atomic levels of an undriven atom.

11.4.3 *Non-resonant Driving,* $\Delta \neq 0$

We have seen that the dressed states of the system for the exact resonance, $\Delta = \omega_L - \omega_0 = 0$, are maximally entangled states. For a detuned field with $\Delta \neq 0$, the diagonalization of the Hamiltonian (11.13) leads to two non-degenerate eigenvalues

$$E_\pm = \hbar n\omega_0 \pm \hbar\sqrt{\Delta^2 + \frac{1}{4}ng^2}, \tag{11.32}$$

and corresponding dressed states[a]

$$|\psi_{n+}\rangle = \sin\theta\,|1, n\rangle + \cos\theta\,|2, n-1\rangle\,,$$
$$|\psi_{n-}\rangle = \cos\theta\,|1, n\rangle - \sin\theta\,|2, n-1\rangle\,, \tag{11.33}$$

where

$$\cos^2\theta = \frac{1}{2} + \frac{\Delta}{2\sqrt{\Delta^2 + \frac{1}{4}ng^2}}. \tag{11.34}$$

It is evident from Eq. (11.33) that for $\Delta \neq 0$ the dressed states of the system are not maximally entangled states, and for $\Delta \gg \sqrt{n}g$, the states reduce to the product states $|2, n-1\rangle$ and $|1, n\rangle$. This fact is

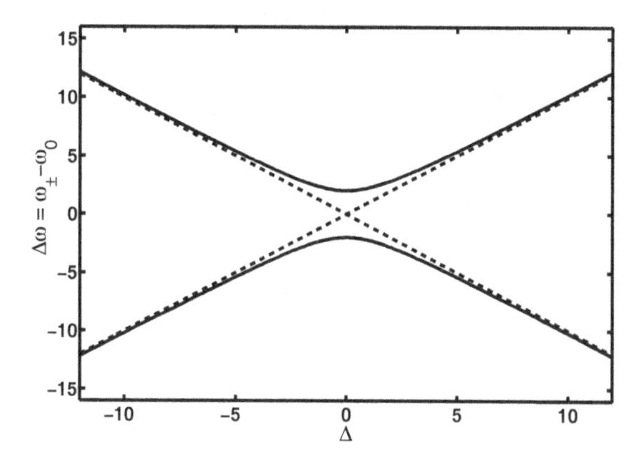

Figure 11.5 Variation of the $n = 1$ dressed-state frequencies $\omega_{\pm}$ with detuning Δ around the unperturbed frequency ω_0 for $g = 4$ (solid lines) and $g = 0$ (dashed lines).

simply related to the distinguishability problem of two systems that for $\Delta \neq 0$ one could distinguish between the atom and the field.

Figure 11.5 shows a variation of the dressed-state frequencies $\omega_{\pm} = E_{\pm}/\hbar$ of the $n = 1$ eigenvectors with the detuning Δ around the unperturbed energy ω_0. We see a crossing of the energy levels in the absence of the coupling at $\Delta = 0$, and the appearance of the avoided crossing effect when $g \neq 0$.

Perhaps the most interesting aspects of the non-maximally entangled dressed states relate to their imbalanced populations. Detailed calculation of the populations of the dressed states of a detuned field is left as an exercise, see Exercise 11.6. The imbalanced populations result in a population inversion between dressed states of two neighbouring manifolds. This implies that a field coupled to the system can be amplified if tuned to resonance with the dressed states for which there exists the population inversion. If the field is a cavity field, one can obtain a lasing action [64]. Since there is no population inversion between the upper $|2\rangle$ and lower $|1\rangle$ atomic bare states, one obtains lasing without population inversion [65].

[a]Details of the derivation of the eigenvalues and the corresponding eigenvectors (dressed states) for a detuned field are left for the reader as a tutorial exercise.

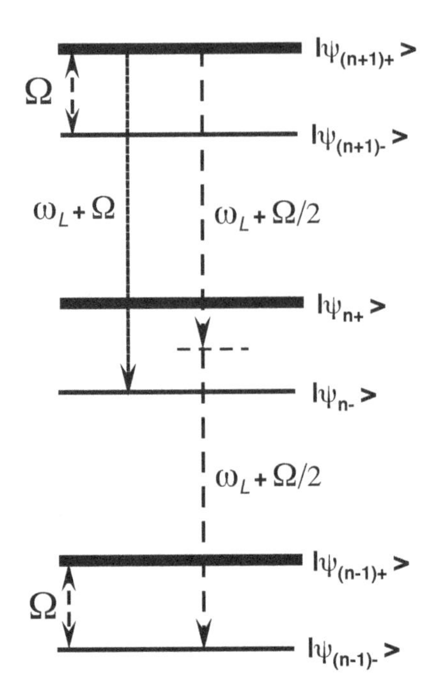

Figure 11.6 Dressed states of a two-level atom driven by an off-resonant strong laser field. The dressed states of each doubled have imbalanced populations, which is indicated by different thicknesses of the lines representing the states. Inverted one-photon transition occurs at frequency $\omega_L + \Omega$ (short dashed line), and two-photon transition occurs at frequency $\omega_L + \Omega/2$ (long dashed line).

It has also been predicted and demonstrated experimentally that a two-photon lasing can be realized between dressed states of two manifolds separated in frequency by $2\omega_L$ for which a population inversion exists [66, 67]. Figure 11.6 shows dressed states of a two-level atom driven by an off-resonant ($\Delta \neq 0$) strong laser field ($n \gg 1$). The energy difference between two dressed states within each manifold is constant and equal to the Rabi frequency Ω of the driving field. The lines representing the dressed states have thicknesses that indicate their populations. One-photon population inversion exists at frequency $\omega_L + \Omega$, and two-photon inversion exists at frequency $\omega_L + \Omega/2$.

11.5 Summary

Let us summarize this chapter by listing the successful steps in an application of the dressed-atom technique:

(1) Consider the fully quantum version of the Hamiltonian of a given system in which both the atomic system and the driving field are quantized.
(2) Write the Hamiltonian in a matrix form using as the basis product states of the atomic and field (photon number) states.
(3) Diagonalise the matrix using the standard diagonalization techniques.
(4) The eigenvectors of the matrix are the dressed states of the system, and the corresponding eigenvalues are the energies of the dressed states. In general, the dressed states group into manifolds.
(5) Draw the dressed states of two neighbouring manifolds, n and $n-1$ or $n+1$ and n, and calculate dipole moments between the dressed states. If the atomic system has no permanent dipole moments, non-vanishing matrix elements of the atomic dipole moment operator occur only between neighbouring manifolds.
(6) The dressed-atom method is most useful when the energy difference between the dressed states of a given manifold are large compared to the damping rates of the bare atomic transitions. It is then possible to make the secular approximation in which we neglect coupling between diagonal and off-diagonal elements of the density matrix of the system.

Further discussion of the dressed-atom technique is left to the exercises.

Exercises

11.1 Show that dressed states of a two-level atom driven by a detuned laser field are not maximally entangled states.

11.2 Calculate damping rates between dressed states of a two-level atom driven by a detuned laser field.

(a) How the damping rates vary with the detuning?
(b) What are the values of the damping rates for $\Delta \to \pm\infty$?

11.3 Calculate dressed states of a three-level atom in the *vee* configuration with two non-degenerate transitions $|1\rangle \to |2\rangle$ and $|3\rangle \to |2\rangle$ driven by a single laser field of frequency $\omega_L = (\omega_1 + \omega_2)/2$, where ω_1 and ω_2 are the frequencies of the $|1\rangle \to |2\rangle$ and $|3\rangle \to |2\rangle$ transitions, respectively.

11.4 Calculate dressed states of a three-level atom in the *vee* configuration with two non-degenerate transitions $|1\rangle \to |2\rangle$ and $|3\rangle \to |2\rangle$ driven by two lasers each coupled to only one of the atomic transitions. Assume that the lasers are on resonance with the atomic transitions to which they are coupled.

11.5 Using equations of motion for the expectation values of the spin operators S^+, S^- and S of a two-level atom driven by a coherent laser field

(a) Find the expectation values of the atomic dipole moments $\langle S^{\pm}\rangle$ and the atomic inversion $\langle S_z\rangle$ in the steady-state limit $(t \to \infty)$.
(b) Show that the magnitude of the average dipole moment is maximal for $\Omega = \Gamma/\sqrt{2}$.
(c) Are the spin oscillations coherent? Comment on the result.
(d) Find for what values of Ω the x-component of the atomic spin (S_x) is squeezed?

11.6 Using the results of Exercise 11.5 for the steady-state solutions for the atomic dipole moments $\langle S^{\pm}\rangle$ and the atomic inversion $\langle S_z\rangle$, show that

(a) $\langle S_z\rangle < 0$ for any values of the parameters involved, that in the bare states basis, there is no population inversion between the upper $|2\rangle$ and lower $|1\rangle$ atomic states.
(b) Find the populations of the dressed-states to show that even there is no population inversion in the bare state basis, there can be a population inversion between the dressed states. Under which condition, the population can be inverted between the dressed states?

11.7 Consider a two-level atom with no permanent dipole moments and driven by a resonant single-mode laser field.

(a) Show that the induced dipole moments are different from zero only between dressed states of two neighbouring manifolds.

(b) There are four possible dipole transitions between dressed states of two neighbouring manifolds. Which pairs of the possible transitions can produce interference fringes in the emitted field?

Chapter 12

Fokker–Planck Equation

12.1 Introduction

In Chapter 9, we derived the master equation for a reduced density operator of a two-level atom coupled to a multi-mode vacuum field. We have seen that the master equation is an operator type equation that, in general, is not easy to solve. We have illustrated the solution of the master equation by representing the density operator in the basis of the two states of the two-level atom. The purpose of the representation of the master equation was to obtain c-number differential equations that are equivalent to the operator equations, but are more readily soluble.

In this chapter, we continue the presentation of approaches to solve problems described by the density operator using as an example the master equation of a harmonic oscillator interacting with a broadband thermal reservoir. As we shall see, with the theory of the photon number and the coherent state P representations developed earlier in this book, it is possible to transform the master equation into c-number differential equations. First, we will consider the photon number representation of the density operator and show that the master equation can be transformed into a hierarchy of first-order differential equations that can be

Quantum Optics for Beginners
Zbigniew Ficek and Mohamed Ridza Wahiddin

ISBN 978-981-4411-75-2 (Hardcover), 978-981-4411-76-9 (eBook)
www.panstanford.com

solved by the standard mathematical methods. Next, we will discuss the coherent state representation, P representation, that transforms the master equation into a c-number differential equation called the Fokker–Planck equation (FPE). We then show how the equation can be viewed as a stochastic equation that, for specific initial conditions, can be solved analytically by a direct integration. The latter part of the chapter illustrates the technique of solving the FPE when direct solutions are not possible. This technique involves stochastic differential equations (SDE) approach and will be illustrated on two examples of typical problems in quantum optics: (i) single-cavity mode driven by a classical coherent field and (ii) effect of two-photon losses on the driven cavity mode. Thus, the reader will have an opportunity to study applications of stochastic methods to quantum optics problems.

12.2 Master Equation of a Harmonic Oscillator

Consider a simple one-mode harmonic oscillator interacting with a multi-mode field (reservoir), whose modes are in vacuum thermal states. An understanding of the dynamics of such system is of great use in quantum optics since, in principle at least, all problems involving bosonic fields can be represented in terms of the harmonic oscillator.

The interaction Hamiltonian of a single-mode harmonic oscillator and a vacuum thermal field is given by

$$\hat{H}_{\text{int}} = -\frac{1}{2} i\hbar \sum_{k} g_k \left[\hat{a}^\dagger \hat{b}_k(t) - \hat{a}\hat{b}_k^\dagger(t)\right], \tag{12.1}$$

where $\hat{a}$ and $\hat{a}^\dagger$ are the annihilation and creation operators of the single-mode harmonic oscillator, and $\hat{b}_k$ and $\hat{b}_k^\dagger$ are the annihilation and creation operators of the multi-mode vacuum field.

The thermal state of the reservoir is characterized by the following correlation functions

$$\begin{aligned}
\langle \hat{b}_k(t)\, \hat{b}_{k'}^\dagger(t')\rangle &= (1+N)\,\delta_{kk'}\delta(t-t')\,, \\
\langle \hat{b}_k^\dagger(t)\, \hat{b}_{k'}(t')\rangle &= N\delta_{kk'}\delta(t-t')\,,
\end{aligned} \tag{12.2}$$

where N is the number of photons in the thermal modes.

Assuming that the reservoir is a broadband thermal field (Markovian), and after tracing with respect to the reservoir field, the master equation of the density operator of the system can be written as

$$\frac{\partial}{\partial t}\rho = -i\omega_0\left[\hat{a}^{\dagger}\hat{a},\,\rho\right] - \frac{1}{2}\Gamma\left(1+N\right)\left(\hat{a}^{\dagger}\hat{a}\rho + \rho\hat{a}^{\dagger}\hat{a} - 2\hat{a}\rho\hat{a}^{\dagger}\right)$$
$$-\frac{1}{2}\Gamma N\left(\hat{a}\hat{a}^{\dagger}\rho + \rho\hat{a}\hat{a}^{\dagger} - 2\hat{a}^{\dagger}\rho\hat{a}\right), \tag{12.3}$$

where Γ is the damping (pumping) rate of the harmonic oscillator. The first term in the above equation leads to the coherent (Hamiltonian) evolution, while the other two terms lead to dissipation of the evolution. The dissipation is enhanced by the presence of thermal photons in the reservoir.

12.3 Photon Number Representation

In many practical problems involving single-mode fields, such us a cavity field, the photon number representation is very useful in finding the density operator of the field. Here, we illustrate the application of the photon number representation to the evolution of the single-mode harmonic oscillator. In the photon number representation, the density operator of the harmonic oscillator can be written as

$$\rho = \sum_{mn}\rho_{nm}|n\rangle\langle m|. \tag{12.4}$$

Using this representation, the master equation (12.3) takes the form

$$\sum_{mn}\frac{\partial}{\partial t}\rho_{nm}|n\rangle\langle m| = -i\omega_0\sum_{mn}\left(\hat{a}^{\dagger}\hat{a}|n\rangle\langle m| - |n\rangle\langle m|\hat{a}^{\dagger}\hat{a}\right)\rho_{nm}$$
$$-\frac{\Gamma}{2}\left(1+N\right)\sum_{mn}\rho_{nm}\left(\hat{a}^{\dagger}\hat{a}|n\rangle\langle m| + |n\rangle\langle m|\hat{a}^{\dagger}\hat{a} - 2\hat{a}|n\rangle\langle m|\hat{a}^{\dagger}\right)$$
$$-\frac{\Gamma}{2}N\sum_{mn}\rho_{nm}\left(\hat{a}\hat{a}^{\dagger}|n\rangle\langle m| + |n\rangle\langle m|\hat{a}\hat{a}^{\dagger} - 2\hat{a}^{\dagger}|n\rangle\langle m|\hat{a}\right). \tag{12.5}$$

After carrying out the operations of the creation and annihilation operators on the number states, we get

$$\sum_{mn} \frac{\partial}{\partial t}\rho_{nm}|n\rangle\langle m| = -i\omega_0 \sum_{mn} (n-m)\,\rho_{nm}|n\rangle\langle m|$$
$$-\frac{\Gamma}{2}(1+N)\sum_{mn}\rho_{nm}\Big(n|n\rangle\langle m| + m|n\rangle\langle m|$$
$$-2\sqrt{nm}|n-1\rangle\langle m-1|\Big)$$
$$-\frac{\Gamma}{2}N\sum_{mn}\rho_{nm}\Big[(n+1)\,|n\rangle\langle m| + (m+1)\,|n\rangle\langle m|$$
$$-2\sqrt{(n+1)(m+1)}|n+1\rangle\langle m+1|\Big]. \qquad (12.6)$$

Now, by comparing the coefficients standing at the same $|n\rangle\langle n|$, we find that the density matrix elements satisfy the following equation of motion

$$\frac{\partial}{\partial t}\rho_{nm} = -i\omega_0\,(n-m)\,\rho_{nm}$$
$$-\frac{\Gamma}{2}(1+N)\left[(n+m)\,\rho_{nm} - 2\sqrt{(n+1)\,(m+1)}\rho_{n+1m+1}\right]$$
$$-\frac{\Gamma}{2}N\left[(n+m+2)\,\rho_{nm} - 2\sqrt{nm}\rho_{n-1m-1}\right]. \qquad (12.7)$$

This differential equation for the density matrix elements gives an infinite hierarchy of differential (c-number) equations. Note that in this simple case, the equations of motion for the off-diagonal ($n \neq m$) matrix elements are decoupled from the equations of motion for the diagonal ($n = m$) matrix elements. This decoupling significantly simplifies the solution of the system of the differential equations.

Example 12.1 (Steady-state solution for the diagonal matrix elements) *Let us solve the set of the differential equations (12.7) for the diagonal matrix elements. The diagonal elements ρ_{nn} represent populations of the number states. Thus, ρ_{nn} given in function of n is the probability distribution of the population among the number states. The diagonal elements satisfy the following equations of motion*

$$\frac{\partial}{\partial t}\rho_{nn} = -\Gamma\,(1+N)\,[n\rho_{nn} - (n+1)\,\rho_{n+1n+1}]$$
$$-\Gamma N\,[(n+1)\,\rho_{nn} - n\rho_{n-1n-1}]. \qquad (12.8)$$

Consider the steady-state population distribution for which $\frac{\partial}{\partial t}\rho_{nn} = 0$*. In this case, we can find the exact expression for* $P_n = \rho_{nn}$*. For* $n = 0$*, we get from Eq. (12.8)*

$$0 = -NP_0 + (1 + N)\,P_1, \tag{12.9}$$

from which we find

$$P_1 = \frac{N}{1+N}P_0. \tag{12.10}$$

For $n = 1$*, Eq. (12.8) yields*

$$0 = NP_0 - (1 + 3N)\,P_1 + 2\,(1 + N)\,P_2, \tag{12.11}$$

from which we find

$$P_2 = \frac{N^2}{(1+N)^2}P_0. \tag{12.12}$$

Finally, by iteration, we find

$$P_n = \frac{N^n}{(1+N)^n}P_0, \tag{12.13}$$

where P_0 *can be found from the normalization* $P_0 = 1/(1+N)$*, which gives*

$$P_n = \frac{N^n}{(1+N)^{n+1}}. \tag{12.14}$$

Thus, the distribution of photons is a thermal distribution, as one could expect.

Hence, in the steady state (thermal equilibrium), the density operator of the harmonic oscillator can be written as

$$\rho = \sum_n P_n\,|n\rangle\langle n| = \sum_n \frac{N^n}{(1+N)^{n+1}}\,|n\rangle\langle n|\,. \tag{12.15}$$

It is interesting to find the stationary density operator of the harmonic oscillator in the coherent state representation when we know the photon number representation. We can find $P(\alpha)$ from Eq. (7.52)

$$P\,(\alpha) = \frac{\mathrm{e}^{|\alpha|^2}}{\pi^2}\int d^2\beta\,\langle\beta|\,\rho\,|\beta\rangle\,\mathrm{e}^{|\beta|^2}\mathrm{e}^{-\alpha\beta^*-\alpha^*\beta}. \tag{12.16}$$

Using the number state representation of the coherent state $|\beta\rangle$ and Eq. (12.14), we obtain

$$\begin{aligned}
\langle\beta|\,\rho\,|\beta\rangle &= \sum_n \frac{N^n}{(1+N)^{n+1}} \sum_{m,k} \frac{\beta^{*m}\beta^k}{\sqrt{m!k!}} \langle m|\,n\rangle\,\langle n|\,k\rangle \mathrm{e}^{-|\beta|^2} \\
&= \sum_n \frac{N^n}{(1+N)^{n+1}} \frac{|\beta|^{2n}}{n!} \mathrm{e}^{-|\beta|^2} \\
&= \frac{1}{1+N} \sum_n \left(\frac{N|\beta|^2}{1+N}\right)^n \frac{1}{n!} \mathrm{e}^{-|\beta|^2} \\
&= \frac{1}{1+N} \exp\left(-\frac{|\beta|^2}{1+N}\right). \qquad (12.17)
\end{aligned}$$

Substituting Eq. (12.17) into Eq. (12.16), and performing the integration, we obtain

$$P\,(\alpha) = \frac{1}{\pi N} \exp\left(-\frac{|\alpha|^2}{N}\right). \qquad (12.18)$$

Thus, $P\,(\alpha)$ obeys the Gaussian distribution. The probability distribution function may be regarded as a statistical distribution superposed by a large number of electromagnetic (EM) modes of random phases.

In the following subsection, we will show how to find the probability distribution function $P\,(\alpha)$ from the master equation of the density operator of the harmonic oscillator in the coherent state representation (FPE).

12.4 *P* Representation: Fokker–Planck Equation

We now turn to a consideration of the coherent state, or P representation of the master equation [68]. We will illustrate the major steps of the derivation of the FPE on a simplified master equation with $N = 0$, and next will extend the derivation to the case of $N \neq 0$. The reader wishing to pursue the theory of the FPE further is referred to the book by Carmichael [69].

Suppose, there exists a time-dependent P distribution $P(\alpha, t)$. Then, using the P representation for the density operator of the harmonic oscillator

$$\hat{\rho} = \int d^2\alpha\, P\,(\alpha)\, |\alpha\rangle\langle\alpha|\,, \qquad (12.19)$$

we can transform the master equation, with $N = 0$, into an integro-differential equation

$$\begin{aligned}\int d^2\alpha \frac{\partial}{\partial t} P(\alpha, t) |\alpha\rangle\langle\alpha| = \int d^2\alpha P(\alpha, t) \\ \times \left\{-i\omega_0 \left(\hat{a}^\dagger \hat{a}|\alpha\rangle\langle\alpha| - |\alpha\rangle\langle\alpha|\hat{a}^\dagger \hat{a}\right)\right. \\ -\frac{\Gamma}{2} \left(\hat{a}^\dagger \hat{a}|\alpha\rangle\langle\alpha| + |\alpha\rangle\right. \\ \left.\left. \times \langle\alpha|\hat{a}^\dagger \hat{a} - 2\hat{a}|\alpha\rangle\langle\alpha|\hat{a}^\dagger\right)\right\}. \qquad (12.20)\end{aligned}$$

We now perform the action of the annihilation and creation operators on the coherent state using the following relations

$$\begin{aligned}\hat{a}^\dagger \hat{a}|\alpha\rangle\langle\alpha| &= \alpha\left(\alpha^* + \frac{\partial}{\partial\alpha}\right)|\alpha\rangle\langle\alpha|, \\ |\alpha\rangle\langle\alpha|\hat{a}^\dagger \hat{a} &= \alpha^*\left(\alpha + \frac{\partial}{\partial\alpha^*}\right)|\alpha\rangle\langle\alpha|, \\ \hat{a}|\alpha\rangle\langle\alpha|\hat{a}^\dagger &= |\alpha|^2|\alpha\rangle\langle\alpha|. \qquad (12.21)\end{aligned}$$

Before we go further with the derivation of the FPE, we stop for a moment to prove the relations (12.21). In fact, in order to prove the above relations, it is enough to show that

$$\hat{a}^\dagger|\alpha\rangle\langle\alpha| = \left(\alpha^* + \frac{\partial}{\partial\alpha}\right)|\alpha\rangle\langle\alpha|. \qquad (12.22)$$

Proof. First, we will show that

$$\hat{a}^\dagger|\alpha\rangle = \left(\frac{1}{2}\alpha^* + \frac{\partial}{\partial\alpha}\right)|\alpha\rangle. \qquad (12.23)$$

In the photon number representation, we can write

$$\begin{aligned}\hat{a}^\dagger|\alpha\rangle &= e^{-\frac{1}{2}|\alpha|^2}\sum_{n=0}^{\infty}\frac{\alpha^n}{\sqrt{n!}}\hat{a}^\dagger|n\rangle = e^{-\frac{1}{2}|\alpha|^2}\sum_{n=0}^{\infty}\frac{\alpha^n\sqrt{n+1}}{\sqrt{n!}}|n+1\rangle \\ &= e^{-\frac{1}{2}|\alpha|^2}\sum_{n=0}^{\infty}\frac{(n+1)\alpha^n}{\sqrt{(n+1)!}}|n+1\rangle \\ &= e^{-\frac{1}{2}|\alpha|^2}\frac{\partial}{\partial\alpha}\sum_{n=0}^{\infty}\frac{\alpha^{n+1}}{\sqrt{(n+1)!}}|n+1\rangle \\ &= e^{-\frac{1}{2}|\alpha|^2}\frac{\partial}{\partial\alpha}\sum_{k=1}^{\infty}\frac{\alpha^k}{\sqrt{k!}}|k\rangle = e^{-\frac{1}{2}|\alpha|^2}\frac{\partial}{\partial\alpha}\sum_{k=0}^{\infty}\frac{\alpha^k}{\sqrt{k!}}|k\rangle. \qquad (12.24)\end{aligned}$$

However,

$$\frac{\partial}{\partial\alpha}|\alpha\rangle = \frac{\partial}{\partial\alpha}\mathrm{e}^{-\frac{1}{2}|\alpha|^2}\sum_{n=0}^{\infty}\frac{\alpha^n}{\sqrt{n!}}|n\rangle$$
$$= -\frac{1}{2}\alpha^*|\alpha\rangle + \mathrm{e}^{-\frac{1}{2}|\alpha|^2}\frac{\partial}{\partial\alpha}\sum_{n=0}^{\infty}\frac{\alpha^n}{\sqrt{n!}}|n\rangle\,. \tag{12.25}$$

Thus,

$$\hat{a}^\dagger|\alpha\rangle = \left(\frac{1}{2}\alpha^* + \frac{\partial}{\partial\alpha}\right)|\alpha\rangle, \tag{12.26}$$

and then

$$\hat{a}^\dagger|\alpha\rangle\langle\alpha| = \left[\left(\frac{1}{2}\alpha^* + \frac{\partial}{\partial\alpha}\right)|\alpha\rangle\right]\langle\alpha|. \tag{12.27}$$

Since

$$\frac{\partial}{\partial\alpha}|\alpha\rangle\langle\alpha| = \left(\frac{\partial}{\partial\alpha}|\alpha\rangle\right)\langle\alpha| + |\alpha\rangle\frac{\partial}{\partial\alpha}\langle\alpha|, \tag{12.28}$$

and

$$\frac{\partial}{\partial\alpha}\langle\alpha| = \frac{\partial}{\partial\alpha}\left(\mathrm{e}^{-\frac{1}{2}|\alpha|^2}\sum_{n=0}^{\infty}\frac{(\alpha^*)^n}{\sqrt{n!}}\langle n|\right) = -\frac{1}{2}\alpha^*\langle\alpha|, \tag{12.29}$$

we finally obtain

$$\hat{a}^\dagger|\alpha\rangle\langle\alpha| = \frac{1}{2}\alpha^*|\alpha\rangle\langle\alpha| + \frac{\partial}{\partial\alpha}\left(|\alpha\rangle\langle\alpha|\right) - |\alpha\rangle\frac{\partial}{\partial\alpha}\langle\alpha|$$
$$= \left(\alpha^* + \frac{\partial}{\partial\alpha}\right)|\alpha\rangle\langle\alpha|, \tag{12.30}$$

as required. □

Using Eq. (12.21), the master equation simplifies to

$$\int d^2\alpha\frac{\partial}{\partial t}P(\alpha,t)|\alpha\rangle\langle\alpha| = \int d^2\alpha P(\alpha,t)\left\{-i\omega_0\left(\alpha\frac{\partial}{\partial\alpha} - \alpha^*\frac{\partial}{\partial\alpha^*}\right)\right.$$
$$\left.-\frac{1}{2}\Gamma\left[\alpha\frac{\partial}{\partial\alpha} + \alpha^*\frac{\partial}{\partial\alpha^*}\right]|\alpha\rangle\langle\alpha|\right\}. \tag{12.31}$$

The partial derivatives that act to the right on $|\alpha\rangle\langle\alpha|$ can be transferred to the distribution $P(\alpha, t)$ by integrating by parts. Since $P(\alpha, t)$ vanishes at $\pm\infty$ the integration by parts gives

$$\int d^2\alpha P(\alpha,t)\,\alpha\left(\frac{\partial}{\partial\alpha}|\alpha\rangle\langle\alpha|\right) = -\int d^2\alpha\left(\frac{\partial}{\partial\alpha}\alpha P(\alpha,t)\right)|\alpha\rangle\langle\alpha|. \tag{12.32}$$

Finally, we get

$$\frac{\partial}{\partial t}P(\alpha, t) = \left[\left(\frac{1}{2}\Gamma + i\omega_0\right)\frac{\partial}{\partial \alpha}\alpha + \left(\frac{1}{2}\Gamma - i\omega_0\right)\frac{\partial}{\partial \alpha^*}\alpha^*\right]P(\alpha, t). \tag{12.33}$$

This is the FPE for the damped harmonic oscillator in the P representation. It contains only the first derivative terms.

We now assume that the reservoir is in a thermal state. In this case the master equation of the density operator of the harmonic oscillator is of the form

$$\begin{aligned}\frac{\partial}{\partial t}\rho = &-i\omega_0\left[\hat{a}^\dagger\hat{a}, \rho\right] - \frac{1}{2}\Gamma(1+N)\left(\hat{a}^\dagger\hat{a}\rho + \rho\hat{a}^\dagger\hat{a} - 2\hat{a}\rho\hat{a}^\dagger\right)\\ &-\frac{1}{2}\Gamma N\left(\hat{a}\hat{a}^\dagger\rho + \rho\hat{a}\hat{a}^\dagger - 2\hat{a}^\dagger\rho\hat{a}\right).\end{aligned} \tag{12.34}$$

The extra term introduced by N will produce the second-order derivative in the FPE:

$$\hat{a}^\dagger|\alpha\rangle\langle\alpha|\hat{a} = \left(\frac{\partial}{\partial\alpha} + \alpha^*\right)\left(\frac{\partial}{\partial\alpha^*} + \alpha\right)|\alpha\rangle\langle\alpha|, \tag{12.35}$$

which results in the FPE of the form

$$\begin{aligned}\frac{\partial}{\partial t}P(\alpha, t) = &\left\{\left(\frac{1}{2}\Gamma + i\omega_0\right)\frac{\partial}{\partial\alpha}\alpha + \left(\frac{1}{2}\Gamma - i\omega_0\right)\frac{\partial}{\partial\alpha^*}\alpha^*\right.\\ &\left.+\Gamma N\frac{\partial^2}{\partial\alpha\partial\alpha^*}\right\}P(\alpha, t).\end{aligned} \tag{12.36}$$

The proof is left to the reader as an exercise.

Thus, the master equation for the system density operator ρ can be transformed into a FPE describing the evolution of the quasi-probability distribution function. The FPEs generally do not have exact solutions, except for linear cases or one-dimensional systems. Despite this, an approximate solution can often be found, especially in cases where nonlinear effects do not arise.

12.5 Drift and Diffusion Coefficients

The FPE, which is the equation of motion for the P distribution is often called the phase space equation of motion for the damped harmonic oscillator. This equation can be written as

$$\frac{\partial}{\partial t}P\left(x_i, x_j, t\right) = \left[-\frac{\partial}{\partial x_i}A(x_i) + \frac{1}{2}\frac{\partial^2}{\partial x_i \partial x_j}D\left(x_i, x_j\right)\right]P\left(x_i, x_j, t\right), \tag{12.37}$$

where $x_1 = \alpha$, $x_2 = \alpha^*$, and

$$A_1 = -\left(\frac{1}{2}\Gamma + i\omega_0\right)\alpha, \quad A_2 = -\left(\frac{1}{2}\Gamma - i\omega_0\right)\alpha^*,$$
$$D_{12} = 2\Gamma N. \tag{12.38}$$

The first derivative term determines the deterministic motion and is called the drift term. The second derivative term will cause a broadening (narrowing) or diffusion of $P(\alpha, t)$ and is called the diffusion term.

In order to show more explicitly that A and D are drift and diffusion coefficients, respectively, we consider the one-dimensional FPE with the variable x and calculate the average value and variance of x, defined as

$$\langle x(t)\rangle = \int_{-\infty}^{\infty} dx\, x P(x, t),$$
$$\sigma^2(t) = \left\langle x^2(t)\right\rangle - \langle x(t)\rangle^2, \tag{12.39}$$

with

$$\left\langle x^2(t)\right\rangle = \int_{-\infty}^{\infty} dx\, x^2 P(x, t), \tag{12.40}$$

where $P(x, t)$ is given by the one-dimensional FPE

$$\frac{\partial}{\partial t}P(x, t) = \left[-\frac{\partial}{\partial x}A(x) + \frac{1}{2}\frac{\partial^2}{\partial x^2}D(x)\right]P(x, t). \tag{12.41}$$

Consider equations of motion for $\langle x(t)\rangle$ and $\sigma^2(t)$:

$$\langle \dot{x}(t)\rangle = \frac{d}{dt}\int_{-\infty}^{\infty} dx\, x P(x, t) = \int_{-\infty}^{\infty} dx\, x\frac{\partial}{\partial t}P(x, t)$$
$$= -\int_{-\infty}^{\infty} dx\, x\frac{\partial}{\partial x}(AP) + \frac{1}{2}\int_{-\infty}^{\infty} dx\, x\frac{\partial^2}{\partial x^2}(DP). \tag{12.42}$$

Integrating by parts, and assuming that P and its derivatives vanish sufficiently fast at infinity, we obtain

$$\langle \dot{x}(t)\rangle = \int_{-\infty}^{\infty} dx A P = \langle A(x)\rangle. \tag{12.43}$$

Similarly,

$$\left\langle \dot{x^2(t)} \right\rangle = -\int_{-\infty}^{\infty} dx\, x^2 \frac{\partial}{\partial x}(AP) + \frac{1}{2}\int_{-\infty}^{\infty} dx\, x^2 \frac{\partial^2}{\partial x^2}(DP)$$
$$= 2\left\langle xA(x)\right\rangle + \left\langle D(x)\right\rangle . \quad (12.44)$$

Hence,

$$(\dot{\sigma^2}) = 2\left\langle xA(x)\right\rangle - 2\left\langle x\right\rangle \left\langle A(x)\right\rangle + \left\langle D(x)\right\rangle . \quad (12.45)$$

If the FPE is linear, with $A(x) = Ax$ and $D(x) = D$, where A and D are constant, the equations of motion for $\langle x\rangle$ and σ^2 become

$$\langle \dot{x}\rangle = A\langle x\rangle ,$$
$$\dot{\sigma}^2 = 2A\sigma^2 + D. \quad (12.46)$$

The solutions of the above equations are

$$\langle x(t)\rangle = \langle x(0)\rangle \, \mathrm{e}^{At},$$
$$\sigma^2(t) = \sigma^2(0)\, \mathrm{e}^{2At} - \left(\frac{D}{2A}\right)\left(1 - \mathrm{e}^{2At}\right). \quad (12.47)$$

The meaning of the coefficients A and D is now clear. The coefficient A determines the motion of the average. Since the maximum of the distribution P coincidences with the average, the coefficient A determines motion (drift) of the distribution. The coefficient D broadens the distribution. For example, for $D > 0$ and $A < 0$ the initial distribution $\sigma^2(0)$ increases in time, thus D acts as a source of fluctuations and, therefore, is called the diffusion term or fluctuation term.

12.6 Solution of the Fokker–Planck Equation

The linear multi-dimensional FPE, Eq. (12.36), is complicated and cannot be solved analytically with an arbitrary initial conditions. However, the FPE can be solved analytically for some specific initial conditions and in this section, we show that the analytical solution is possible with the initial condition that at $t = 0$ the harmonic oscillator was in a coherent state $|\alpha_0\rangle$, for which

$$P(\alpha, 0) = \delta^2(\alpha - \alpha_0). \quad (12.48)$$

Before trying to solve the FPE, we first simplify the FPE by transforming the variables α and α^* into a frame rotating at frequency ω_0:

$$\tilde{\alpha} = \alpha e^{i\omega_0 t}, \qquad \tilde{\alpha}^* = \alpha^* e^{-i\omega_0 t}, \tag{12.49}$$

and

$$\tilde{P}(\tilde{\alpha}, t) = P(\alpha, t). \tag{12.50}$$

With this transformation, we have

$$\begin{aligned}\frac{\partial}{\partial t}\tilde{P} &= \frac{\partial P}{\partial t} + \frac{\partial P}{\partial \alpha}\frac{\partial \alpha}{\partial t} + \frac{\partial P}{\partial \alpha^*}\frac{\partial \alpha^*}{\partial t} \\ &= \frac{\partial P}{\partial t} - i\omega_0\left(\alpha\frac{\partial P}{\partial \alpha} - \alpha^*\frac{\partial P}{\partial \alpha^*}\right) \\ &= \frac{\partial P}{\partial t} - i\omega_0\left(\frac{\partial}{\partial \alpha}\alpha - \frac{\partial}{\partial \alpha^*}\alpha^*\right)P. \end{aligned}\tag{12.51}$$

Substituting this into the FPE, and employing the results

$$\begin{aligned}\frac{\partial}{\partial \alpha} &= \frac{\partial}{\partial \tilde{\alpha}}\frac{\partial \tilde{\alpha}}{\partial \alpha} = \frac{\partial}{\partial \tilde{\alpha}}e^{i\omega_0 t}, \\ \frac{\partial}{\partial \alpha^*} &= \frac{\partial}{\partial \tilde{\alpha}^*}\frac{\partial \tilde{\alpha}^*}{\partial \alpha^*} = \frac{\partial}{\partial \tilde{\alpha}^*}e^{-i\omega_0 t}, \end{aligned}\tag{12.52}$$

we obtain

$$\frac{\partial}{\partial t}\tilde{P}(\tilde{\alpha}, t) = \left[\frac{\Gamma}{2}\left(\frac{\partial}{\partial \tilde{\alpha}}\tilde{\alpha} + \frac{\partial}{\partial \tilde{\alpha}^*}\tilde{\alpha}^*\right) + \Gamma N\frac{\partial^2}{\partial \tilde{\alpha}\partial \tilde{\alpha}^*}\right]\tilde{P}(\tilde{\alpha}, t). \tag{12.53}$$

We now change the variables from $\tilde{\alpha}$ and $\tilde{\alpha}^*$ to their real and imaginary parts

$$\tilde{\alpha} = x + iy, \qquad \tilde{\alpha}^* = x - iy, \tag{12.54}$$

and obtain

$$\begin{aligned}\frac{\partial}{\partial \tilde{\alpha}}\tilde{\alpha} &= \frac{1}{2}\left(\frac{\partial}{\partial x}x + \frac{\partial}{\partial y}y\right), \\ \frac{\partial}{\partial \tilde{\alpha}^*}\tilde{\alpha}^* &= \frac{1}{2}\left(\frac{\partial}{\partial x}x + \frac{\partial}{\partial y}y\right), \\ \frac{\partial^2}{\partial \tilde{\alpha}\partial \tilde{\alpha}^*} &= \frac{1}{4}\left(\frac{\partial^2}{\partial x^2} + \frac{\partial^2}{\partial y^2}\right). \end{aligned}\tag{12.55}$$

Substituting Eq. (12.55) into Eq. (12.53), we obtain the following FPE

$$\frac{\partial}{\partial t}\tilde{P}(x, y, t) = \left[\frac{\Gamma}{2}\left(\frac{\partial}{\partial x}x + \frac{\partial}{\partial y}y\right) + \frac{\Gamma N}{4}\left(\frac{\partial^2}{\partial x^2} + \frac{\partial^2}{\partial y^2}\right)\right]\tilde{P}(x, y, t). \tag{12.56}$$

Since the terms dependent on x and y are separated from each other, we look for a solution of the equation of the form

$$\tilde{P}(x, y, t) = X(x, t)\, Y(y, t)\,. \tag{12.57}$$

From Eq. (12.56), we see that the functions X and Y satisfy the independent equations

$$\frac{\partial}{\partial t}X = \left(\frac{\Gamma}{2}\frac{\partial}{\partial x}x + \frac{\Gamma N}{4}\frac{\partial^2}{\partial x^2}\right)X, \tag{12.58}$$

$$\frac{\partial}{\partial t}Y = \left(\frac{\Gamma}{2}\frac{\partial}{\partial y}y + \frac{\Gamma N}{4}\frac{\partial^2}{\partial y^2}\right)Y. \tag{12.59}$$

The right-hand sides of these equations contain first- and second-order derivatives. We can simplify these to a equation containing only first-order derivatives by taking Fourier transform on both sides of the equations. Let

$$F(u, t) = \int_{-\infty}^{\infty} dx X(x, t)\, \mathrm{e}^{ixu}, \tag{12.60}$$

then

$$\begin{aligned}\frac{\partial F}{\partial t} &= \int_{-\infty}^{\infty} dx \frac{\partial}{\partial t} X(x, t)\, \mathrm{e}^{ixu} \\ &= \frac{\Gamma}{2}\int_{-\infty}^{\infty} dx \left(\frac{\partial}{\partial x}xX\right)\mathrm{e}^{ixu} + \frac{\Gamma N}{4}\int_{-\infty}^{\infty} dx \frac{\partial^2 X}{\partial x^2}\mathrm{e}^{ixu}.\end{aligned} \tag{12.61}$$

Integrating by parts and assuming that X and its derivatives vanish sufficiently fast at infinity, we obtain

$$\frac{\partial F}{\partial t} = -\left(\frac{\Gamma}{2}u\frac{\partial}{\partial u} + \frac{\Gamma N}{4}u^2\right)F. \tag{12.62}$$

We will solve Eq. (12.62) by the method of characteristics. The subsidiary (characteristic) equation is

$$\frac{dt}{1} = \frac{du}{(\Gamma/2)u} = -\frac{dF}{(\Gamma N/4)u^2 F}. \tag{12.63}$$

Its solutions are

$$\begin{aligned} u \exp\left(-\frac{\Gamma}{2}t\right) &= \text{const.}, \\ F \exp\left(\frac{N}{4}u^2\right) &= \text{const.} \end{aligned} \tag{12.64}$$

Thus, F must have the general form

$$F(u,t) = \Phi\left(u\mathrm{e}^{-\frac{\Gamma}{2}t}\right)\mathrm{e}^{-\frac{N}{4}u^2}, \tag{12.65}$$

where Φ is an arbitrary function. We find Φ from the initial conditions that

$$\begin{aligned} F(u,0) &= \int_{-\infty}^{\infty} dx X(x,0)\,\mathrm{e}^{ixu} \\ &= \int_{-\infty}^{\infty} dx \mathrm{e}^{ixu}\delta(x-x_0) = \mathrm{e}^{ix_0u}, \end{aligned} \tag{12.66}$$

which gives

$$\Phi(u) = \mathrm{e}^{ix_0u}\mathrm{e}^{\frac{N}{4}u^2}. \tag{12.67}$$

Hence,

$$F(u,t) = \exp\left[ix_0u\mathrm{e}^{-\frac{\Gamma}{2}t}\right]\exp\left[-\frac{N}{4}u^2\left(1-\mathrm{e}^{-\Gamma t}\right)\right]. \tag{12.68}$$

Taking the inverse Fourier transform, we get

$$\begin{aligned} X(x,t) &= \frac{1}{2\pi}\int_{-\infty}^{\infty} du\, F(u,t)\,\mathrm{e}^{-ixu} \\ &= \frac{1}{\sqrt{\pi N\left(1-\mathrm{e}^{-\Gamma t}\right)}}\exp\left[-\frac{\left(x-x_0\mathrm{e}^{-\frac{\Gamma}{2}t}\right)^2}{N\left(1-\mathrm{e}^{-\Gamma t}\right)}\right]. \end{aligned} \tag{12.69}$$

Equation (12.59) for $Y(y,t)$ can be solved in the similar fashion, and then

$$\begin{aligned} P(x,y,t) &= \frac{1}{\pi N\left(1-\mathrm{e}^{-\Gamma t}\right)} \\ &\times \exp\left[-\frac{\left(x-x_0\mathrm{e}^{-\frac{\Gamma}{2}t}\right)^2+\left(y-y_0\mathrm{e}^{-\frac{\Gamma}{2}t}\right)^2}{N\left(1-\mathrm{e}^{-\Gamma t}\right)}\right], \end{aligned} \tag{12.70}$$

or in terms of α and α^*:

$$P(\alpha,t) = \frac{1}{\pi N\left(1-\mathrm{e}^{-\Gamma t}\right)}\exp\left[-\frac{\left|\alpha-\alpha_0\mathrm{e}^{-\frac{\Gamma}{2}t}\mathrm{e}^{-i\omega_0 t}\right|^2}{N\left(1-\mathrm{e}^{-\Gamma t}\right)}\right]. \tag{12.71}$$

Thus, $P(\alpha,t)$ is a two-dimensional Gaussian distribution. For $t\to\infty$, the distribution reduces to its steady-state value

$$P(\alpha) = \frac{1}{\pi N}\mathrm{e}^{-\frac{|\alpha|^2}{N}}, \tag{12.72}$$

which is the same we have obtained before using the photon number distribution (see Eq. (12.18)).

For $t = 0$ the distribution is a delta function, the width of the distribution increases with time and reaches the value N for $t \to \infty$.

We have seen in the above example, that a linear FPE can be solved analytically even when it is multi-dimensional. Nonlinear FPEs are difficult to solve analytically, but the one-dimensional case can be easily solved in the steady state. To illustrate this, consider the one-dimensional FPE, which in the steady state can be written as

$$\frac{d}{dx}\left(-A(x)P_{\mathrm{ss}}(x) + \frac{1}{2}\frac{d}{dx}D(x)P_{\mathrm{ss}}(x)\right) = 0. \tag{12.73}$$

Integrating over x, we get the first-order differential equation

$$\frac{d}{dx}\left(D(x)P_{\mathrm{ss}}(x)\right) = 2A(x)P_{\mathrm{ss}}(x) + \mathrm{const.} \tag{12.74}$$

Since $P_{\mathrm{ss}}(x)$ and $(d/dx)P_{\mathrm{ss}}(x)$ vanish at infinity, the constant is zero. Therefore, we get

$$\frac{1}{D(x)P_{\mathrm{ss}}(x)}\frac{d}{dx}\left(D(x)P_{\mathrm{ss}}(x)\right) = 2\frac{A(x)}{D(x)}. \tag{12.75}$$

Its solution is

$$P_{\mathrm{ss}}(x) = \frac{1}{\mathcal{N}}\frac{1}{D(x)}\exp\left(2\int dx\frac{A(x)}{D(x)}\right), \tag{12.76}$$

where $\mathcal{N}$ is a normalization constant.

In summary, we point out that the analytical solution (12.76) is valid for any form of $A(x)$ and $D(x)$.

12.7 Stochastic Differential Equations

We have shown that in the special case of initial coherent state, a direct solution of the multi-dimensional FPE can be found. In cases, where direct solutions are not possible, one can alternatively employ the SDE approach [70]. This approach is based on the fact that for a FPE with positive diffusion matrix, there exists a set of equivalent SDE. The positive defined diffusion matrix $D_{ij}(\alpha)$ can always be factorized into the form

$$D_{ij}(\alpha) = B_{ij}(\alpha)\,B_{ij}^{T}(\alpha). \tag{12.77}$$

Then, according to the Ito rule, a set of SDE equivalent to the FPE can be written in the following (Ito) form

$$\frac{d\vec{\alpha}}{dt} = -A\left(\vec{\alpha}\right) + B\left(\vec{\alpha}\right)\vec{\psi}\left(t\right), \tag{12.78}$$

where $\vec{\alpha}$ is a column vector $(\alpha_1, \ldots, \alpha_N, \alpha_1^*, \ldots, \alpha_N^*)$, and $\vec{\psi}$ is a column vector. Its components $\psi_i\left(t\right)$ are real *independent* Gaussian white noise terms with zero mean value $\langle\psi_i\left(t\right)\rangle = 0$ and delta-δ correlated in time

$$\langle\psi_i\left(t\right)\psi_j\left(t'\right)\rangle = \delta_{ij}\delta\left(t - t'\right). \tag{12.79}$$

The above SDE can be treated using direct numerical simulation techniques or analytical methods.

Example 12.2 (Single-cavity mode driven by a classical coherent field) *Consider the simplest and the most well-known example in quantum optics of a single mode cavity driven by a classical coherent laser field coupled to the external environment.*

The dynamics of the cavity mode are represented by the master equation

$$\frac{\partial}{\partial t}\rho = -\frac{i}{\hbar}\left[\hat{H}_{sys}, \rho\right] - \frac{1}{2}\Gamma\left(\hat{a}^\dagger\hat{a}\rho + \rho\hat{a}^\dagger\hat{a} - 2\hat{a}\rho\hat{a}^\dagger\right), \tag{12.80}$$

where Γ is the damping rate of the cavity mode resulting from the coupling of the cavity into the environment (reservoir), and $\hat{H}_{sys}$ is the system Hamiltonian of the form

$$\hat{H}_{sys} = \hbar\omega\hat{a}^\dagger\hat{a} + i\hbar\left(Ee^{-i\omega_{\mathrm{L}}t}\hat{a}^\dagger - E^*e^{i\omega_{\mathrm{L}}t}\hat{a}\right), \tag{12.81}$$

where E is the amplitude (Rabi frequency) and ω_{L} is the angular (or carrier) frequency of the driving field, respectively.

The master equation contains time-dependent coefficient, and it is straightforward to transform the equation into a type interaction picture (rotating frame) in which the coefficients are time independent. This can be done introducing a Hamiltonian

$$\hat{H}_0 = \hbar\omega_{\mathrm{L}}\hat{a}^\dagger\hat{a}, \tag{12.82}$$

and the transformed density operator $\tilde{\rho} = \exp(-i\hat{H}_0t/\hbar)\rho\exp(i\hat{H}_0t/\hbar)$, giving master equation of the interaction picture

$$\frac{\partial}{\partial t}\tilde{\rho} = \left[-i\Delta\hat{a}^\dagger\hat{a} + E\hat{a}^\dagger - E^*\hat{a}, \tilde{\rho}\right] - \frac{1}{2}\Gamma\left(\hat{a}^\dagger\hat{a}\tilde{\rho} + \tilde{\rho}\hat{a}^\dagger\hat{a} - 2\hat{a}\tilde{\rho}\hat{a}^\dagger\right), \tag{12.83}$$

where $\Delta = \omega - \omega_{\mathrm{L}}$ is the detuning between the cavity mode and the laser frequency. In practical terms, the detuning must be much smaller that the cavity mode spacing for the single-mode approximation to be applicable.

Next step of the calculations is to transform the master equation (12.83) into a FPE for the P function, giving the result

$$\frac{\partial}{\partial t} P(\alpha, t) = \left\{ \frac{\partial}{\partial \alpha} \left[\left(\frac{1}{2}\Gamma + i\Delta \right) \alpha - E \right] + \frac{\partial}{\partial \alpha^*} \left[\left(\frac{1}{2}\Gamma - i\Delta \right) \alpha^* - E^* \right] \right\} P(\alpha, t). \quad (12.84)$$

The FPE can be transformed using the Ito rule to obtain the corresponding stochastic equations. In the case considered here, however, since there is no diffusion term, there is no noise term. Hence, the SDE are of the form

$$\frac{d\alpha}{dt} = -\left(\frac{1}{2}\Gamma + i\Delta \right) \alpha + E,$$
$$\frac{d\alpha^*}{dt} = -\left(\frac{1}{2}\Gamma - i\Delta \right) \alpha^* + E^*. \quad (12.85)$$

The steady-state solution of Eqs. (12.85) is

$$\alpha_{\mathrm{ss}} = \frac{E}{\left(\frac{1}{2}\Gamma + i\Delta\right)}, \quad (12.86)$$

and using the operator correspondences, this results in

$$\langle \hat{a} \rangle_{\mathrm{ss}} = \frac{E}{\left(\frac{1}{2}\Gamma + i\Delta\right)}. \quad (12.87)$$

As a consequence, the intra-cavity photon number is given by

$$\langle \hat{a}^\dagger \hat{a} \rangle = \left| \frac{E}{\left(\frac{1}{2}\Gamma + i\Delta\right)} \right|^2 = \frac{E^2}{\frac{1}{4}\Gamma + \Delta^2}. \quad (12.88)$$

This demonstrates the usual, expected classical behaviour of a Fabry–Perot interferometer, that the intensity of the cavity field has a peak at the cavity resonance frequency, with a Lorentzian line-shape.

The equations of motion for α and α^* are not coupled to each other, thus can be solved exactly for an arbitrary time t by a simple integration, giving the result

$$\alpha(t) = \alpha(0)\mathrm{e}^{-\left(\frac{1}{2}\Gamma + i\Delta\right)t} + E \int_0^t dt' \mathrm{e}^{-\left(\frac{1}{2}\Gamma + i\Delta\right)(t-t')}. \quad (12.89)$$

Equation (12.89) shows that $\alpha(t)$ is a deterministic quantity, so that $\langle\hat{a}^n\rangle = \langle\alpha^n(t)\rangle = \alpha^n(t)$. This means that operators moments simply factorize to result in

$$\begin{aligned}\langle(\hat{a}^\dagger)^m\hat{a}^n\rangle &= \langle(\alpha^*(t))^m\,(\alpha(t))^n\rangle = (\alpha^*(t))^m\,(\alpha(t))^n \\ &= \langle\hat{a}^\dagger\rangle^m\langle\hat{a}\rangle^n. \end{aligned} \tag{12.90}$$

The above relation is only true for coherent states, and can even be used as alternative definition of the coherent states. Therefore, the coherently driven and damped cavity preserves the coherence, so that the cavity field is always in the coherent state. This preservation of coherence under damping is one of the remarkable properties of coherent states, that have made them a very universal and basic entity of laser physics.

However, the coherence is not preserved if a nonlinear cavity damping is included. We will illustrate it in the next example, where we will include a two-photon damping of the cavity mode. Similar conclusions can be obtained with one-photon damping.

Example 12.3 (Effect of two-photon losses on the driven cavity mode) *Suppose, apart from the ordinary damping, considered in the Example 12.2, the cavity mode is also damped by two-photon losses, for example, due to a two-photon absorption. Then, the master equation of the system can be written as*

$$\begin{aligned}\frac{\partial}{\partial t}\rho = &-\frac{i}{\hbar}\left[\hat{H}_{sys}, \rho\right] - \frac{1}{2}\Gamma\left(\hat{a}^\dagger\hat{a}\rho + \rho\hat{a}^\dagger\hat{a} - 2\hat{a}\rho\hat{a}^\dagger\right) \\ &-\frac{1}{2}\kappa\left(\hat{a}^{\dagger 2}\hat{a}^2\rho + \rho\hat{a}^{\dagger 2}\hat{a}^2 - 2\hat{a}^2\rho\hat{a}^{\dagger 2}\right),\end{aligned} \tag{12.91}$$

where κ is the two-photon loss coefficient, and $\hat{H}_{sys}$ is given in Eq. (12.81). Following the standard procedure, we first transform the master equation to an interaction picture and next into a FPE, which is of the form

$$\begin{aligned}\frac{\partial}{\partial t}P(\alpha, t) = &\left\{\frac{\partial}{\partial\alpha}\left[\left(\frac{1}{2}\Gamma + i\Delta\right)\alpha + \kappa\alpha^2\alpha^* - E\right]\right. \\ &+\frac{\partial}{\partial\alpha^*}\left[\left(\frac{1}{2}\Gamma - i\Delta\right)\alpha^* + \kappa\alpha^{*2}\alpha - E^*\right] \\ &\left.+\frac{1}{2}\frac{\partial^2}{\partial\alpha^2}\left(-\kappa\alpha^2\right) + \frac{1}{2}\frac{\partial^2}{\partial\alpha^{*2}}\left(-\kappa\alpha^{*2}\right)\right\}P(\alpha, t).\end{aligned} \tag{12.92}$$

Using the Ito rule, we turn the FPE into the following SDE

$$\frac{d\alpha}{dt} = -\left(\frac{1}{2}\Gamma + i\Delta\right)\alpha - \kappa\alpha^2\alpha^* + E + B_{11}\psi_1(t),$$
$$\frac{d\alpha^*}{dt} = -\left(\frac{1}{2}\Gamma - i\Delta\right)\alpha^* - \kappa\alpha^{*2}\alpha + E^* + B_{22}\psi_2(t), \quad (12.93)$$

where $\psi_i(t)$ are independent Gaussian noise terms with zero means and the following non-zero correlations

$$\langle\psi_1(t)\,\psi_1(t')\rangle = \delta\left(t - t'\right),$$
$$\langle\psi_2(t)\,\psi_2(t')\rangle = \delta\left(t - t'\right). \quad (12.94)$$

The parameters B_{11} and B_{22} appearing in Eq. (12.93), are the diagonal matrix elements of the matrix B, which can be found from the diffusion matrix $D(D = BB^T)$:

$$D = \begin{pmatrix} -\kappa\alpha^2 & 0 \\ 0 & -\kappa\alpha^{*2} \end{pmatrix}. \quad (12.95)$$

Then, the matrix B is of the form:

$$B = \begin{pmatrix} i\sqrt{\kappa}\alpha & 0 \\ 0 & -i\sqrt{\kappa}\alpha^* \end{pmatrix}. \quad (12.96)$$

However, there is a problem with the equations (12.93). Since ψ_1 and ψ_2 are independent stochastic processes, the equation of motion for α^* is not the complex conjugate of α. This example allows us to understand why we have to employ the positive P representation in this type of problems rather than the Glauber–Sudarshan P representation.

If we use the positive P representation, the FPE becomes

$$\frac{\partial}{\partial t}P(\alpha, \beta, t) = \left\{\frac{\partial}{\partial\alpha}\left[\left(\frac{1}{2}\Gamma + i\Delta\right)\alpha + \kappa\alpha^2\beta^* - E\right]\right.$$
$$+\frac{\partial}{\partial\beta^*}\left[\left(\frac{1}{2}\Gamma - i\Delta\right)\alpha^* + \kappa\beta^{*2}\alpha - E^*\right]$$
$$\left.+\frac{1}{2}\frac{\partial^2}{\partial\alpha^2}\left(-\kappa\alpha^2\right) + \frac{1}{2}\frac{\partial^2}{\partial\beta^{*2}}\left(-\kappa\beta^{*2}\right)\right\}P(\alpha, \beta, t), \quad (12.97)$$

from which, we find the following Ito SDE

$$\frac{d\alpha}{dt} = -\left(\frac{1}{2}\Gamma + i\Delta\right)\alpha - \kappa\alpha^2\beta^* + E + i\sqrt{\kappa}\alpha\psi_1(t),$$
$$\frac{d\beta^*}{dt} = -\left(\frac{1}{2}\Gamma - i\Delta\right)\beta^* - \kappa\beta^{*2}\alpha + E^* - i\sqrt{\kappa}\beta^*\psi_2(t). \quad (12.98)$$

Hence, we see that employing the positive P representation resolves the problem since β^* is independent of α.

Exercises

12.1 Using the general formalism for the derivation of the master equation in the Born–Markov approximations (Chapter 9), derive the master equation for the density operator of the harmonic oscillator coupled to a broadband thermal reservoir

$$\frac{\partial}{\partial t}\rho = -i\omega_0\left[\hat{a}^\dagger\hat{a}, \rho\right] - \frac{1}{2}\Gamma(1+N)\left(\hat{a}^\dagger\hat{a}\rho + \rho\hat{a}^\dagger\hat{a} - 2\hat{a}\rho\hat{a}^\dagger\right)$$
$$-\frac{1}{2}\Gamma N\left(\hat{a}\hat{a}^\dagger\rho + \rho\hat{a}\hat{a}^\dagger - 2\hat{a}^\dagger\rho\hat{a}\right),$$

where N is the number of photons in the reservoir modes.

12.2 Consider the master equation for the density operator of the harmonic oscillator in a thermal reservoir derived in Exercise 12.1.

(a) What is the physical interpretation of the terms in the master equation? Give a graphical sketch of the processes involved.

(b) Calculate the diagonal elements of the density operator, ρ_{nn}, which give the probability of finding n photons in the harmonic oscillator mode.

(c) What is the condition for the steady state? Find the photon number distribution in this case.

12.3 Calculate the time evolution of a coherent state of the harmonic oscillator, $\rho(0) = |\alpha\rangle\langle\alpha|$, when the oscillator evolves in a zero-temperature ($T = 0$) reservoir.

12.4 Show that the master equation of the density operator of the harmonic oscillator in a thermal reservoir can be converted into the FPE of the form

$$\frac{\partial}{\partial t}P(\alpha, t) = \left\{\left(\frac{1}{2}\Gamma + i\omega_0\right)\frac{\partial}{\partial\alpha}\alpha + \left(\frac{1}{2}\Gamma - i\omega_0\right)\frac{\partial}{\partial\alpha^*}\alpha^*\right.$$
$$\left.+\Gamma N\frac{\partial^2}{\partial\alpha\partial\alpha^*}\right\}P(\alpha, t).$$

Chapter 13

Quantum Trajectory Theory

13.1 Introduction

We have already introduced major techniques used in quantum optics to solve various problems. Those techniques involve either integrating differential equations or transforming the master equation into c-number differential equations. But, there is yet another way to solve problems in quantum optics, in particular those involving non-Markovian systems, a powerful procedure based on the statistical and photodetection theories.

Most quantum optical systems are open quantum systems, that is, are systems that interact with their surroundings via energy exchange. Hence, one may resort to approximate perturbation techniques to analyse these interactions. On the other hand, the master equation approach describes the behaviour of statistical ensemble rather than the individual behaviour of the element of the ensemble. However, new theories, which describe the evolution of single-quantum systems, are being developed. These include the quantum trajectory theory [71, 72], Monte Carlo wavefunction method [73], waiting-time distributions and quantum stochastic equations. Here, we will briefly describe the bare essentials of the quantum trajectory theory based on direct photoelectric detection.

Quantum Optics for Beginners
Zbigniew Ficek and Mohamed Ridza Wahiddin

ISBN 978-981-4411-75-2 (Hardcover), 978-981-4411-76-9 (eBook)
www.panstanford.com

13.2 Quantum Trajectories

The quantum trajectory theory, first introduced by Carmichael [71], is constructed around the standard theory of photoelectric detection and the master equation theory of a photoemissive source. It therefore links the statistics of photoelectron emissions to a dynamical process involving photon emissions taking place at the source. The quantum trajectory approach also provides a powerful computational method. It is quite easy to implement on a personal computer. The computer simulation can generate *trajectories* for a stochastic wavefunction that describes the current state of the quantum mechanical source, conditioned on a particular past history of coherent evolution and collapses. These would clarify the physical interpretation, since they can give the intuition of what is going on with respect to the source in a visible form; the standard master equation approach does not allow this concrete visualization.

The connection between the conditioned wavefunction and the master equation is that an ensemble average taken over a large number of trajectories, with respect to the conditioned wavefunction, reproduces the results of a master equation calculation. The *unravelling* of the master equation has been applied to various quantum optical systems including the driven Jaynes–Cummings model, field-quadrature measurements and optical second harmonic generation. Note that different unravellings can be constructed for different measurement schemes to give complimentary pictures of a quantized source. The quantum trajectory theory has essentially the following two main features.

The dynamics of an individual quantum system: the dynamics of an open system is composed of (i) continuous coherent evolution and (ii) discrete quantum jumps (emission/absorption of energy quanta). Randomness is introduced in the jump processes, an inherent property of quantum systems as opposed to the deterministic evolution of classical systems.

The path in the configuration space (Hilbert space) that the individual system takes in the course of time evolution, the dynamics of the corresponding wavefunction $|\Psi\rangle$, is regarded as a quantum trajectory.

13.2.1 *Formulation of the Quantum Trajectory Theory*

The formulation of the quantum trajectory treatment starts with a master equation. A typical master equation is of the form

$$\dot{\rho} = \frac{1}{i\hbar}\left[\hat{H}_{\mathrm{S}}, \rho\right] + \frac{1}{2}\left(2\hat{C}\rho\hat{C}^{+} - \hat{C}^{+}\hat{C}\rho - \rho\hat{C}^{+}\hat{C}\right), \qquad (13.1)$$

where H_{S} is the Hamiltonian of the closed system and $\hat{C}$ is the collapse operator that determines the way the wavefunction $|\Psi\rangle$ changes in the course of energy exchange. The master equation of the form (13.1), normally called the Lindblad form, describes many systems coupled to different type of reservoirs. We now determine the quantum trajectory method, also called the Monte Carlo wavefunction method.

The quantum trajectory theory determines whether an element of an ensemble described by Eq. (13.1) is subjected to either one of the following dynamics:

(a) *Coherent dynamics* (between jumps)

$$i\hbar\frac{d}{dt}|\Psi\rangle = \tilde{H}\,|\Psi\rangle, \qquad (13.2)$$

where $\tilde{H} = \hat{H}_{\mathrm{S}-\frac{1}{2}i\hbar\hat{C}^{+}\hat{C}}$ is a non-Hermitian Hamiltonian.

(b) *Jumps*

$$|\Psi\rangle \longleftarrow \hat{C}\,|\Psi\rangle. \qquad (13.3)$$

We see from Eq. (13.2) that the norm of $|\Psi\rangle$ $(= \langle\Psi|\Psi\rangle)$ is not conserved since $\tilde{H}$ is not Hermitian. Thus, we call $|\Psi\rangle$ in Eq. (13.2) an un-normalized wavefunction and write it as $\left|\bar{\Psi}\right\rangle$. Let us rewrite Eq. (13.2) as

$$i\hbar\frac{d}{dt}\left|\bar{\Psi}\right\rangle = \tilde{H}\left|\bar{\Psi}\right\rangle. \qquad (13.4)$$

The question now is: When do the quantum jumps occur? The answer is: Quantum trajectory theory prescribes that the event of a jump is determined by the following rule.

Rule: A jump occurs during the time interval $[t, t + \Delta t)$.

In order to decide whether a quantum jump has occurred, we define a machine-generated random number $R \in [0, 1)$ and compare it to

$$\Delta p(t) = \Delta t \frac{\langle \bar{\Psi} | \hat{C}^{+} \hat{C} | \bar{\Psi} \rangle}{\langle \bar{\Psi} | \bar{\Psi} \rangle}. \tag{13.5}$$

If $\Delta p(t)$ is greater than or equal to a machine-generated random number $R \in [0, 1)$ then a jump occurs. Since $\Delta p(t)$ is determined by $|\bar{\Psi}\rangle$, it follows that $|\bar{\Psi}(t + \Delta t)\rangle$ is determined by $|\bar{\Psi}\rangle$. In quantum trajectory language we say $|\bar{\Psi}\rangle$ is *conditioned* on its dynamics history, and therefore we call the wavefunction $|\bar{\Psi}\rangle$ a *conditioned wavefunction*. We will denote it as $|\bar{\Psi}_C\rangle$.

In summary, we give a typical algorithm to highlight the main features of the quantum trajectory theory.

$$\begin{aligned}
& t = 0 \\
& (*)\ \text{If}\ \Delta p(t) = \Delta t \frac{\langle \bar{\Psi}_C | \hat{C}^{+} \hat{C} | \bar{\Psi}_C \rangle}{\langle \bar{\Psi}_C | \bar{\Psi}_C \rangle} \geq R \in [0, 1) \\
& \text{then}\ |\bar{\Psi}_C\rangle \leftarrow \hat{C}\, |\bar{\Psi}_C\rangle \\
& \text{else}\ i\hbar \tfrac{d}{dt} |\bar{\Psi}_C\rangle = \tilde{H}\, |\bar{\Psi}_C\rangle \\
& \text{end if} \\
& t \leftarrow t + \Delta t \\
& \text{Go to}\ (*)
\end{aligned} \tag{13.6}$$

Hence, one needs to find only $\tilde{H}$ and $\hat{C}$. It is also important to choose Δt so that $\Delta p(t) \ll 1$. But we need to compromise so that the computation time is not too long.

13.3 Cavity QED Laser

The quantum trajectory approach was born more or less for the needs in the field of cavity QED (quantum electrodynamics). For some cavity QED problems the standard methods are either invalid or difficult to apply, but the quantum trajectory approach provides a new and useful way to proceed. We now illustrate the theory to study the photon statistics of a cavity QED laser. Theoretically, enhancing the generation of photons in the laser cavity mode may decrease the threshold-pumping rate in a laser. The fraction of pumped photons, which enter the cavity mode β may be enhanced by using a laser cavity with a width on the order of the wavelength of the laser light. In such a cavity, spontaneous decay out of the laser mode is inhibited by the boundary conditions of the cavity. This cavity QED effect causes a reduction in threshold [74]. The ideal cavity QED laser ($\beta = 1$) would have a number of laser-mode photons proportional to pump power P, and consequently this device also does not have a well-defined pump threshold [75], or even could work as a thresholdless laser [76].

Consider a single-mode laser theory based on a three-level homogeneously broadened gain medium. The laser mode and lasing transition are assumed to be exactly resonant. The lower level of the lasing transition is rapidly depleted so that to a good approximation it remains empty. We give below the corresponding rate equations, which generalize the semiclassical theory by including spontaneous emission into the laser mode

$$\gamma^{-1}\dot{n} = -\lambda n + \beta n N + \beta N,$$
$$\gamma^{-1}\dot{N} = -N + P - \beta n N. \qquad (13.7)$$

Here, γ is the modified spontaneous emission rate to modes other than the laser mode, n is the photon number in the laser mode, N is the carrier number (atoms in the upper level of the lasing transition), λ is the cavity decay rate and P is the pumping rate, measured in units of the spontaneous emission rate.

Note that the rate equations tell us about the intensity of the emitted light. They cannot, however, tell us what kind of light is emitted. Further information can be obtained by considering an associated birth–death model, which includes some probability treatment. The model is essentially a translation of the Einstein rate equation theory into probabilistic language for a field with uncertain energy density proportional to n. The pumping is included in a form that produces Poisson fluctuations in the carrier number. The model is mathematically represented by the master equation for the probability $p_{n,N}$ of finding n photons in the laser mode and N carriers:

$$\begin{aligned}\gamma^{-1}\dot{p}_{n,N} = &-\lambda[np_{n,N} - (n+1)p_{n+1,N}]\\ &-\beta[nNp_{n,N} - (n-1)(N+1)p_{n-1,N+1}]\\ &-\beta[Np_{n,N} - (N+1)p_{n-1,N+1}] + P[p_{n,N-1} - p_{n,N}]\\ &-(1-\beta)[Np_{n,N} - (N+1)p_{n,N+1}]. \qquad (13.8)\end{aligned}$$

The average photon and carrier numbers in this case satisfy

$$\gamma^{-1}\langle\dot{n}\rangle = -\lambda\langle n\rangle + \beta\langle nN\rangle + \beta\langle N\rangle,$$
$$\gamma^{-1}\langle\dot{N}\rangle = -\langle N\rangle + P - \beta\langle nN\rangle. \tag{13.9}$$

Equations (13.9) reduce to Eq. (13.7) if we make the factorization

$$\langle nN\rangle = \langle n\rangle\langle N\rangle. \tag{13.10}$$

We now apply the quantum trajectory method to solve the birth–death master equation (13.8) for the case of a thresholdless cavity QED laser. The following is an outline of the scheme.

For a given state at time t:

(1) Construct three jump probabilities.
(2) Get three uniformly distributed random numbers.
(3) Compare and decide which jump to make or no jump at all.

Three processes are involved here. Namely, pumping process, cavity losses and stimulated/spontaneous emissions into the laser mode. We therefore get the following three types of probabilities:

(1) Probability of pumping process, $\text{Prob}_{\text{pump}}(t_i) = P\Delta t$.
(2) Probability of losses, $\text{Prob}_{\text{loss}}(t_i) = n(t_i)\lambda\Delta t$.
(3) Probability of emission into the laser mode,
$\text{Prob}_{\text{emission}}(t_i) = N(t_i)[n(t_i) + 1]\Delta t$.

One must also generate three uniformly distributed random numbers, $r_1(t_i)$, $r_2(t_i)$ and $r_3(t_i) \in (0, 1)$ respectively to be compared with the probabilities above.

If $\text{Prob}_{\text{pump}}(t_i) \geq r_1(t_i)$, then the number of carriers is increased by one. Otherwise, if $\text{Prob}_{\text{loss}}(t_i) \geq r_2(t_i)$, then the losses process will cause the number of photons to be reduced by 1. Furthermore, if $\text{Prob}_{\text{emission}}(t_i) \geq r_3(t_i)$ then the stimulated/spontaneous emission will cause the number of carriers to decrease by 1 and simultaneously the number of photons increases by 1.

One can then easily investigate the photon statistics of the laser by calculating the photon number n and the ensemble averages $\langle n\rangle$ and $\langle n^2\rangle$. Figures 13.1 and 13.2 show the photon and carrier numbers vs. dimensionless time τ for the case of $P = 1.0$, $\lambda = 0.1$ and $\beta = 1.0$.

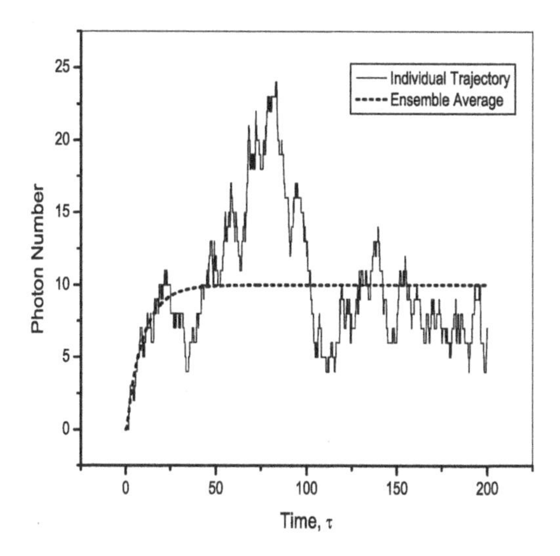

Figure 13.1 Individual trajectory and ensemble average of the photon number versus dimensionless time for $P = 1.0, \lambda = 0.1$, and $\beta = 1.0$.

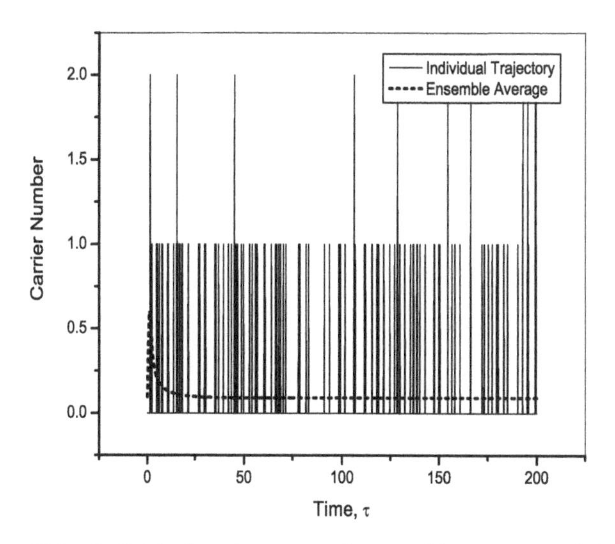

Figure 13.2 Individual trajectory and ensemble average of the carrier number vs. dimensionless time for $P = 1, \lambda = 0.1$ and $\beta = 1$.

Exercises

13.1 Determine the $\tilde{H}$ and the collapse operators $\hat{C}_1$ and $\hat{C}_2$ for a two-level atom in thermal equilibrium.

13.2 Explain why the master equation (13.8) is insufficient to calculate the linewidth of the laser.

13.3 Sketch the flowchart to implement the quantum trajectory method to determine $\langle N \rangle$, $\langle n \rangle$ and $\langle n^2 \rangle$ for the above cavity QED model.

Chapter 14

Interaction-Free Measurements

14.1 Introduction

In Chapter 3, we have discussed the direct measurement scheme in which a measured field is directly absorbed by photodetectors. In the direct detection process, the field is destroyed as the detector absorbs all the field interacting with it. From the point of view of quantum physics, the measurement destroys the state vector of the field such that the state collapses to the appropriate eigenstate of the measured field. For example, if prior to the measurement the state vector of the field is $|\alpha\rangle + |\beta\rangle$, at the measurement the state vector collapses to $|\alpha\rangle$ if the value α is obtained, or $|\beta\rangle$ if β is obtained.

Apart from the collapse of the wave function, the measurement disturbs the field (system) such that it is impossible to predict the future development of the system. This is the major difference between the quantum and classical physics that in the quantum physics the measurement disturbs the system to an extent that cannot be made arbitrary small. This fact is clear from the Heisenberg uncertainty relation that with an observation of the position of a particle within an accuracy Δx disturbs the momentum of the particle with an uncertainty Δp such that $\Delta x \Delta p \geq \hbar/2$. We may call it a *disturbance interpretation* of quantum

Quantum Optics for Beginners
Zbigniew Ficek and Mohamed Ridza Wahiddin

ISBN 978-981-4411-75-2 (Hardcover), 978-981-4411-76-9 (eBook)
www.panstanford.com

theory, that the physical act of detection disturbs the state of the observed system uncontrollably, so that we cannot have complete knowledge of the system, and we cannot therefore predict its future development. Some experiments called negative-result experiments and interaction-free measurements will help us to understand how one can detect a photon without destroying it and how one could detect an object without interacting with it.

14.2 Negative-Result Measurements

In negative-result experiments, the result of a photon detection is obtained not through the occurrence of a physical event, detection of a photoelectron, as for a normal measurement, but by the absence of such an event.

Consider an experiment, shown in Fig. 14.1, which was proposed by Epstein [77], and involves a Mach–Zehnder interferometer. In the interferometer a beam of light is split by a 50/50 beam splitter into two beam travelling in arms S_1 and S_2. The mirrors M_1 and M_2 can be fixed or can move (recoil) under the impact of a photon, but the mirrors do not absorb (destroy) the photon. Thus, we can detect a photon by observing which mirror recoiled. Hence, the

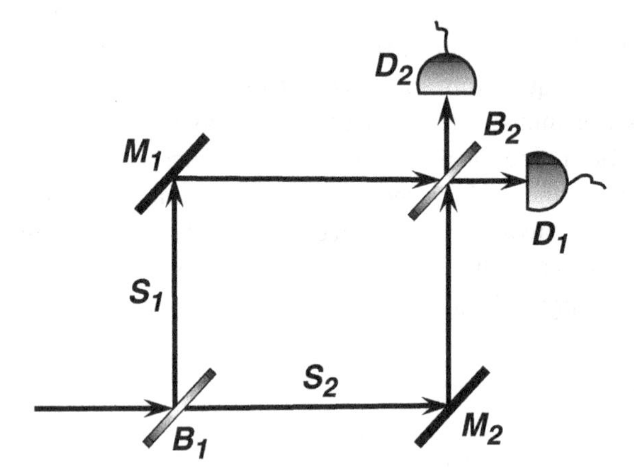

Figure 14.1 Scheme to demonstrate negative-result measurement.

measurements can be achieved seemingly without an interaction between the complete measuring setup (detectors) and the photon.

Assume for a moment, that the arm S_2 is blocked. Then, according to quantum mechanics, if the mirror M_1 recoiled we can say that the collapse of the state vector of the photon to the state $|\phi\rangle$, corresponding to the arm S_1 have been achieved without any interaction. Now, assume that both arms are opened and both mirrors are made moveable. Then, one of the mirrors will recoil as each photon passes through the system. This tells us along which arm the photon is passing, and as a result the coherence of the two beams is lost and the photon has an equal probability to be detected at D_1 or D_2.

Assume now that M_1 is moveable and M_2 is fixed. If M_1 moves, we deduce that the photon is in arm S_1, if it does not move, we deduce that the photon is in arm S_2. In this case we have gain exactly the same information as in the previous cases. However, in this case we have seemingly performed a measurement and achieved a collapse of the wave function without any direct detection of the photon. Again, there is an equal probability of the photon being detected at D_1 or D_2.

Finally, assume that both mirrors are fixed. In this case, the two arms S_1 and S_2 are coherent, and the geometry of the interferometer may be arranged such that every photon will be detected at one of the detectors, say D_1, with none detected at D_2. In this case, we observe photons only by the direct detection.

14.3 Experimental Schemes of Interaction-Free Measurements

In this section we discuss an effect, referred to as interaction-free measurement, which is based mostly on the same type of physical situation as the negative-result measurement. However, the arrangement is carefully constructed to make it natural to claim that one is able to detect the presence of an object without interacting with it at all. This type of measurement is not possible under classical theory, but using the concepts of quantum theory the claim can be justified. The method we use makes particular use of the fact that light consists of photons, and where a beam of light has a choice of paths, we will talk of an individual photon taking one pass or another. One can notice, that such an idea would be totally inappropriate under a classical model of light.

We will analyse in details two schemes of the interaction-free measurement, one proposed by Elitzur and Vaidman [78] and the other proposed by Kwiat *et al.* [79].

14.3.1 *The Elitzur and Vaidman Scheme*

The Elitzur and Vaidman scheme uses a single Mach–Zehnder interferometer, in a similar way as in the Epstein proposal. The difference is that the present scheme assumes that the mirrors are fixed and there is an object in the arm S_1, as shown in Fig. 14.2. Without the object in the arm S_1, one can arrange the system such that all the photons will be detected by D_1, none by D_2.

However, if there is an object in the arm S_1, photons will reach the second beam splitter travelling only through the arm S_2, and then the probabilities of detecting the photons by the detectors D_1 and D_2 are both equal to 1/2. Thus, in this arrangement, the detector D_2 detects photons only if the blocking object is in one of the arms.

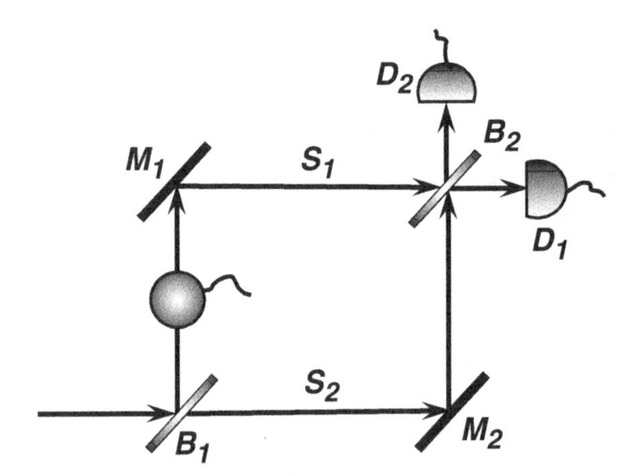

Figure 14.2 The scheme of Elitzur and Vaidman to demonstrate interaction-free measurement.

We can explain this method theoretically in the following way. We know from classical optics that a reflection changes the phase of the wave by $\pi/2$, or alternatively in terms of quantum states of the wave, multiplies the state vector by i. Let us indicate the state of a photon moving to the right by $|1\rangle$, and that of a photon moving upwards by $|2\rangle$. At the beam splitter B_1, the state $|1\rangle$ will change to

$$|1\rangle \rightarrow \frac{1}{\sqrt{2}}\left(|1\rangle + i|2\rangle\right). \tag{14.1}$$

At the mirrors M_1 and M_2, the states will change into

$$|1\rangle \rightarrow i|2\rangle, \qquad |2\rangle \rightarrow i|1\rangle. \tag{14.2}$$

Thus, if the object is absent, the evolution of the initial state $|1\rangle$ will be as follows:

$$\begin{aligned} |1\rangle &\rightarrow \frac{1}{\sqrt{2}}\left(|1\rangle + i|2\rangle\right) \rightarrow \frac{1}{\sqrt{2}}\left(i|2\rangle - |1\rangle\right) \\ &\rightarrow \frac{1}{2}\left(i|2\rangle - |1\rangle\right) - \frac{1}{2}\left(|1\rangle + i|2\rangle\right) = -|1\rangle. \end{aligned} \tag{14.3}$$

Hence, if the object is absent, the photon leaves B_2 moving towards the right and is detected by the detector D_1. On the other hand, if the object is present, the evolution of the initial state $|1\rangle$ is as follows:

$$\begin{aligned} |1\rangle &\rightarrow \frac{1}{\sqrt{2}}\left(|1\rangle + i|2\rangle\right) \rightarrow \frac{1}{\sqrt{2}}\left(i|2\rangle + i|s\rangle\right) \\ &\rightarrow \frac{1}{2}\left(i|2\rangle - |1\rangle\right) + \frac{i}{\sqrt{2}}|s\rangle, \end{aligned} \tag{14.4}$$

where $|s\rangle$ is the state of a photon scattered by the object. This equation shows that the detectors D_1 and D_2 will each click with probability 1/4, and there is probability 1/2 that there is no detection.

In summary of this section, if the detector D_1 clicks, no information is obtained about the object. This could happen whether the object is present in the arm S_1 or not. If there is no click in both D_1 and D_2, we have discovered that there is the object, but our measurement has not been interaction free as the photon has been scattered or absorbed by the object. However, if the detector D_2 clicks, we find that the object is present in the system and the photon has not interacted with it, otherwise the photon would have been scattered or absorbed.

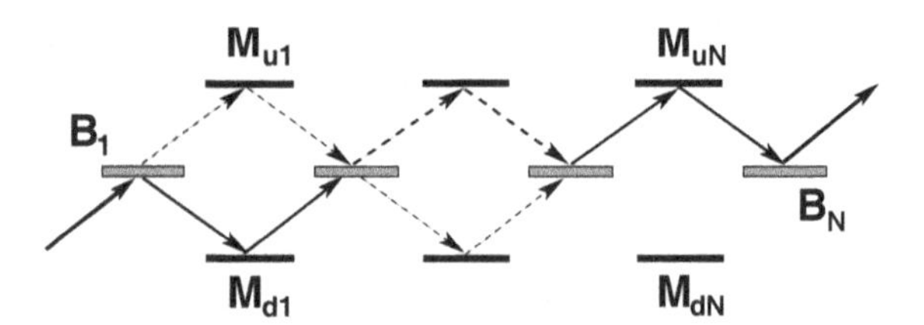

Figure 14.3 The scheme of Kwiat *et al.* to demonstrate interaction-free measurements.

14.3.2 *The Kwiat et al. Scheme*

In the scheme proposed by Kwiat *et al.* [79], instead of a single Mach–Zehnder interferometer, a series of N interferometers is used, the amplitude reflectivity r of each beam splitter being equal to $\cos(\pi/2N)$, and the amplitude transmissivity $t = \sin(\pi/2N)$. The scheme is shown in Fig. 14.3. A photon entering at the lower part of the interferometer and passing through all N interferometers will be transferred into the upper part of the interferometer. Thus, if there is no object in one of the interferometers, each photon is certain to exit via the *up* port of the last beam splitter. If there is the object, there is a non-zero probability that the photon will leave through the *down* port of the last beam splitter.

This experiment can be explained theoretically as follows. Let us consider the state of the photon after leaving down the second beam splitter, that is, the first interferometer. There are two contributions to the state of the photon. The first one is from the pass $B_{r1} \to M_{d1} \to B_{r2}$ and the second one is from the pass $B_{t1} \to M_{u1} \to B_{t2}$, where B_{ri} and B_{ti} mean reflection and transmission at the ith mirror, respectively.

The first beam has been reflected twice at beam splitters and once at the mirror M_{d1}, three reflections giving a phase factor i^3, and the amplitude

$$\alpha_1 = -i\cos^2\left(\frac{\pi}{2N}\right). \tag{14.5}$$

The second beam has been transmitted twice at the beam splitters and reflected once at the mirror M_{u1}, giving the total transmission coefficient multiplied by the factor i as

$$\alpha_2 = i\sin^2\left(\frac{\pi}{2N}\right). \tag{14.6}$$

Thus, after passing the first interferometer (two beam splitters), the intensity of the light going downwards is given by

$$I_2 = |\alpha_1 + \alpha_2|^2 = \cos^2\left(\frac{\pi}{N}\right). \tag{14.7}$$

Hence, after passing m beam splitters, the intensity downwards will be

$$I_m = \cos^2\left(\frac{m\pi}{2N}\right). \tag{14.8}$$

If the photon passes $m = N$ beam splitters, intensity exiting the whole system at the *down* port is equal to zero.

However, if there is an object in one of the arms of the interferometer, the intensity I_N will be different from zero, as in this case $m < N$. Then, a detector located in the *down* port after the last beam splitter will click. Following our previous analysis, if there is an object in one of the upper arms of the interferometer, the amplitude α_2 is suppressed and then the intensity of the light going downwards at the end of the first interferometer is given by

$$\tilde{I}_1 = |\alpha_1|^2 = \cos^4\left(\frac{\pi}{2N}\right). \tag{14.9}$$

Hence, if there are objects in the upper arms and after passing m beam splitters, the intensity downwards will be

$$\tilde{I}_m = \cos^{2m}\left(\frac{\pi}{2N}\right). \tag{14.10}$$

Thus, for $m = N$ we find that the probability of the photon leaving the system via the lower port is equal to $\cos^{2N}(\pi/2N)$, which becomes very close to unity when N is large.

The detection of the photon leaving the interferometers through the lower port is equivalent to the observation that there is an object inside the interferometer. This is measurement-free detection as the photon was not scattered or absorbed by the object, and the probability of obtaining this interaction-free measurement is greater than the Elitzur and Vaidman maximum of $1/2$, and tends to unity for large N. The probability of the photon *not* leaving by the lower port is equal to $1 - \cos^{2N}(\pi/2N)$, and corresponds to the probability that the photon was scattered by the object. Naturally, this also corresponds to an observation that there is the object inside the interferometer, but in this case is not, of course, interaction free.

Exercises

14.1 A single photon is injected into a Mach–Zehnder interferometer. The two beam splitters in the interferometer are identical but not the 50/50 beam splitters.

(a) Write an expression for the state of one of the output arms.

(b) What is the expectation value of the number of photons in this output arm, expressed in the beam splitters reflection and transmission coefficients?

14.2 Prove that in the Kwiat *et al.* experiment, the intensity downwards after passing two interferometers (three beam splitters) is

$$I_3 = \cos^2\left(\frac{3\pi}{2N}\right).$$

14.3 Show that, if after each beam splitter in the Kwiat *et al.* experiment, a detector is placed in the path of one of the two modes, the probability that the photon leaves the last beam splitter in the other mode is converging to 1 for large N. Explain why this corresponds to an "interaction-free" measurement of the presence of the detectors.

14.4 Consider a single-mode field incident on a beam splitter of the amplitude reflectivity r and the amplitude transmissivity t. The incident beam splits into reflected and transmitted beams.

(a) Show that in the classical treatment of the process the three fields, the incident i, reflected r and transmitted t beams, satisfy the energy conservation.

(b) Show that in the quantum treatment of the process, where the complex classical amplitudes are replaced by the annihilation and creations operators, the commutation relations for the three fields

$$[\hat{a}_j, \hat{a}_j^\dagger] = 1 \qquad \text{and} \qquad [\hat{a}_r, \hat{a}_t^\dagger] = 0, \qquad j = i, r, t$$

are satisfied only if one includes the fourth beam splitter input port vertical to the incident beam and being in the vacuum state.

14.5 The evolution of a state as it passes a beam splitter is equivalent to a rotation in Cartesian coordinates that it can be represented by a unitary matrix

$$\mathcal{R}(\theta) = \begin{pmatrix} \cos\theta & i\sin\theta \\ i\sin\theta & \cos\theta \end{pmatrix}.$$

The operator $\mathcal{R}(\theta)$ is often called as rotational operator. Show that rotational operators multiply like exponentials, that is, $\mathcal{R}(\theta)\mathcal{R}(\phi) = \mathcal{R}(\theta + \phi)$.

Chapter 15

Classical and Quantum Interference

15.1 Introduction

Optical interference is regarded as a classical phenomenon and is usually completely described in classical terms, in which optical fields are represented by classical waves. Does it mean that optical interference, which is fundamentally ascribable to the phenomenon of a superposition of wave amplitudes, cannot be applied to test quantum phenomena? The superposition principle is at the heart of quantum physics and one could expect that it should distinguish the quantum nature of light from the wave nature. Our experience based on extensive theoretical analysis and experimental observations shows that classical and quantum theories of optical interference readily explain the presence of an interference pattern resulting from the first-order coherence. However, there are interference effects involving the higher order (second-order) coherence that distinguish the quantum (photon) nature of light from the wave nature. Quantum interference has recently returned to prominence because of its utility in manipulating spontaneous emission and other radiative properties of atomic systems. In this chapter, we present elementary concepts and definitions of both the classical

Quantum Optics for Beginners
Zbigniew Ficek and Mohamed Ridza Wahiddin

ISBN 978-981-4411-75-2 (Hardcover), 978-981-4411-76-9 (eBook)
www.panstanford.com

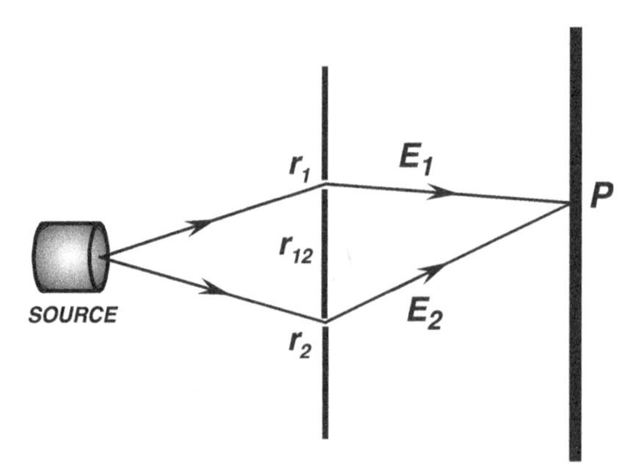

Figure 15.1 Schematic diagram of the Young's double-slit experiment.

and quantum theory of optical interference and discuss similarities and differences between these two approaches [80].

15.2 First-Order Coherence

The elementary model for a demonstration of the first-order coherence is the Young's double-slit experiment, shown in Fig. 15.1. In this experiment, a light beam from a single source passes through two slits, located at $\vec{r}_1$ and $\vec{r}_2$, where it undergoes splitting into two beams of amplitudes $\vec{E}_1$ and $\vec{E}_2$, respectively. These two beams, which act as if they came from two sources, are then detected on a distant screen S. The average intensity of the beams measured at a point P on the screen can be written as

$$I(\vec{R}, t) = \langle \vec{E}^*(\vec{R}, t) \cdot \vec{E}(\vec{R}, t)\rangle = \langle (\vec{E}_1^* + \vec{E}_2^*) \cdot (\vec{E}_1 + \vec{E}_2)\rangle$$
$$= I_1 + I_2 + 2\text{Re}\left\langle E_1^* E_2 \right\rangle, \tag{15.1}$$

where $I_i = \left\langle E_i^* E_i \right\rangle$ is the average intensity of the ith beam, and $2\text{Re}\left\langle E_1^* E_2 \right\rangle$ is the superposition term of the two amplitudes. This term is responsible for the interference effect.[a]

[a]In the derivation of Eq. (15.1), we have used the superposition principle for the electric field that at any instance, the field at any point in space arising from several sources is the vector sum of the contributions that each source would have produced if it were acting alone. Moreover, we have assumed that the vectors $\vec{E}_1$ and $\vec{E}_2$ lie along the same line. This is a good approximation if the observation point P is very

Let us analyse in details the dependence of the superposition term in Eq. (15.1) on the geometry of the experiment. Since the observation point P lies in the far-field zone of the radiation emitted by the slits, $R \gg r_{12}$, that is, the separation between the slits is very small compared to the distance to the point P, the fields at the observation point can be approximated by plane waves for which we can write

$$\begin{aligned} E(\vec{R}_i, t - t_i) &\approx E(\vec{R}_i, t)\mathrm{e}^{-i(\omega_i t_i + \phi_i)} \\ &= E(\vec{R}_i, t)\mathrm{e}^{-i(\omega_i R_i/c + \phi_i)}, \quad i = 1, 2, \end{aligned} \tag{15.2}$$

where ω_i is the angular frequency of the ith field and ϕ_i is its initial phase which, in general, can depend on time. Since the observation point lies in far-field zone of the radiation emitted by the slits, we can write approximately

$$R_i = |\vec{R} - \vec{r}_i| \approx R - \hat{R} \cdot \vec{r}_i, \tag{15.3}$$

where $\hat{R} = \vec{R}/R$ is the unit vector in the direction $\vec{R}$.

Using the plane-wave approximation, we can write the intensity as

$$\begin{aligned} I(\vec{R}, t) &= I_1 + I_2 + 2\mathrm{Re}\left\{\langle E_1^* E_2\rangle \mathrm{e}^{ik_0 \hat{R}\cdot\vec{r}_{12}}\right\} \\ &= I_1 + I_2 + 2\langle E_1^* E_2\rangle \cos(k_0 \hat{R} \cdot \vec{r}_{12}) \\ &= I_1 + I_2 + 2\sqrt{I_1 I_2}\, g^{(1)}(\vec{R}) \cos(k_0 \hat{R} \cdot \vec{r}_{12}), \end{aligned} \tag{15.4}$$

where $\vec{r}_{12} = \vec{r}_2 - \vec{r}_1$ is the distance between the slits, and

$$g^{(1)}(\vec{R}) = \frac{\langle E_1^* E_2\rangle}{\sqrt{I_1 I_2}} \tag{15.5}$$

is the normalized first-order coherence function. The coherence function is a measure of correlation between the two beams relative to the intensity of the beams. Note that in the derivation of Eq. (15.4) we have assumed that the two waves have the same frequency ($\omega_1 = \omega_2$) and phases ($\phi_1 = \phi_2$).

Equation (15.4) shows that the average intensity detected at the screen depends on the coherence between the beams. In the case of perfectly correlated fields, $|g^{(1)}| = 1$, and then the intensity can vary from $(\sqrt{I_1} - \sqrt{I_2})^2$ to $(\sqrt{I_1} + \sqrt{I_2})^2$, giving the so-called *interference*

---

far from the slits, that is, the point P lies in the far-field zone of the radiation emitted by the slits.

pattern. When $I_1 = I_2 = I_0$, the total average intensity varies from $\langle I\rangle_{\text{min}} = 0$ to $\langle I\rangle_{\text{max}} = 4\langle I_0\rangle$, giving perfect interference pattern. For two independent fields, $g^{(1)} = 0$, and than the resulting intensity is just a sum of the intensities of the two fields, which does not vary with the position of P.

In the quantum description of the electromagnetic (EM) field, the field amplitude is represented by the field operators, and the coherence effects are given in terms of the positive and negative frequency parts of the field operator $\hat{E}$. In this case, the average intensity in the Young's experiment can be written as

$$I(\vec{R}, t) = \langle E^{(-)}(\vec{R}, t)E^{(+)}(\vec{R}, t)\rangle = \langle (E_1^{(-)} + E_2^{(-)})(E_1^{(+)} + E_2^{(+)})\rangle$$
$$= I_1 + I_2 + 2\text{Re}\langle E_1^{(-)} E_2^{(+)}\rangle. \tag{15.6}$$

and the normalized first-order coherence takes the form

$$g^{(1)}(\vec{R}) = \frac{\langle E_1^{(-)} E_2^{(+)}\rangle}{\sqrt{I_1 I_2}}. \tag{15.7}$$

The coherence function (15.7) described by the field operators is formally similar to the coherence function (15.5) described by the classical field amplitudes. The similarity arises from the fact that an experiment cannot distinguish between classical and quantum effects described by first-order correlation functions.

The usual measure of the depth of modulation (sharpness) of interference fringes is a visibility in an interference pattern defined as

$$\mathcal{V} = \frac{\langle I(\vec{R}, t)\rangle_{\text{max}} - \langle I(\vec{R}, t)\rangle_{\text{min}}}{\langle I(\vec{R}, t)\rangle_{\text{max}} + \langle I(\vec{R}, t)\rangle_{\text{min}}}, \tag{15.8}$$

where $\langle I(\vec{R}, t)\rangle_{\text{max}}$ and $\langle I(\vec{R}, t)\rangle_{\text{min}}$ represent the intensity maxima and minima at the point P, respectively.

Since,

$$\langle I\rangle_{\text{max}} = \langle I_1\rangle + \langle I_2\rangle + 2\sqrt{I_1 I_2}\,|g^{(1)}|, \tag{15.9}$$

and

$$\langle I\rangle_{\text{min}} = \langle I_1\rangle + \langle I_2\rangle - 2\sqrt{I_1 I_2}\,|g^{(1)}|, \tag{15.10}$$

we obtain

$$\mathcal{V} = \frac{2\sqrt{I_1 I_2}}{(I_1 + I_2)}|g^{(1)}|. \tag{15.11}$$

Thus, $|g^{(1)}|$ determines the visibility of the interference fringes. In the special case of equal intensities of the two fields ($I_1 = I_2$), Eq. (15.11) reduces to $\mathcal{V} = |g^{(1)}|$, that is, $|g^{(1)}|$ is then simply equal to the visibility. For perfectly correlated fields $|g^{(1)}| = 1$, and then $\mathcal{V} = 1$, while $\mathcal{V} = 0$ for uncorrelated fields.

15.3 Welcher Weg Problem

One may notice from Eq. (15.11) that in the case of $I_1 \neq I_2$, the visibility is always smaller than one even for perfectly correlated fields. This fact is related to the problem of extracting which-way (Welcher weg) information has been transferred through the slits into the point P. The observation of an interference pattern and the acquisition of which-way information has been transmitted are mutually exclusive. We can introduce an inequality according to which the fringe visibility $\mathcal{V}$ displayed at the point P and an absolute upper bound on the amount of which-way information $\mathcal{D}$ that can be detected at the point P are related by

$$\mathcal{D}^2 + \mathcal{V}^2 \leq 1. \qquad (15.12)$$

Hence, the extreme situations characterized by perfect fringe visibility ($\mathcal{V} = 1$) or full knowledge of which-way information has been transmitted ($\mathcal{D} = 1$) are mutually exclusive. In order to distinguish which-way information has been transmitted, one can locate an intensity detector at the point P and adjust it to measure a field of a particular intensity I_{d}. When the fields coming from the slits have the same intensities, the detector cannot distinguish which-way the detected field came to the point P, so there is no which-way information available ($\mathcal{D} = 0$) resulting in perfect fringe visibility ($\mathcal{V} = 1$). On the other hand, when the intensities of the fields are different ($I_1 \neq I_2$), the detector adjusted to measure a particular intensity can distinguish which way the field came to the point P resulting in the disappearance of the interference fringes. This is clearly seen from Eq. (15.11), if $I_1 \gg I_2$ or $I_1 \ll I_2$, the visibility $\mathcal{V} \approx 0$ even for $|g^{(1)}| = 1$.

The first-order correlation function is very sensitive to the frequency and phase of the detected fields. Suppose, the fields have different frequencies ($\omega_1 \neq \omega_2$) and the phases ($\phi_1 \neq \phi_2$). We can centre the frequencies around the average frequency of the two fields as

$$\omega_1 = \omega_0 + \frac{1}{2}\Delta, \quad \omega_2 = \omega_0 - \frac{1}{2}\Delta, \tag{15.13}$$

where $\omega_0 = (\omega_1 + \omega_2)/2$ is the average frequency of the fields and $\Delta = \omega_1 - \omega_2$ is the frequency difference (detuning) between the fields. Substituting Eq. (15.2) with Eqs. (15.13) and (15.3) into Eq. (15.5), we obtain

$$g^{(1)}(\vec{R}) = |g^{(1)}(\vec{R})| \exp\left(ik_0 \hat{R} \cdot \vec{r}_{12}\right) \exp\left[i\left(k_0 \tilde{R}\frac{\Delta}{\omega_0} + \delta\phi\right)\right], \tag{15.14}$$

where $\tilde{R} = R + \frac{1}{2}\hat{R} \cdot (\vec{r}_1 + \vec{r}_2)$, $\delta\phi = \phi_1 - \phi_2$, $k_0 = \omega_0/c = 2\pi/\lambda_0$, and λ_0 represents the mean wavelength of the fields.

Let us analyse the physical meaning of the exponents appearing on the right-hand side of Eq. (15.14). The first exponent depends on the separation between the slits and the position $\vec{R}$ of the point P. For small separations the exponent slowly changes with the position $\vec{R}$ and leads to minima and maxima in the interference pattern. The minima appear whenever

$$k_0 \hat{R} \cdot \vec{r}_{12} = (2l + 1)\pi, \quad l = 0, \pm 1, \pm 2, \ldots. \tag{15.15}$$

The second exponent, appearing in Eq. (15.14), depends on the sum of the position of the slits, the ratio Δ/ω_0 and the difference $\delta\phi$ between the initial phases of the fields. This term introduces limits on the visibility of the interference pattern and can affect the pattern only if the frequencies and the initial phases of the fields are different. Even for equal and well stabilized phases, but significantly different frequencies of the fields such that $\Delta/\omega_0 \approx 1$, the exponent oscillates rapidly with $\vec{R}$ leading to the disappearance of the interference pattern. Thus, in order to observe an interference pattern it is important to have two fields of well-stabilized phases and equal or nearly equal frequencies. Otherwise, no interference pattern can be observed even if the fields are perfectly correlated.

The dependence of the interference pattern on the frequencies and phases of the fields is related to the problem of extracting which-way information has been transferred to the observation point P.

For perfectly correlated fields with equal frequencies ($\Delta = 0$) and equal initial phases ($\phi_1 = \phi_2$), the total intensity at the point P is

$$\langle I(\vec{R})\rangle = 2\langle I_0\rangle \left[1 + \cos\left(k_0\hat{R}\cdot\vec{r}_{12}\right)\right], \qquad (15.16)$$

giving maximum possible interference pattern with the maximum visibility of 100%. When $\Delta \neq 0$ and/or $\phi_1 \neq \phi_2$, the total intensity at the point P is given by

$$\langle I(\vec{R})\rangle = 2\langle I_0\rangle \left\{1 + \cos(k_0\vec{R}\cdot\vec{r}_{12})\cos\left(k_0\tilde{R}\frac{\Delta}{\omega_0} + \delta\phi\right)\right.$$
$$\left. - \sin\left(k_0\hat{R}\cdot\vec{r}_{12}\right)\sin\left(k_0\tilde{R}\frac{\Delta}{\omega_0} + \delta\phi\right)\right\}. \qquad (15.17)$$

In this case the intensity exhibits additional cosine and sine modulations, and at the minima the intensity is different from zero indicating that the maximum depth of modulation of 100% is not possible for two fields of different frequencies and/or initial phases. Moreover, for large differences between the frequencies of the fields ($\Delta/\omega_0 \gg 1$), the terms $\cos(k_0\tilde{R}\frac{\Delta}{\omega_0} + \delta\phi)$ and $\sin(k_0\tilde{R}\frac{\Delta}{\omega_0} + \delta\phi)$ rapidly oscillate with $\vec{R}$ and average to zero, which washes out the interference pattern. In terms of which-way information has been transferred, a detector located at the point P and adjusted to measure a particular frequency or phase could distinguish the frequency or the phase of the two fields. Clearly, one could tell which way the detected field came to the point P. Thus, whether which-way information is available or not depends on the intensities as well as frequencies and phases of the interfering fields. Maximum possible which-way information results in the lack of the interference pattern, and vice versa, the lack of which-way information results in maximum interference pattern.

15.4 Second-Order Coherence

We can extend the analysis of interference phenomenon to higher order correlation functions, which involve intensities of the measured fields. Here, we illustrate some properties of the second-order correlation function of a classical field. In the following section, we will extend the analysis to the second-order correlation function of quantum fields.

The second-order (intensity) correlation function of a classical field of a complex amplitude $E(\vec{R}, t)$ is defined as

$$G^{(2)}(\vec{R}_1, t_1; \vec{R}_2, t_2) = \langle E^*(\vec{R}_1, t_1)E^*(\vec{R}_2, t_2)E(\vec{R}_2, t_2)E(\vec{R}_1, t_1)\rangle$$
$$= \langle I(\vec{R}_1, t_1)I(\vec{R}_2, t_2)\rangle, \qquad (15.18)$$

where $I(\vec{R}_1, t_1)$ and $I(\vec{R}_2, t_2)$ are the instantaneous intensities of the field detected at a point $\vec{R}_1$ at time t_1 and at a point $\vec{R}_2$ at time t_2, respectively.

In the plane-wave approximation, the second-order correlation function can be written as

$$G^{(2)}(\vec{R}_1, t_1; \vec{R}_2, t_2) = \sum_{i,j,k,l=1}^{2} \langle E_i^*(t_1)\, E_k^*(t_2)E_l(t_2)E_j(t_1)\rangle$$
$$\times e^{ik\left(\hat{R}_1\cdot\vec{r}_{ij}+\hat{R}_2\cdot\vec{r}_{kl}\right)} e^{i\left(\phi_i+\phi_k-\phi_l-\phi_j\right)}, \qquad (15.19)$$

where $k = 2\pi/\lambda$ and λ is the wavelength of the field. There are 16 correlation functions contributing to the right-hand side of Eq. (15.19), each accompanied by an phase factor that depends on the relative phase of the fields.

The second-order correlation function has completely different coherence properties than the first-order correlation function.

1. Interference pattern can be observed in the second-order correlation function, but in contrast to the first-order correlation function, the interference appears between two points located at $\vec{R}_1$ and $\vec{R}_2$.
2. Interference pattern can be observed even if the fields are produced by two independent sources for which the phase difference $\phi_1 - \phi_2$ is completely random. It is easy to verify that in this case the second-order correlation function (15.19) takes the form

$$G^{(2)}(\vec{R}_1, t_1; \vec{R}_2, t_2) = \langle I_1^2\rangle + \langle I_2^2\rangle + 2\langle I_1 I_2\rangle$$
$$+ 2\langle I_1 I_2\rangle \cos\left[k\vec{r}_{12}\cdot\left(\hat{R}_1 - \hat{R}_2\right)\right], \qquad (15.20)$$

where $I_i = I_i(t_i)$ $(i = 1, 2)$.

We see that the second-order correlation function exhibits a cosine modulation with the separation $\vec{R}_1 - \vec{R}_2$ of the two detectors. This is an interference although it involves intensities of two independent fields. Thus, an interference pattern can be observed even for two completely independent fields. Similar to the first-order correlation function, the sharpness of the fringes depends on the relative intensities of the fields and degrades with an increasing difference between I_1 and I_2. For equal intensities, $I_1 = I_2 = I_0$, the correlation function (15.20) reduces to

$$G^{(2)}(\vec{R}_1, t; \vec{R}_2, t) = 4\langle I_0^2 \rangle \left\{1 + \frac{1}{2} \cos\left[k\vec{r}_{12} \cdot \left(\hat{R}_1 - \hat{R}_2\right)\right]\right\}. \quad (15.21)$$

In analogy to the visibility in the first-order correlation function, we can define the visibility of the interference pattern of the intensity correlations as

$$\mathcal{V}_2 = \frac{G^{(2)}_{\mathrm{max}} - G^{(2)}_{\mathrm{min}}}{G^{(2)}_{\mathrm{max}} + G^{(2)}_{\mathrm{min}}}. \quad (15.22)$$

It can be verified from Eqs. (15.21) and (15.22) that in the case of a classical field, an interference pattern can be observed with the maximum possible visibility of $\mathcal{V}_2 = \frac{1}{2}$. Hence, we can conclude that two independent fields of random and uncorrelated phases can exhibit an interference pattern in the intensity correlation with a maximum visibility of 50%. This is the classical limit for the second-order interference. In the next section, we will discuss the second-order correlations involving quantum fields. We will see that with the quantum fields, visibilities larger than 50% may be observed.

15.5 Two-Photon Interference and Quantum Non-locality

The second-order correlation function can be completely different if one considers the quantum description of the field. In quantum optics, the most important quantity is the electric field, which is represented by the field operator $\hat{E}(\vec{R}, t)$. The correlation and coherence properties are discussed in terms of the positive and negative frequency parts $\hat{E}^{(+)}(\vec{R}, t)$ and $\hat{E}^{(-)}(\vec{R}, t)$. Here, we discuss separately the spatial and temporal non-classical interference effects in the two-photon correlations.

15.5.1 *Spatial Non-classical Two-Photon Interference*

In the case of the quantum description of the field, the second-order correlation function is defined in terms of the normally ordered field operators $\hat{E}^{(+)}$ and $\hat{E}^{(-)}$ as

$$G^{(2)}(\vec{R}_1, t_1; \vec{R}_2, t_2) = \langle \hat{E}^{(-)}(\vec{R}_1, t_1)\hat{E}^{(-)}(\vec{R}_2, t_2) \\ \times \hat{E}^{(+)}(\vec{R}_2, t_2)\hat{E}^{(+)}(\vec{R}_1, t_1)\rangle, \quad (15.23)$$

where the average is taken over a state $|i\rangle$ of the field. Usually, the state $|i\rangle$ is taken as an initial state of the field.

If we know the density operator ρ for the field, we can calculate the second-order correlation functions as

$$G^{(2)}(\vec{R}_1, t_1; \vec{R}_2, t_2) = \mathrm{Tr}\left\{\rho \hat{E}^{(-)}(\vec{R}_1, t_1)\hat{E}^{(-)}(\vec{R}_2, t_2) \\ \times \hat{E}^{(+)}(\vec{R}_2, t_2)\hat{E}^{(+)}(\vec{R}_1, t_1)\right\}, \quad (15.24)$$

where the trace is taken over the initial state $|i\rangle$.

The correlation functions described by the field operators are similar to the correlation functions of the classical field. A closer look at the first- and second-order correlation functions could suggest that the only difference between the correlation functions is the classical amplitudes $E^*(\vec{R}, t)$ and $E(\vec{R}, t)$ are replaced by the field operators $\hat{E}^{(-)}(\vec{R}, t)$ and $\hat{E}^{(+)}(\vec{R}, t)$. This is true as long as the first-order correlation functions are considered, where the interference effects do not distinguish between the quantum and classical theories of the EM field. However, there are significant differences between the classical and quantum descriptions of the field in the properties of the second-order correlation function.

To illustrate this, consider two independent single-mode fields of equal frequencies and polarizations. Suppose there are initially n photons in the field 1 and m photons in the field 2, and the state vectors of the fields are the Fock states $|\psi_1\rangle = |n\rangle$ and $|\psi_2\rangle = |m\rangle$. The initial state of the two fields is the direct product of the single-field states, $|\psi\rangle = |n\rangle \otimes |m\rangle$. Using the explicit form of $E^{(\pm)}(\vec{R}, t)$, Eq. (1.56) simplified to a single-mode field and the expectation value with respect to the initial state of the fields, we find that the equal time $t_1 = t_2 = t$ two-photon correlation function (15.23) takes the form

$$G^{(2)}(\vec{R}_1, t; \vec{R}_2, t) = \left(\frac{\hbar\omega}{2\epsilon_0 V}\right)^2 \{n(n-1) + m(m-1) \\ + 2nm\left(1 + \cos\left[k\vec{r}_{12} \cdot \left(\hat{R}_1 - \hat{R}_2\right)\right]\right)\}. \quad (15.25)$$

We note that the first two terms on the right-hand side of Eq. (15.25) vanish when the number of photons in each field is smaller than two, that is, when $n < 2$ and $m < 2$. In this limit the correlation function (15.25) reduces to

$$G^{(2)}\left(\vec{R}_1, t; \vec{R}_2, t\right) = 2\left(\frac{\hbar\omega}{2\epsilon_0 V}\right)^2 \left\{1 + \cos\left[k\vec{r}_{12} \cdot \left(\hat{R}_1 - \hat{R}_2\right)\right]\right\}. \quad (15.26)$$

In order to examine the visibility of the interference pattern, we substitute Eq. (15.26) into Eq. (15.22) and find that perfect interference pattern with the visibility $\mathcal{V}_2 = 1$ can be observed in the second-order correlation function of two quantum fields each containing only one photon. As we have noted, the classical theory predicts only a visibility of $\mathcal{V}_2 = 0.5$. Thus, a visibility $\mathcal{V}_2 > 0.5$ can be regarded as a non-classical effect. For $n, m \gg 1$, the first two terms on the right-hand side of Eq. (15.25) are large, $m(m-1) \approx n(n-1) \approx n^2$, and then the quantum correlation function (15.25) reduces to that of the classical field.

It follows from Eq. (15.26) that the second-order correlation function vanishes when

$$k\vec{r}_{12} \cdot \left(\hat{R}_1 - \hat{R}_2\right) = (2l + 1)\pi, \quad l = 0, \pm 1, \pm 2, \ldots \quad (15.27)$$

This shows that two photons can never be detected at two points separated by an odd number of $\lambda/2r_{12}$, despite the fact that one photon can be detected anywhere.

These spatial non-classical correlations were observed experimentally by Ou and Mandel [81], who measured the variation of the correlations between two output beams from a beam splitter with the relative position of the detectors. In the experiment, two photons produced by a degenerate parametric oscillator (DPO) fall on a beam splitter BS from opposite sides, as illustrated in Fig. 15.2. The beam splitter outputs are received and measured by two photodetectors D1 and D2 located at different points and their positions can be

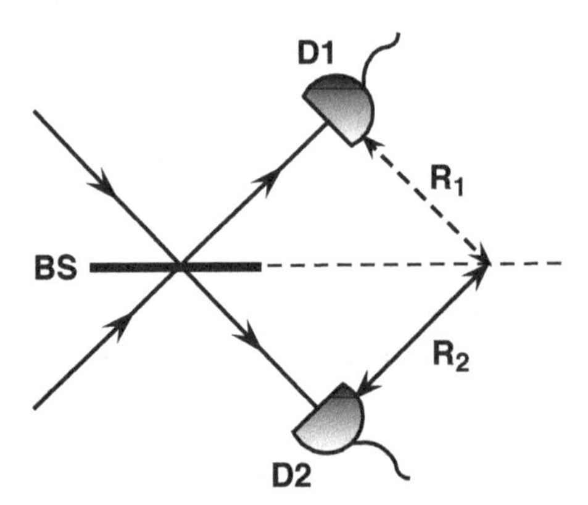

Figure 15.2 Schematic diagram of the Ou and Mandel experiment to demonstrate the spatial variation of the two-photon correlations with relative position of the two photodetectors.

varied transversely to the incident beams. The detected signals are then multiplied at a coincidence counter which gives the joint probability $G^{(2)}(\vec{R}_1, t; \vec{R}_2, t)$ as the function of the relative position of the detectors. In the experiment, the detector D2 was fixed at a constant position R_2, and the relative distance $R = R_1 - R_2$ was varied by moving the detector D1.

In Fig. 15.3, we plot the joint probability $G^{(2)}(\vec{R}_1, t; \vec{R}_2, t)$ as a function of $R_1 - R_2 = R$. Clearly, for some relative positions of the two detectors the joint probability vanishes indicating the non-classical two-photon correlations between the beams.

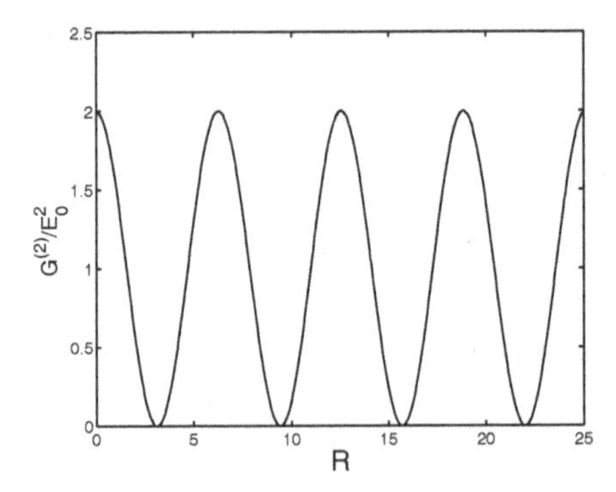

Figure 15.3 The joint probability $G^{(2)}(\vec{R}_1, t; \vec{R}_2, t)$ as a function of the relative position of the detectors for $kr_{12} = 1$ and $E_0^2 = 2\left(\hbar\omega/(2\epsilon_0 V)\right)^2$.

The vanishing of $G^{(2)}(\vec{R}_1, t; \vec{R}_2, t)$ for two photons at widely separated points $\vec{R}_1$ and $\vec{R}_2$ is an example of quantum-mechanical non-locality, that the outcome of a detection measurement at $\vec{R}_1$ appears to be influenced by where we have chosen to locate the $\vec{R}_2$ detector. At certain positions $\vec{R}_2$, we can never detect a photon at $\vec{R}_1$ when there is a photon detected at $\vec{R}_2$, whereas at other position $\vec{R}_2$ it is possible. The photon correlation argument shows clearly that quantum theory does not in general describe an objective physical reality independent of observation.

Einstein–Podolsky–Rosen (EPR) took the view that local realism must be valid [82]. They therefore argued that quantum mechanics must be incomplete. One would have to assume the existence of 'hidden variables', that are not part of quantum theory, in order to describe the localized sub-systems consistently with the quantum predictions. The argument is perhaps best viewed as a demonstration of the inconsistency between quantum mechanics as we know it (that is without *completion*) and local realism. However, the EPR paradox was refuted by an experimental demonstration of the violation of the Bell inequalities [83]. Bell derived a set of inequalities [84] that can be violated only if a given theory, in our case the quantum mechanics, is inconsistent with any local realism [85, 86].

15.5.2 *Temporal Non-classical Two-Photon Interference*

In the preceding section, we have shown that spatial correlations between two photons can lead to non-classical interference effects in the two-photon correlations. Here, we consider *temporal* correlations between photons produced by the same source. As a detector of the time correlations of photons, we consider a simple situation of two-photon interference at a beam splitter, shown in Fig. 15.4.

The photons of the modes a and b incident on a beam splitter of the reflectivity η and produce output modes c and d. The amplitudes of the output modes are related to the amplitudes of the input modes by

$$\hat{c} = i\sqrt{\eta}\,\hat{a} + \sqrt{1-\eta}\,\hat{b},$$
$$\hat{d} = i\sqrt{\eta}\,\hat{b} + \sqrt{1-\eta}\,\hat{a}, \tag{15.28}$$

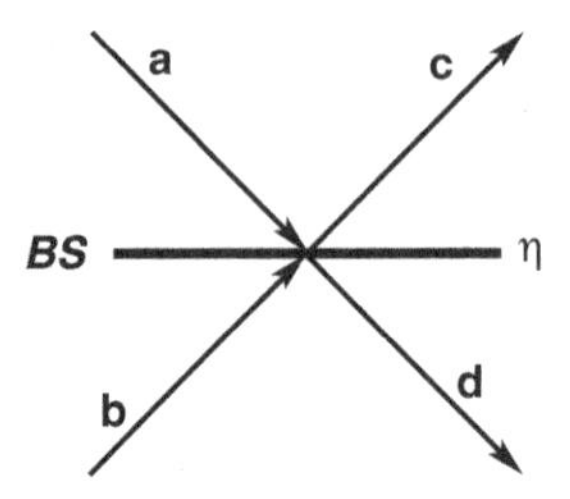

Figure 15.4 The input (a, b) and output (c, d) beams at a beam splitter BS of reflectivity η.

where the factor i indicates a $\pi/2$ phase shift between the reflected and transmitted fields.

The joint (coincidence) probability that a photon is detected in the arm c at time t and another one in the arm d at time $t + \tau$ is proportional to the second-order correlation function

$$P_{cd}(\tau) = \langle\Psi|\hat{E}_c^{(-)}(t)\,\hat{E}_d^{(-)}(t+\tau)\,\hat{E}_d^{(+)}(t+\tau)\,\hat{E}_c^{(+)}(t)\,|\Psi\rangle, \quad (15.29)$$

where $|\Psi\rangle$ is the state of the input fields, $\hat{E}_u^{(+)}(t) = \lambda\hat{u}(t)$, $(\hat{u} = \hat{c}, \hat{d})$ is the field amplitude of the output n-th mode and λ is a constant.

For an arbitrary state of the input fields, and in the limit of a long time, the coincidence probability takes the form

$$\begin{aligned}P_{cd}(\tau) = |\lambda|^4 \{&\eta(1-\eta)\langle\hat{a}^\dagger\hat{a}^\dagger(\tau)\hat{a}(\tau)\hat{a}\rangle + \eta(1-\eta)\langle\hat{b}^\dagger\hat{b}^\dagger(\tau)\hat{b}(\tau)\hat{b}\rangle \\ &-\eta(1-\eta)\langle\hat{a}^\dagger\hat{b}^\dagger(\tau)\hat{a}(\tau)\hat{b}\rangle + (1-\eta)^2\langle\hat{b}^\dagger\hat{a}^\dagger(\tau)\hat{a}(\tau)\hat{b}\rangle \\ &+\eta^2\langle\hat{a}^\dagger\hat{b}^\dagger(\tau)\hat{b}(\tau)\hat{a}\rangle - \eta(1-\eta)\langle\hat{b}^\dagger\hat{a}^\dagger(\tau)\hat{b}(\tau)\hat{a}\rangle\}\,. \quad (15.30)\end{aligned}$$

The first two terms on the right-hand side of Eq. (15.30) describe correlations between reflected and transmitted photons of the same input beam. These correlations vanish if there is only one photon in each of the input beams. The third and fourth term describe correlations between the amplitudes of the reflected–reflected and transmitted–transmitted photons. The last two terms arise from the interference between the amplitudes of the reflected–transmitted and transmitted–reflected photons of the two beams mixed at the beam splitter, and are the real quantum interference contributions to the coincidence probability. If the state of the input fields is $|\Psi\rangle = |1\rangle_a|1\rangle_b$, that each of the input fields contains only one photon, and $\tau = 0$, the coincidence probability takes the form

$$P_{cd}(0) = (1 - 2\eta)^2\,. \quad (15.31)$$

Thus, if the beam splitter is either fully reflecting ($\eta = 1$) or fully transmissive ($\eta = 0$), the probability that a coincidence will be detected is 1. In other words, there is a photon in each of the output ports. This agrees with our intuition: if the beam splitter is fully reflective (transmissive), the photon in the mode a will be reflected (transmitted) into mode c (d), and vice versa for the photon in the mode b.

An interesting quantum interference effect arises when a 50/50 ($\eta = 1/2$) beam splitter is used. In this case, the probability of detecting a coincidence goes to zero, indicating that both photons are always found together in either c or d. This effect results from quantum interference that the two paths are indistinguishable as the detected photons have the same frequency and can come from either of the two input modes. This effect is called in the literature as the Hong–Ou–Mandel (HOM) dip.

Where the name HOM dip came from?

In the experiment, Hong, Ou and Mandel [87] measured time separations between two photons by interference at a beam splitter. The experimental setup is shown in Fig. 15.5. Two photons of the same frequency are produced by a degenerate parametric down-conversion process (DPO) and fall on the beam splitter BS from opposite sides. In order to introduce a time delay between the photons, the beam splitter can be translated slightly in the vertical

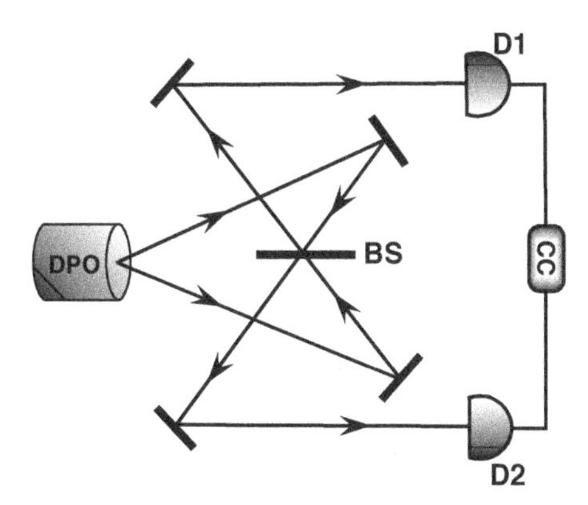

Figure 15.5 The experimental setup to measure two-photon coincidence for different delay times.

direction. This shortens the path for one photon relative to the other. The coincidences were detected by two photodetectors D1 and D2 and multiplied at the coincidence counter CC.

In the parametric down-conversion, photon pairs are randomly produced with the cross-correlations between the idler and signal modes given by a Gaussian distribution in time

$$\langle \hat{a}^{\dagger}\hat{b}^{\dagger}(\tau)\hat{a}(\tau)\hat{b}\rangle = \langle \hat{a}^{\dagger}\hat{b}\rangle\langle \hat{b}^{\dagger}(\tau)\hat{a}(\tau)\rangle = \exp\left[-\left(\Delta\omega\tau\right)^{2}\right], \qquad (15.32)$$

where $\Delta\omega$ is the bandwidth of the down-converted beam.

Hence, for a 50/50 beam splitter and quantum description of the fields

$$P_{cd}(\tau) = \frac{|\lambda|^{4}}{2}\left(1 - \mathrm{e}^{-\Delta\omega^{2}\tau^{2}}\right). \qquad (15.33)$$

Evidently, the coincidence probability vanishes at $\tau = 0$. In the case of classical description of the fields, the coincidence probability (15.30) reduces to

$$P_{cd}(\tau) = \frac{|\lambda|^{4}I^{2}}{2}\left(1 - \frac{1}{2}\mathrm{e}^{-\Delta\omega^{2}\tau^{2}}\right), \qquad (15.34)$$

where I is the intensity of the input fields (assumed to be equal for both fields). Thus, for classical fields the coincidence probability can be reduced maximally to 1/2.

The coincidence probability of a quantum field is plotted in Fig. 15.6. The figure clearly shows the presence of the HOM dip, that the coincidence probability $P_{cd}(\tau)$ vanishes at $\tau = 0$ and approaches the classical limit of $P_{cd}(\tau) = 1/2$ as $\tau \to \infty$.

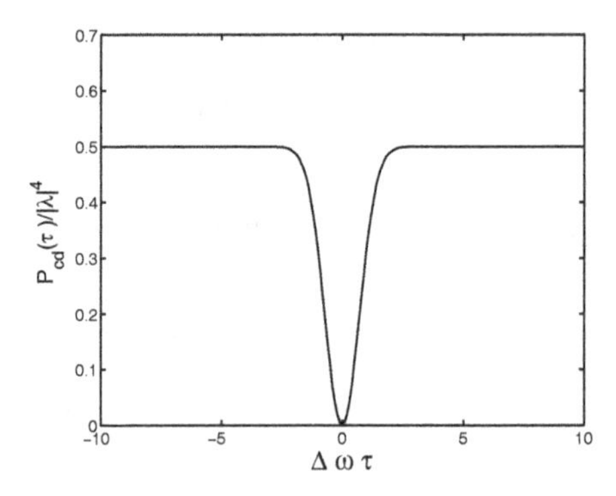

Figure 15.6 The time dependence of the two-photon coincidence probability.

15.6 Summary

Quantum interference, in particular the two-photon interference, has played a major role in the development of new fields in quantum optics called quantum information, quantum cryptography, quantum teleportation and quantum computation. The fundamental features of quantum physics, such as EPR paradox, entanglement, Bell inequalities and photon polarization correlations have been tested in quantum optics experiments. The reader wishing to learn more about various applications of quantum interference and its impact on the development of new fields of quantum optics is referred to recent books [88–90].

Exercises

15.1 Explain the physical reason for which in the interference between two beams of different intensities, $I_1 \neq I_2$, the visibility of the interference fringes $\mathcal{V}$ is not equal to the first-order coherence $|g^{(1)}|$.

15.2 Consider a three-level atom in two configurations: $\vee$ and Λ. For the $\vee$-type atom the ground state is $|2\rangle$ and the upper states are $|1\rangle$ and $|3\rangle$, whereas for the Λ-type atom the upper state is $|2\rangle$ and the two ground states are $|1\rangle$ and $|3\rangle$. In both atoms the only allowed transitions are $|2\rangle \leftrightarrow |1\rangle$ and $|2\rangle \leftrightarrow |3\rangle$.

(a) If we treat each transition as a source of light, find which of these two systems has no coherence (interference) terms in the total intensity of the emitted light.

(b) Show that the other system has non-zero coherence terms only if it is in a linear superposition of the states $|1\rangle$ and $|3\rangle$, that is,

$$|\Psi\rangle = a|1\rangle + b|3\rangle, \qquad |a|^2 + |b|^2 = 1.$$

(c) Assuming that a and b are real numbers, for what values of a and b does the emitted intensity equal to zero?

15.3 Show that the transformation

$$\begin{pmatrix} c \\ d \end{pmatrix} = \begin{pmatrix} t & -r \\ r & t \end{pmatrix} \begin{pmatrix} a \\ b \end{pmatrix},$$

between annihilation operators of the input (a, b) and output (c, d) beams at a beam splitter of reflectivity r and transmittivity t can be associated with the interaction Hamiltonian

$$\hat{H} = i\hbar g(ab^{\dagger} - ba^{\dagger}),$$

where $\sin(gt) = r$.

15.4 *Beam-splitting a single photon.* A transformation between annihilation operators of the input (a, b) and output (c, d) beams at a 50/50 beam splitter is of the form

$$\begin{pmatrix} c \\ d \end{pmatrix} = \frac{1}{\sqrt{2}} \begin{pmatrix} 1 & -1 \\ 1 & 1 \end{pmatrix} \begin{pmatrix} a \\ b \end{pmatrix}.$$

(a) What is the physical interpretation of this transformation?

(b) What is the output state of the beam splitter if the input state is a single photon state $|\Psi\rangle = |1\rangle_a |0\rangle_b$?

15.5 *Beam-splitting a coherent state.* What is the output state of a 50/50 beam splitter if the input state is $|\Psi\rangle = |\alpha\rangle_a |0\rangle_b$, where $|\alpha\rangle$ is a coherent state of amplitude α.

15.6 *Beam-splitting two photons.* Consider the case when in total two photons are present in the input beams exciting two arms in a 50/50 beam splitter.

(a) What would be the output state if the two photons are arranged such that each excite one of the two input arms, $|\Psi\rangle = |1\rangle_a |1\rangle_b$?

(b) Calculate the coincidence probability $P_{cd}(0)$, given by Eq. (15.30), for the state of the input beams $|\Psi\rangle = |0\rangle_a |2\rangle_b$.

(c) What would be the coincidence probability if the state of the input beams was a *noon* state $|\Psi\rangle = (|2\rangle_a |0\rangle_b + |0\rangle_a |2\rangle_b) / \sqrt{2}$?

15.7 *Beam-splitting multi-photon states.* The input state of a beam splitter with n photons in input arm a and m photons in input arm b can be written in terms of the zero photon state $|0, 0\rangle$ as

$$|n, m\rangle = \frac{\left(\hat{a}_a^{\dagger}\right)^n}{\sqrt{n}} \frac{\left(\hat{a}_b^{\dagger}\right)^m}{\sqrt{m}} |0, 0\rangle.$$

(a) If the input state is $|2, 0\rangle$, what would be the output state of the beam splitter in terms of the $|0, 0\rangle$ state?

(b) What would be the output state in terms of the $|0, 0\rangle$ state if the input state was changed to $|1, 1\rangle$?

Chapter 16

Atom–Atom Entanglement

16.1 Introduction

In this chapter, we will discuss the central topic in the current quantum optics studies, that is, entanglement. We focus mostly on the problem of creation of entanglement between two atoms and address the question how one could create multi-atom entangled states. The term entanglement, one of the most intriguing properties of multi-particle systems, was introduced by Schrödinger in his discussions of the foundations of quantum mechanics. It describes a multi-particle system that has the astonishing property that the results of a measurement on one particle cannot be specified independently of the results of measurements on the other particles. Although entangled systems can be physically separated, they can no longer be considered as independent, even when they are very far from one another. Entanglement is not only at the heart of the distinction between quantum and classical mechanics, but is now regarded to be a resource central to the development of quantum technologies ranging from quantum information, quantum cryptography, teleportation and quantum computation to atomic and molecular spectroscopy. These practical implementations all stem from the realization that we may control and manipulate

Quantum Optics for Beginners
Zbigniew Ficek and Mohamed Ridza Wahiddin

ISBN 978-981-4411-75-2 (Hardcover), 978-981-4411-76-9 (eBook)
www.panstanford.com

quantum systems at the level of single atoms and photons to store information and transfer the information between distant systems in a controlled way and with high fidelity. This field of study is a relatively new one, and considerable research activity is taking place at present. The most active ones are the studies of practical schemes for creation of entanglement between trapped atoms or ions [91, 94].

We first illustrate how one can create entanglement in a simple system of two two-level atoms (two qubits). Next, we will illustrate how the entanglement can be related to quantum interference. Although the entangling procedure is illustrated for two-level atoms, it can be extended into multi-atom systems or multi-level atoms (qutrits).

16.2 Two-Atom Systems

We start with the analysis of the energy states of a two-atom systems. In the absence of the interatomic interactions and the driving laser field, the space of the two-atom system is spanned by four product states

$$|g_1\rangle|g_2\rangle, \quad |e_1\rangle|g_2\rangle, \quad |g_1\rangle|e_2\rangle, \quad |e_1\rangle|e_2\rangle, \tag{16.1}$$

with corresponding energies

$$E_{gg} = -\hbar\omega_0, \quad E_{eg} = -\hbar\Delta, \quad E_{ge} = \hbar\Delta, \quad E_{ee} = \hbar\omega_0, \tag{16.2}$$

where $\omega_0 = (\omega_1 + \omega_2)/2$ is the average frequency of the atomic transition frequencies, $\Delta = (\omega_2 - \omega_1)/2$ is the detuning of the atomic frequencies, and $|g_i\rangle$ and $|e_i\rangle$ denote the ground and excited states of the ith atom, respectively.

From the energy eigenvalues, we see that the product states $|e_1\rangle|g_2\rangle$ and $|g_1\rangle|e_2\rangle$ form a pair of nearly degenerated states differing in energy by $2\hbar\Delta$, whereas the ground and the upper states are separated by $2\hbar\omega_0$.

Suppose, the atoms can interact with each other through the vacuum field by a coherent exchange of photons. The interaction is represented by the dipole–dipole interaction, which depends on the separation between the atoms and the orientation of the atomic dipole moments in respect to the interatomic axis [92–94]. The explicit form of the dipole–dipole interaction strength is given by

$$\Omega_{12} = \frac{3}{4}\Gamma\left\{-\left[1-(\hat{\mu}\cdot\hat{r}_{12})^2\right]\frac{\cos(kr_{12})}{kr_{12}}\right.$$
$$\left.+\left[1-3(\hat{\mu}\cdot\hat{r}_{12})^2\right]\left[\frac{\sin(kr_{12})}{(kr_{12})^2}+\frac{\cos(kr_{12})}{(kr_{12})^3}\right]\right\}, \quad (16.3)$$

where $\hat{\mu}$ is the unit vector along the dipole moments of the atoms, which we have assumed to be parallel ($\hat{\mu} = \hat{\mu}_1 = \hat{\mu}_2$), $\hat{r}_{12}$ is the unit vector in the direction of $\vec{r}_{12}$, $k = \omega_0/c$ is the wave number of the atomic transition and Γ is the spontaneous emission damping rate of the atoms.

When we include the dipole–dipole interaction into the Hamiltonian of the two-atom system, the product states combine into two linear superpositions (entangled states), with their energies shifted from $\pm\hbar\Delta$ by the dipole–dipole interaction energy. To check this explicitly, we begin with writing the Hamiltonian of two atoms that includes the dipole–dipole interaction

$$\hat{H}_{aa} = \sum_{i=1}^{2}\hbar\omega_i S_i^z + \hbar\sum_{i\neq j}\Omega_{ij}S_i^+S_j^-. \quad (16.4)$$

Next, we write the Hamiltonian in the basis of the product states (16.1) and arrive to a matrix of the form

$$\hat{H}_{aa} = \hbar\begin{pmatrix} -\omega_0 & 0 & 0 & 0 \\ 0 & -\Delta & \Omega_{12} & 0 \\ 0 & \Omega_{12} & \Delta & 0 \\ 0 & 0 & 0 & \omega_0 \end{pmatrix}. \quad (16.5)$$

Evidently, in the presence of the dipole–dipole interaction the matrix (16.5) is not diagonal, which indicates that the product states (16.1) are not the eigenstates of the two-atom system. The actual energy states of the system are readily found by the diagonalization of the matrix (16.5). We will diagonalize the matrix (16.5) separately for the case of identical ($\Delta = 0$) and non-identical ($\Delta \neq 0$) atoms to find eigenstates of the systems and their energies.

16.3 Entangled States of Two Identical Atoms

Consider first a system of two identical atoms ($\Delta = 0$). As we have already mentioned, in order to find energies and corresponding energy states of the system, we have to diagonalize the matrix (16.5). This is a relatively simple matrix to diagonalize and the resulting energies and corresponding eigenstates are of the form

$$\begin{aligned} E_g &= -\hbar\omega_0, & |g\rangle &= |g_1\rangle|g_2\rangle, \\ E_s &= \hbar\Omega_{12}, & |s\rangle &= \frac{1}{\sqrt{2}}\left(|e_1\rangle|g_2\rangle + |g_1\rangle|e_2\rangle\right), \\ E_a &= -\hbar\Omega_{12}, & |a\rangle &= \frac{1}{\sqrt{2}}\left(|e_1\rangle|g_2\rangle - |g_1\rangle|e_2\rangle\right), \\ E_e &= \hbar\omega_0, & |e\rangle &= |e_1\rangle|e_2\rangle. \end{aligned} \tag{16.6}$$

The eigenstates (16.6), first introduced by Dicke [95] are known as the collective states of two interacting atoms. They have few interesting properties. Firstly, the energies of the ground state $|g\rangle$ and the upper state $|e\rangle$ are not affected by the dipole–dipole interaction. Secondly, the states $|s\rangle$ and $|a\rangle$ are shifted from their unperturbed energies by the amounts $\pm\Omega_{12}$, the dipole–dipole energy. Finally, the most important property of the collective states is that the states $|s\rangle$ and $|a\rangle$ are an example of maximally entangled states of the two-atom system. The states are equally weighted linear superpositions of the product states which cannot be separated into product states of the individual atoms.

We show the collective states of two identical atoms in Fig. 16.1. It is seen that in the collective states representation, the two-atom system behaves as a single four-level system, with the ground state $|g\rangle$, the upper state $|e\rangle$, and two intermediate states: the symmetric state $|s\rangle$ and the antisymmetric state $|a\rangle$. The energies of the intermediate states depend on the dipole–dipole interaction and these states suffer a large shift when the interatomic separation is small. There are two transition channels $|e\rangle \to |s\rangle \to |g\rangle$ and $|e\rangle \to |a\rangle \to |g\rangle$, each with two cascade non-degenerate transitions. For two identical atoms, these two channels are uncorrelated, but a detailed analysis involving the master equation of the two-atom system shows that the transitions in these channels are damped with significantly different rates. The transitions through the symmetric

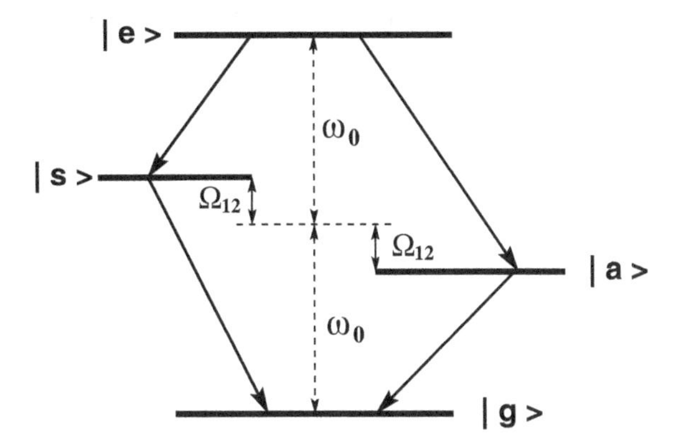

Figure 16.1 Collective states of two identical atoms. The energies of the symmetric and antisymmetric states are shifted by the dipole–dipole interaction Ω_{12}. The arrows indicate possible one-photon transitions.

state are damped with an enhanced (superradiant) rate, while the transitions through the antisymmetric state are damped with a reduced (subradiant) rate. The details of the calculations of the transition rates are left to the reader as a tutorial exercise.

16.4 Entangled States of Two Non-identical Atoms

As we have already seen, in the case of two identical atoms, the dipole–dipole interaction leads to the maximally entangled symmetric and antisymmetric states. These states decay independently with different damping rates. Furthermore, in the case of the small sample model of two atoms the antisymmetric state decouples from the external coherent field and the environment, and consequently does not decay. The decoupling of the antisymmetric state from the coherent field prevents the state from the external coherent interactions. This indicates that the initially non-populated antisymmetric state remains unpopulated for all times and the population is distributed only between three collective states.

When the atoms are non-identical with different transition frequencies, the states (16.6) are no longer the eigenstates of the Hamiltonian (16.2). However, we can still find the new eigenstates simply by the diagonalization of the matrix (16.5) with $\Delta \neq 0$. It results in the following energies and corresponding eigenstates

$$\begin{aligned}
E_g &= -\hbar\omega_0, & |g\rangle &= |g_1\rangle|g_2\rangle,\\
E_{s'} &= \hbar w, & |s'\rangle &= \beta|e_1\rangle|g_2\rangle + \alpha|g_1\rangle|e_2\rangle,\\
E_{a'} &= -\hbar w, & |a'\rangle &= \alpha|e_1\rangle|g_2\rangle - \beta|g_1\rangle|e_2\rangle,\\
E_e &= \hbar\omega_0, & |e\rangle &= |e_1\rangle|e_2\rangle,
\end{aligned} \tag{16.7}$$

where

$$\alpha = \frac{d}{\sqrt{d^2 + \Omega_{12}^2}}, \quad \beta = \frac{\Omega_{12}}{\sqrt{d^2 + \Omega_{12}^2}}, \quad w = \sqrt{\Omega_{12}^2 + \Delta^2}, \tag{16.8}$$

and $d = \Delta + \sqrt{\Omega_{12}^2 + \Delta^2}$.

The energy level structure of the collective system of two non-identical atoms is similar to that of the identical atoms, with the ground state $|g\rangle$, the upper state $|e\rangle$ and two intermediate states $|s'\rangle$ and $|a'\rangle$. The effect of the frequency difference Δ on the collective atomic states is to increase the splitting between the intermediate levels, which now is equal to $w = \sqrt{\Omega_{12}^2 + \Delta^2}$. However, the most dramatic effect of the detuning Δ is on the degree of entanglement of the intermediate states $|s'\rangle$ and $|a'\rangle$ that in the case of non-identical atoms the states are no longer maximally entangled states. For $\Delta = 0$ the states $|s'\rangle$ and $|a'\rangle$ reduce to the maximally entangled states $|s\rangle$ and $|a\rangle$, whereas for $\Delta \gg \Omega_{12}$ the entangled states reduce to the product states $|e_1\rangle|g_2\rangle$ and $|g_1\rangle|e_2\rangle$, respectively.

16.5 Creation of Entanglement between Two Atoms

We now consider excitation processes that can lead to a preparation of the two-atom system in only one of the collective states. In particular, we will focus on processes that can prepare the two-atom system in the entangled symmetric state $|s\rangle$. Our main interest, however, is in the preparation of the system in the maximally entangled antisymmetric state $|a\rangle$, which is known as a decoherence-free state [94]. The central idea is to choose the distance between the atoms such that the resulting level shift is large enough to consider the possible transitions between the collective states separately. This will allow to make a selective excitation of the symmetric and antisymmetric states and, therefore, to create controlled entanglement between the atoms.

16.5.1 *Preparation of Atoms in Entangled Symmetric State*

A system of two identical two-level atoms may be prepared in the symmetric state $|s\rangle$ by a short laser pulse. The conditions for a selective excitation of the collective atomic states can be analysed from the interaction Hamiltonian of the laser field with the two-atom system.

We start with the Hamiltonian of two interacting atoms driven by an external (classical) laser field

$$\hat{H} = \hbar\omega_0 \sum_{i=1}^{2} S_i^z + \hbar \sum_{i \neq j}^{2} \Omega_{ij} S_i^+ S_j^- + \hat{H}_{\mathrm{L}}, \tag{16.9}$$

where

$$\hat{H}_{\mathrm{L}} = -\frac{1}{2}\hbar \sum_{i=1}^{2} \left[\Omega\left(\vec{r}_i\right) S_i^+ \mathrm{e}^{-i(\omega_{\mathrm{L}} t + \phi_{\mathrm{L}})} + \mathrm{H.c.}\right] \tag{16.10}$$

is the interaction Hamiltonian of the atoms with the laser field of the Rabi frequency $\Omega\left(\vec{r}_i\right)$, the angular frequency ω_{L} and phase ϕ_{L}.

Note that the Rabi frequencies of the driving field are evaluated at the positions of the atoms and are defined as

$$\Omega\left(\vec{r}_i\right) \equiv \Omega_i = \vec{\mu}_i \cdot \vec{E}_{\mathrm{L}} \mathrm{e}^{i\vec{k}_{\mathrm{L}} \cdot \vec{r}_i}/\hbar, \tag{16.11}$$

where $\vec{E}_{\mathrm{L}}$ is the amplitude and $\vec{k}_{\mathrm{L}}$ is the wave vector of the driving field, respectively. The Rabi frequencies depend on the positions of the atoms and can be different for the atoms located at different points.

In the interaction picture and in the basis of the collective states (16.6), the Hamiltonian (16.9) can be written as

$$\hat{H} = \hat{H}_a + \hat{H}_{\mathrm{L}}, \tag{16.12}$$

where

$$\hat{H}_a = \hbar\{\Delta_{\mathrm{L}}\left(|e\rangle\langle e| - |g\rangle\langle g|\right) + \left(\Delta_{\mathrm{L}} + \Omega_{12}\right)|s\rangle\langle s| + \left(\Delta_{\mathrm{L}} - \Omega_{12}\right)|a\rangle\langle a|\}, \tag{16.13}$$

$$\hat{H}_{\mathrm{L}} = -\frac{1}{2}\hbar \sum_{i=1}^{2} \left[\Omega\left(\vec{r}_i\right) S_i^+ + \mathrm{H.c.}\right], \tag{16.14}$$

and $\Delta_{\rm L} = \omega_{\rm L} - \omega_0$ is the detuning of the laser frequency from the atomic transition frequencies ω_0.

We make the unitary transformation

$$\tilde{H}_{\rm L} = {\rm e}^{i\hat{H}_a t/\hbar}\hat{H}_{\rm L}{\rm e}^{-i\hat{H}_a t/\hbar}, \tag{16.15}$$

and find that the transformed interaction Hamiltonian $\tilde{H}_{\rm L}$ is given by

$$\begin{aligned}\tilde{H}_{\rm L} = -\frac{\hbar}{2\sqrt{2}}\Big\{(\Omega_1+\Omega_2)\left(S_{es}^{+}{\rm e}^{i(\Delta_{\rm L}+\Omega_{12})t} + S_{sg}^{+}{\rm e}^{i(\Delta_{\rm L}-\Omega_{12})t}\right) \\ + (\Omega_2-\Omega_1)\left(S_{ag}^{+}{\rm e}^{i(\Delta_{\rm L}+\Omega_{12})t} + S_{ea}^{+}{\rm e}^{i(\Delta_{\rm L}-\Omega_{12})t}\right) + {\rm H.c.}\Big\}.\end{aligned} \tag{16.16}$$

The Hamiltonian (16.16) represents the interaction of the laser field with the collective two-atom system, and in the transformed form contains terms oscillating at frequencies $(\Delta_{\rm L} \pm \Omega_{12})$, which correspond to the two separate groups of transitions between the collective atomic states at frequencies $\omega_{\rm L} = \omega_0 + \Omega_{12}$ and $\omega_{\rm L} = \omega_0 - \Omega_{12}$. The $\Delta_{\rm L} + \Omega_{12}$ frequencies are separated from $\Delta_{\rm L} - \Omega_{12}$ frequencies by $2\Omega_{12}$, and hence the two groups of the transitions evolve separately when $\Omega_{12} \gg \Gamma$. Depending on the frequency, the laser can be selectively tuned to one of the two groups of the transitions. When $\omega_{\rm L} = \omega_0 + \Omega_{12}$, the detuning $\Delta_{\rm L} = \Omega_{12}$, so that the laser is then tuned to exact resonance with the $|e\rangle - |a\rangle$ and $|g\rangle - |s\rangle$ transitions. In this case, the terms in the Hamiltonian (16.16) corresponding to these transitions have no explicit time dependence. In contrast, the $|g\rangle - |a\rangle$ and $|e\rangle - |s\rangle$ transitions are off-resonant and the terms corresponding to these transitions have an explicit time dependence $\exp(\pm 2i\Omega_{12}t)$. If $\Omega_{12} \gg \Gamma$, the off-resonant terms rapidly oscillate with the frequency $2\Omega_{12}$, and then we can make a secular approximation in which we neglect all those rapidly oscillating terms. The interaction Hamiltonian can then be written in the simplified form

$$\tilde{H}_{\rm L} = -\frac{\hbar}{2\sqrt{2}}\left[(\Omega_1+\Omega_2)\,S_{sg}^{+} + (\Omega_2-\Omega_1)\,S_{ea}^{+} + {\rm H.c.}\right]. \tag{16.17}$$

It is seen that the laser field couples to the transitions with significantly different Rabi frequencies. The coupling strength of the laser to the $|g\rangle - |s\rangle$ transition is proportional to the sum of the Rabi frequencies $\Omega_1 + \Omega_2$, whereas the coupling strength of the laser to the $|a\rangle - |e\rangle$ transition is proportional to the difference of the Rabi frequencies $\Omega_1 - \Omega_2$. According to Eq. (16.11) the Rabi frequencies Ω_1 and Ω_2 of two identical atoms differ only by the phase factor $\exp(i\vec{k}_L \cdot \vec{r}_{12})$. Thus, in order to selectively excite the $|g\rangle - |s\rangle$ transition, the driving laser field should be in phase with both atoms, that is, $\Omega_1 = \Omega_2$. This can be achieved by choosing the propagation vector $\vec{k}_L$ of the laser orthogonal to the line joining the atoms. Under this condition we can make a further simplification and truncate the state vector of the system into two states $|g\rangle$ and $|s\rangle$. In this two-state approximation we find from the Schrödinger equation the time evolution of the population $P_s(t)$ of the state $|s\rangle$ as

$$P_s(t) = \sin^2\left(\frac{1}{\sqrt{2}}\Omega t\right), \tag{16.18}$$

where $\Omega = \Omega_1 = \Omega_2$. The population oscillates with the Rabi frequency of the $|s\rangle - |g\rangle$ transition and at certain times $P_s(t) = 1$ indicating that all the population is in the symmetric state. This happens at times

$$T_n = (2n+1)\frac{\pi}{\sqrt{2}\Omega}, \quad n = 0, 1, \ldots. \tag{16.19}$$

Hence, the system can be prepared in the state $|s\rangle$ by simply applying a laser pulse, for example, with the duration T_0, which is a standard π pulse.

The two-state approximation is of course an idealization, and a possibility that all the transitions can be driven by the laser imposes significant limits on the Rabi frequency and the duration of the pulse. Namely, the Rabi frequency cannot be too strong in order to avoid the coupling of the laser to the $|s\rangle - |e\rangle$ transition, which could lead to a slight pumping of the population to the state $|e\rangle$. On the other hand, the Rabi frequency cannot be too small as for a small Ω the duration of the pulse, required for the complete transfer of the population into the state $|s\rangle$, becomes longer and then spontaneous emission can occur during the excitation process. Therefore, the transfer of the population to the state $|s\rangle$ cannot be made arbitrarily fast and, in addition, requires a careful estimation of the optimal Rabi frequency, which could be difficult to achieve in a real experimental situation.

16.5.2 *Preparation of Atoms in Entangled Antisymmetric State*

If we choose the laser frequency such that $\Delta_{\mathrm{L}} + \Omega_{12} = 0$, the laser field is then resonant to the $|a\rangle - |g\rangle$ and $|e\rangle - |s\rangle$ transitions and, after the secular approximation, the Hamiltonian (16.16) reduces to

$$\tilde{H}_{\mathrm{L}} = -\frac{\hbar}{2\sqrt{2}}\left[(\Omega_2 - \Omega_1)\, S_{ag}^{+} + (\Omega_1 + \Omega_2)\, S_{es}^{+} + \mathrm{H.c.}\right]. \tag{16.20}$$

Clearly, for $\Omega_1 = -\Omega_2$ the laser couples only to the $|a\rangle - |g\rangle$ transition. Thus, in order to selectively excite the $|g\rangle - |a\rangle$ transition, the atoms should experience opposite phases of the laser field. This can be achieved by choosing the propagation vector $\vec{k}_{\mathrm{L}}$ of the laser along the interatomic axis, and the atomic separations such that

$$\vec{k}_{\mathrm{L}} \cdot \vec{r}_{12} = (2n+1)\,\pi, \quad n = 0, 1, 2, \ldots, \tag{16.21}$$

which corresponds to a situation where the atoms are separated by a distance $r_{12} = (2n+1)\lambda/2$.

One can notice that the smallest distance at which the atoms could experience opposite phases corresponds to $r_{12} = \lambda/2$. However, it can be verified from Eq. (16.3) that at the separation $r_{12} = \lambda/2$ the dipole–dipole interaction Ω_{12} is small, and then all of the transitions between the collective states occur at approximately the same frequency. In this case the secular approximation is not valid and we cannot separate the transitions at $\Delta_{\mathrm{L}} + \Omega_{12}$ from the transitions at $\Delta_{\mathrm{L}} - \Omega_{12}$.

One possible solution to the problem of the selective excitation with opposite phases is to use a standing laser field instead of the running-wave field. If the laser amplitudes differ by the sign, that is, $\vec{E}_{L_1} = -\vec{E}_{L_2} = \vec{E}_0$ and $\vec{k}_{L_1} \cdot \vec{r}_1 = -\vec{k}_{L_2} \cdot \vec{r}_2$, the Rabi frequencies experienced by the atoms are

$$\Omega_1 = \frac{2i}{\hbar}\vec{\mu}_1 \cdot \vec{E}_0 \sin\left(\frac{1}{2}\vec{k}_{\mathrm{L}} \cdot \vec{r}_{12}\right),$$

$$\Omega_2 = -\frac{2i}{\hbar}\vec{\mu}_2 \cdot \vec{E}_0 \sin\left(\frac{1}{2}\vec{k}_{\mathrm{L}} \cdot \vec{r}_{12}\right), \tag{16.22}$$

where $\vec{k}_{\mathrm{L}} = \vec{k}_{L_1} = \vec{k}_{L_2}$, and we have chosen the reference frame such that $\vec{r}_1 = \frac{1}{2}\vec{r}_{12}$ and $\vec{r}_2 = -\frac{1}{2}\vec{r}_{12}$. It follows from Eq. (16.22) that the Rabi frequencies oscillate with opposite phases independent of the separation between the atoms. However, the magnitude of the Rabi frequencies decreases with decreasing r_{12}.

16.5.3 *Creation of Two-Photon Entangled States*

In the previous subsection, we have discussed different excitation processes that can prepare two atoms in the symmetric and antisymmetric entangled states. However, apart from the symmetric and antisymmetric states, there are two other collective states of the two-atom system: the ground state $|g\rangle = |g_1\rangle|g_2\rangle$ and the upper state $|e\rangle = |e_1\rangle|e_2\rangle$. These states are not entangled states. They are product states of the individual atomic states and, the most interesting, their energies are not affected by the dipole–dipole interaction Ω_{12}.

As we shall demonstrate, one can create entangled states involving those two product states. However, this requires the application of an external coherent field, resonant or near resonant with the two-photon frequency $2\omega_0$ separating these states. Moreover, more conditions should be satisfied. The problem is that a coherent field resonant to the $|g\rangle \leftrightarrow |e\rangle$ two-photon transition couples also to one-photon transitions, $|g\rangle \leftrightarrow |s\rangle$ and $|g\rangle \leftrightarrow |a\rangle$. Thus, the field populates not only the upper state $|e\rangle$ but also the intermediate states $|s\rangle$ and $|a\rangle$. The two-photon entangled states are superpositions of the collective ground and excited states with no contribution from the intermediate collective states $|s\rangle$ and $|a\rangle$. Therefore, some arrangements should be done to excite the two-photon transition and at the same time eliminate the coupling of the field to the one-photon transitions.

Let us now investigate how this could be arranged. Suppose, the density matrix of the two-atom system is given by

$$\hat{\rho} = \begin{pmatrix} \rho_{gg} & 0 & 0 & \rho_{ge} \\ 0 & \rho_{ss} & 0 & 0 \\ 0 & 0 & \rho_{aa} & 0 \\ \rho_{eg} & 0 & 0 & \rho_{ee} \end{pmatrix}. \tag{16.23}$$

where ρ_{ij} are the non-zero density matrix elements.

It is seen from Eq. (16.23) that the density matrix of the system is not diagonal due to the presence of the two-photon coherencies ρ_{ge} and ρ_{eg}. This indicates that in the presence of the two-photon coherences, the collective states $|g\rangle$, $|s\rangle$ and $|e\rangle$ are no longer eigenstates of the system. The density matrix can be re-diagonalized by including ρ_{eg} and ρ_{ge} to give the new (entangled) states

$$|u\rangle = \frac{(P_u - \rho_{ee})|g\rangle + \rho_{eg}|e\rangle}{\sqrt{(P_u - \rho_{ee})^2 + |\rho_{eg}|^2}}, \quad |d\rangle = \frac{\rho_{ge}|g\rangle + (P_d - \rho_{gg})|e\rangle}{\sqrt{(P_d - \rho_{gg})^2 + |\rho_{eg}|^2}}, \tag{16.24}$$

and the collective states $|s\rangle$ and $|a\rangle$ remain unchanged. The probabilities (eigenvalues) of the diagonal states are

$$\begin{aligned} P_u &= \frac{1}{2}\left(\rho_{gg} + \rho_{ee}\right) + \frac{1}{2}\sqrt{\left(\rho_{gg} - \rho_{ee}\right)^2 + 4\left|\rho_{eg}\right|^2}, \\ P_d &= \frac{1}{2}\left(\rho_{gg} + \rho_{ee}\right) - \frac{1}{2}\sqrt{\left(\rho_{gg} - \rho_{ee}\right)^2 + 4\left|\rho_{eg}\right|^2}, \\ P_s &= \rho_{ss}, \\ P_a &= \rho_{aa}. \end{aligned} \tag{16.25}$$

The two-photon behaviour of the entangled states (16.24) suggests that the simplest technique for generating the two-photon excitation (TPE) states would be by applying a TPE process. An obvious candidate is a squeezed vacuum field which, as we have seen in Chapter 6, is characterized by strong two-photon correlations, which would enable the transition $|g\rangle \rightarrow |e\rangle$ to occur effectively in a single step without populating the intermediate states. It has been demonstrated that the entangled states (16.24) are analogous to the pairwise atomic states [96–98], or the multi-atom squeezed states predicted in the small sample (Dicke) model of two interacting atoms [99].

16.6 Quantum Interference of the Field Radiated by Two-Atom Systems

In Chapter 15, we investigated quantum interference effects involving classical and quantum electromagnetic (EM) fields. We were not interested in sources that could generate these fields in practice. In this chapter, we focus on practical sources of the field, two-level atoms, and discuss quantum interference effects involving radiation fields emitted by the atoms. We consider both the Young's and the Hanbury Brown and Twiss (HBT) experiments in which we determine, respectively, the first and second-order correlations in the field emitted by two two-level atoms. The advantage of using two-level atoms instead of slits is that at a given time each atom cannot emit more than one photon. Therefore, the two-level atoms can be regarded as sources of single photon fields. The atoms interacting with the vacuum field may decay spontaneously from their upper energy states to the ground states emitting their energy to the surrounding vacuum modes.

16.6.1 *First-Order Interference of the Field Radiated by a Two-Atom System*

Consider a Young's type experiment with the slits replaced by two-level atoms. The atoms are sources of the field registered in the far-field zone by a detector P, as illustrated in Fig. 16.2. We assume that (a) initially only one atom was in the excited state and the other was in its ground state, and (b) initially both atoms were excited.

Since the electric field emitted by an atom is proportional to the radiating dipole moment

$$\hat{E}_i^{(-)} \sim \hat{\mu}_i = |\mu_i| S_i^+, \quad \hat{E}_i^{(+)} \sim \hat{\mu}_i^\dagger = |\mu_i| S_i^-, \tag{16.26}$$

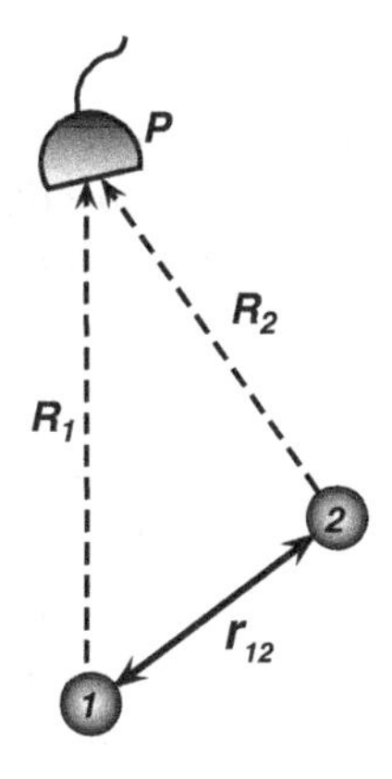

Figure 16.2 Schematic diagram of a Young's type experiment in which slits are replaced by atoms distance r_{12} and radiating fields registered by a detector P located in the far-field zone of the radiating atoms, $R_1 \approx R_2 = R \gg r_{12}$.

we can write the average intensity at the detector P in terms of the atomic spin operators as

$$I(\vec{R}, t) = \Gamma \left[\langle S_1^+ S_1^- \rangle + \langle S_2^+ S_2^- \rangle + 2 \langle S_1^+ S_2^- \rangle \cos\left(k\hat{R} \cdot \vec{r}_{12}\right) \right], \tag{16.27}$$

where Γ is the spontaneous emission rate, equal for both atoms. In the basis of the collective states, the radiation intensity can be written as

$$\begin{aligned} I(\vec{R}, t)/\Gamma &= [\rho_{ee}(t) + \rho_{ss}(t)] \left[1 + \cos\left(k\hat{R} \cdot \vec{r}_{12}\right)\right] \\ &+ [\rho_{ee}(t) + \rho_{aa}(t)] \left[1 - \cos\left(k\hat{R} \cdot \vec{r}_{12}\right)\right] \\ &+ i \left[\rho_{sa}(t) - \rho_{as}(t)\right] \sin\left(k\hat{R} \cdot \vec{r}_{12}\right), \end{aligned} \tag{16.28}$$

where ρ_{ss} and ρ_{aa} are the populations of the entangled states $|s\rangle$ and $|a\rangle$, respectively, and ρ_{sa}, ρ_{as} are coherencies between them.

Example 16.1 (Initially one atom excited) *Assume that only one atom was initially excited. In this case*

$$\rho_{ss}(0) = \rho_{aa}(0) = \rho_{sa}(0) = \rho_{as}(0) = \frac{1}{2}, \quad \rho_{ee}(0) = 0. \tag{16.29}$$

Then, the radiation intensity (16.28) reduces to

$$I(\vec{R}, t)/\Gamma = \rho_{ss}(t) + \rho_{aa}(t) + [\rho_{ss}(t) - \rho_{aa}(t)] \cos\left(k\hat{R} \cdot \vec{r}_{12}\right). \tag{16.30}$$

It follows that an interference can be observed only if $\rho_{ss}(t) \neq \rho_{aa}(t)$. This happens when $\rho_{12}(t)$ and $\rho_{21}(t)$ are different from zero, that is, when there is a coherence between the atoms.

At first thought, the result (16.30) can be surprising. One could argue that with only one atom excited initially, say the atom 1, and the path of the emitted photon is known, the photon travels to the detector along the path R_1, so this should rule out the interference effect. This is true as long as the atoms are independent and in this case one can readily show that $\rho_{ss}(t) = \rho_{aa}(t)$ for all times t. However, when the atoms are coupled to each other, which could happen when the atoms are located in a common reservoir, then the photon emitted by the excited atom could travel to the detector along the path R_1 but also could be absorbed by the second atom and then travel to the detector along the path R_2. Since we do not know which path the emitted photon travelled to the detector, an interference effect occurs.

Example 16.2 (Initially both atoms excited) *If initially both atoms were prepared in their excited states, then*

$$\rho_{ee}(0) = 1, \quad \rho_{ss}(0) = \rho_{aa}(0) = \rho_{sa}(0) = \rho_{as}(0) = 0. \qquad (16.31)$$

In this case, the radiation intensity (16.28) simplifies to

$$I(\vec{R}, t) = 2\Gamma\rho_{ee}(t). \qquad (16.32)$$

Clearly, there are no terms involved that depend on the position of the detector $\vec{R}$. Thus, no variation of the intensity with $\vec{R}$, which on the other hand means no quantum interference can be observed in this case.

The result of the example 16.2 can be understood by referring to Fig. 16.1 When both atoms are initially excited, the population is initially in the upper state $|e\rangle$ and due to spontaneous emission the population can decay to the intermediate states $|s\rangle$ and $|a\rangle$. There are two paths the population decays, $|e\rangle \rightarrow |s\rangle$ and $|e\rangle \rightarrow |a\rangle$. Then, by measuring the population of either $|s\rangle$ or $|a\rangle$, one could recognize which path the population decayed to the lower energy states. Knowledge of the path eliminates the interference fringes.

Example 16.3 (Two non-identical atoms) *When the atoms are not identical with different transition frequencies, $\omega_1 \neq \omega_2$, the radiation intensity can be written as*

$$I(\vec{R}, t)/\Gamma = I_1(t) + I_2(t) + 2\langle\mathcal{E}_1^{(-)}\mathcal{E}_2^{(+)}\rangle \cos\left(k\hat{R}\cdot\vec{r}_{12} + \Delta t\right), \qquad (16.33)$$

where $\Delta = \omega_2 - \omega_1$. In this case for any initial conditions, the interference term depends on time. The magnitude of the term decreases with increasing Δ and the observation time t, and vanishes for long observation times, $t \gg 0$.

The reason for the vanishing of the interference term when the atoms are not identical is that a long observation time allows to determine the frequency of the detected photon, so that it can be known from which atom the photon came to the detector.

16.6.2 *Two-Photon Interference in a Two-Atom System*

We have seen that one-photon interference is sensitive to the one-photon coherences. In contrast, the second-order correlation function can exhibit an interference pattern independent of the one-photon coherences. This type of interference results from the detection process that a detector does not distinguish between two simultaneously detected photons.

We illustrate the temporal and spatial properties of the second-order correlation function with few examples.

Example 16.4 (One-time second-order correlation function) *As a first example, consider the one-time second-order correlation function*

$$\begin{aligned} G^{(2)}(\vec{R}, t) &= \langle E^{(-)}(\vec{R}, t) E^{(-)}(\vec{R}, t) E^{(+)}(\vec{R}, t) E^{(+)}(\vec{R}, t) \rangle \\ &= \Gamma^2 \rho_{ee}(t) \left[1 + \cos\left(k\hat{R} \cdot \vec{r}_{12} \right) \right], \end{aligned} \tag{16.34}$$

where, for simplicity, we have assumed that $\vec{R}_1 = \vec{R}_2$. We see that interference takes place only when the population of the upper state $\rho_{ee}(t) \neq 0$, and no first-order coherence is required to see interference. This type of interference can be achieved by preparing both atoms in their upper states.

Now consider a more general two-photon correlation function and also assume that the atoms are not identical.

Example 16.5 (Two initially excited non-identical atoms) *Consider now two initially excited non-identical atoms, $\rho_{ee}(0) = 1$. If both photons were detected at the same time t, the second-order correlation function takes the form*

$$G^{(2)}(\vec{R}_1, t; \vec{R}_2, t) = \Gamma^2 \left[1 + \cos\left(k\hat{R} \cdot \vec{r}_{12} \right) \right] e^{-4\Gamma t}. \tag{16.35}$$

This shows that even for two significantly different (distinguishable) photons detected at the same time, the second-order correlation function exhibits perfect interference pattern. This surprising result arises from the fact that a detector cannot distinguish between two simultaneously detected photons, even if the photons have significantly different frequencies.

Example 16.6 (Photons detected at different times) *When initially both atoms were excited ($\rho_{ee}(0) = 1$) and the two emitted* *photons were detected at different times, the first at t and the other at* $t+\tau$*, the two-time second-order correlation function takes the form*

$$G^{(2)}(\vec{R}_1, t; \vec{R}_2, t+\tau) = \Gamma^2\left[1+\cos\left(k\hat{R}\cdot\vec{r}_{12}+\Delta\tau\right)\right]e^{-2\Gamma(2t+\tau)}. \tag{16.36}$$

We see that at the interference term is independent of t. It depends on the time difference τ*, but only when the atoms are not identical,* $\Delta \neq 0$*. Thus, for identical atoms perfect interference is observed even if the photons are detected at significantly different times that could even go to infinity. For non-identical atoms, the interference degrades with an increasing* τ*, and is expected to vanish for* $\tau \gg 0$*. To put it another way, when photons emitted from non-identical atoms are well separated in time, it is possible to determine from which atom the photon came to the detector by examining its frequency. This rules out any interference, which is always a manifestation of the intrinsic indistinguishability of possible paths of the detected photons.*

16.7 Summary

In summary of this chapter, we can thus state: Any attempt to *interfere* with the interference phenomenon to find out from which atom the detected photon was emitted leads to a degradation or even elimination of the interference fringes. The observation of an interference pattern and the acquisition of from which atom the photon came to a photodetector are mutually exclusive.

Exercises

16.1 Show that

$$\left\langle S_1^+S_1^-\right\rangle + \left\langle S_2^+S_2^-\right\rangle = \rho_{ss}+\rho_{aa}+2\rho_{ee},$$
$$\left\langle S_1^+S_2^-\right\rangle = \frac{1}{2}\left(\rho_{ss}-\rho_{aa}+\rho_{as}-\rho_{sa}\right).$$

16.2 Write the purity condition $\mathrm{Tr}\rho^2$ in terms of the populations and coherences of the collective (Dicke) states.

16.3 The master equation for two interacting atoms in the ordinary vacuum has the same form as the master equation for the V system, given in the Exercise 9.5 of Chapter 9. The only difference is that now $S_1^+ = |e_1\rangle\langle g_1|$ and $S_2^+ = |e_2\rangle\langle g_2|$.

(a) Assuming that initially only one atom was excited, calculate the time evolution of the populations $\rho_{ss}(t)$, $\rho_{aa}(t)$ and the coherence $\rho_{sa}(t)$.

(b) Under what condition $\rho_{ss}(t) \neq \rho_{aa}(t)$?

(c) Find the time evolution of the intensity $I(t)$.

(d) Analyse the dependence of the intensity on time t and the damping rate Γ_{12} and find conditions for t and Γ_{12} at which one can observe interference pattern.

16.4 Consider two identical two-level atoms interacting with a multi-mode broadband reservoir whose modes are in a squeezed vacuum state. If the atoms are confined to a region much smaller than the atomic transition wavelength (the Dicke model), the density matrix elements in the basis of the collective states satisfy the following equations of motion

$$\begin{aligned}
\dot{\rho}_{ee} &= -2\Gamma\left(N+1\right)\rho_{ee} + 2N\Gamma\rho_{ss} + \Gamma|M|\rho_u, \\
\dot{\rho}_{ss} &= 2\Gamma\left[N - \left(3N+1\right)\rho_{ss} + \rho_{ee} - |M|\rho_u\right], \\
\dot{\rho}_{aa} &= 0, \\
\dot{\rho}_{u} &= 2\Gamma|M| - \Gamma\left(2N+1\right)\rho_u - 6\Gamma|M|\rho_{ss},
\end{aligned}$$

where N is the number of photons in the modes of the squeezed field, M is the two-photon correlation between the modes, see Eq. (6.69), and $\rho_u = \rho_{eg} + \rho_{ge}$.

(a) Show that the steady-state values of the density matrix elements are

$$\begin{aligned}
\rho_{ee} &= \frac{N^2\left(2N+1\right) - \left(2N-1\right)|M|^2}{\left(2N+1\right)\left(3N^2+3N+1-3|M|^2\right)}, \\
\rho_{ss} &= \frac{N\left(N+1\right) - |M|^2}{3N^2+3N+1-3|M|^2}, \\
\rho_{u} &= \frac{2|M|}{\left(2N+1\right)\left(3N^2+3N+1-3|M|^2\right)}.
\end{aligned}$$

(b) Show that in the case of a thermal field with $|M| = 0$ and a classical squeezed field, with the maximal correlations

$|M| = N$, the stationary populations obey a Boltzmann distribution with $\rho_{gg} > \rho_{ss} > \rho_{ee}$.

(c) Show that in the case of a quantum squeezed field with the maximal correlations $|M|^2 = N(N+1)$, the Boltzmann distribution of the populations is violated, that $\rho_{ee} > \rho_{ss}$.

16.5 Using the results of Exercise 16.4, show that in the case of a quantum squeezed field with the maximal correlations $|M|^2 = N(N+1)$, the stationary state of the system is a pure entangled state. Is it maximally or non-maximally entangled state? Under which condition, the state becomes maximally entangled?

Chapter 17

Classical and Quantum Lithography

17.1 Introduction

We have learnt in Chapter 5 how to reduce fluctuations of an electromagnetic (EM) field below the quantum (Heisenberg) limit imposed by the quantum nature of light. It has been shown that the quantum limit can be beaten using non-classical squeezed light of reduced fluctuations. Another issue of significant interest in quantum optics is to beat the diffraction limit imposed on the resolution of measured objects by the wave nature of light. In this chapter, we will illustrate how one can beat the diffraction limit by using the quantum nature of entangled light beams. We shall see that entangled light can lead us to a new domain of quantum optics in which detectors can resolve two closely spaced objects or spectral lines with the minimal resolvable limit significantly reduced or even completely suppressed. This subject is generally known as quantum optical lithography and could be described in short as the ability to print patterns onto certain materials using non-classical light. The developments in this area are of fundamental interest, and they hold promise for advances in optical interferometry and in applications such as quantum metrology and gravitational wave studies.

Quantum Optics for Beginners
Zbigniew Ficek and Mohamed Ridza Wahiddin

ISBN 978-981-4411-75-2 (Hardcover), 978-981-4411-76-9 (eBook)
www.panstanford.com

17.2 Classical Optical Lithography

Optical lithography is a technology for writing features of very small size onto substrates using coherent optical fields. However, the resolution is limited by the Rayleigh diffraction criterion, which states that the minimal resolvable feature size occurs at a spacing corresponding to the distance between an intensity maximum and an adjacent intensity minimum of the diffraction pattern. The Rayleigh diffraction limit is $\Delta x_{\mathrm{min}} = \lambda/2$, where Δx_{min} is the fringe separation and λ is the optical wavelength. This is the best resolution that can be achieved with classical fields. Hence, it has become necessary to use optical fields of very short wavelengths to fabricate smaller objects.

A modification of optical lithography, called *classical interferometric lithography*, involves two coherent plane waves of laser radiation intersecting at an angle 2θ, as shown in Fig. 17.1. A phase shifter (PS) located in one of the two arms introduces a phase difference between the two coherent optical paths producing an interference pattern on the screen S (lithographic plate), with the resolution given by the diffraction formula

$$\Delta x = \frac{\lambda}{2\sin\theta}. \tag{17.1}$$

Then, in the grazing incidence limit of $\theta \to \pi/2$, the minimum resolution is $\Delta x_{\mathrm{min}} = \lambda/2$.

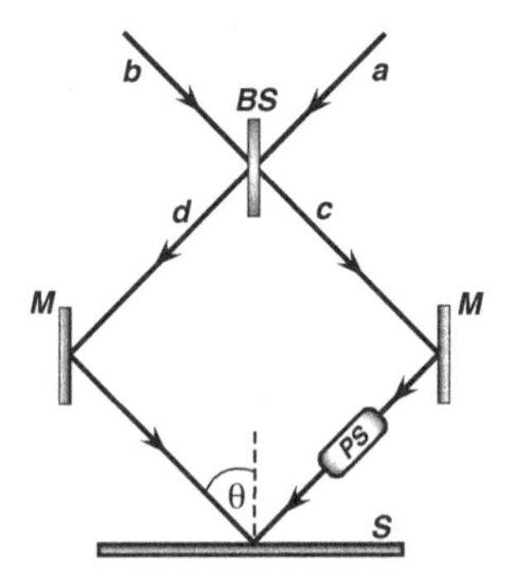

Figure 17.1 Schematic diagram of an interferometric lithography experiment.

17.3 Quantum Lithography

Consider now a quantum description of interference lithography in which we use the photon picture of the incident fields mixed at the 50/50 beam splitter BS. The photons are reflected by the mirrors M onto the screen. We introduce a phase shifter placed in the upper arm which produces a phase shift $\phi = ks$, where $k = 2\pi/\lambda$ and s is the difference between the two optical paths. The input fields, ports a and b, are represented by the annihilation operators $\hat{a}$ and $\hat{b}$ that obey the usual commutation relations

$$[\hat{a}, \hat{a}^{\dagger}] = [\hat{b}, \hat{b}^{\dagger}] = 1, \qquad [\hat{a}, \hat{b}] = 0. \tag{17.2}$$

The output fields c and d are represented by two operators $\hat{c}$ and $\hat{d}$, which are linear combinations of the reflected and transmitted input-field operators as

$$\hat{c} = \frac{1}{\sqrt{2}}\left(i\hat{a} + \hat{b}\right)\mathrm{e}^{i\phi}, \quad \hat{d} = \frac{1}{\sqrt{2}}\left(\hat{a} + i\hat{b}\right), \tag{17.3}$$

where $\exp(i\phi)$ represents the phase difference between the fields impinging on the screen S. Hence, the annihilation operator of the total field detected at the screen S is given by

$$\hat{u} = \hat{c} + \hat{d} = \frac{1}{\sqrt{2}}\left(1 + i\mathrm{e}^{i\phi}\right)\hat{a} + \frac{1}{\sqrt{2}}\left(i + \mathrm{e}^{i\phi}\right)\hat{b}. \tag{17.4}$$

The quantum lithography approach to sub-wavelength resolution is based on n-photon absorption process. The n-photon absorption rate, corresponding to the deposition rate of n photons on the screen, is proportional to the n-order correlation function of the total field operators as

$$A_n = \frac{1}{n!}\left\langle \left(\hat{u}^{\dagger}\right)^n (\hat{u})^n \right\rangle, \tag{17.5}$$

where $\hat{u} = \hat{c} + \hat{d}$, and the average is taken over the initial state of the field.

Let us now illustrate with few examples on how the absorption rate, and then the resulting resolution, depend on the state of the input field.

Example 17.1 (One-photon absorption rate) *Consider first the one-photon absorption rate for an input field in the state*

$$|\Psi_1\rangle = |1\rangle_a|0\rangle_b. \tag{17.6}$$

It is easy to show that in this state the one-photon absorption rate is given by

$$A_1 = \langle \Psi_1 | \hat{u}^\dagger \hat{u} | \Psi_1 \rangle = 1 - \sin\phi, \tag{17.7}$$

which represents an interference pattern that varies with the phase ϕ.

The Rayleigh criterion demands $\phi_{\min} = \pi$ *for the minimal distance between a maximum and an adjacent minimum of the interference pattern, from which we find the minimum fringe spacing* $x_{\min} = \lambda/2$ *(for* $\theta \simeq \pi/2$*). This is the usual classical result, called the single-photon diffraction limit.*

Example 17.2 (Two-photon absorption rate with an input classical state) *Consider now the two-photon absorption rate with an input state*

$$|\Psi_{2c}\rangle = |2\rangle_a |0\rangle_b. \tag{17.8}$$

This state is an example of a so-called two-photon classical state. It gets this name because it is possible to distinguish through which channel the incident photons arrive. We then find

$$\begin{aligned} A_{2c} &= \frac{1}{2!} \langle \Psi_{2c} | \hat{u}^\dagger \hat{u}^\dagger \hat{u} \hat{u} | \Psi_{2c} \rangle \\ &= (1 - \sin\phi)^2 = \frac{3}{2} - 2\sin\phi - \frac{1}{2}\cos 2\phi. \end{aligned} \tag{17.9}$$

We see that there are three terms in the absorption rate. The first is the spatially uniform term 3/2, which is three times larger than desired. The effect of this term is to reduce the contrast of the fringe pattern. The second term is the same as in the one-photon absorption rate. This is an unwanted term since it will mask the effect of the third term $\cos 2\phi$ *that oscillates in space with twice the frequency as the one-photon absorption rate. This dependence leads to narrower interference fringes than that for the one-photon absorption rate.*

The two-photon absorption rate can exhibit even narrower features with the fringe spacing reduced by a factor of two if the input state is a quantum state.

Example 17.3 (Two-photon absorption rate with an input quantum state) *Consider the following a two-photon input state*

$$|\Psi_{2q}\rangle = |1\rangle_a |1\rangle_b. \tag{17.10}$$

The state $|\Psi_{2q}\rangle$ *is an example of a* quantum *state—because there it possesses a photon in each channel, so that when a photon is detected, one cannot say from which channel it originated. With the input state* $|\Psi_{2q}\rangle$*, the two-photon absorption rate takes the form*

$$A_{2q} = \frac{1}{2!}\langle\Psi_{2q}|\hat{u}^{\dagger}\hat{u}^{\dagger}\hat{u}\hat{u}|\Psi_{2q}\rangle = 1 + \cos 2\phi. \qquad (17.11)$$

Comparing Eq. (17.11) with Eq. (17.9), we see that in the case of the input quantum field, the slowly oscillating term $\sin\phi$ *has been eliminated, and we are left with only the* $\cos 2\phi$ *term, giving the minimum resolution* $x_{\min} = \lambda/4$.

In Fig. 17.2, we plot the absorption rates as a function of the phase shift ϕ. It is evident that the classical two-photon excitation pattern has the same period but is narrower than the one-photon excitation pattern, and the quantum two-photon excitation pattern has a

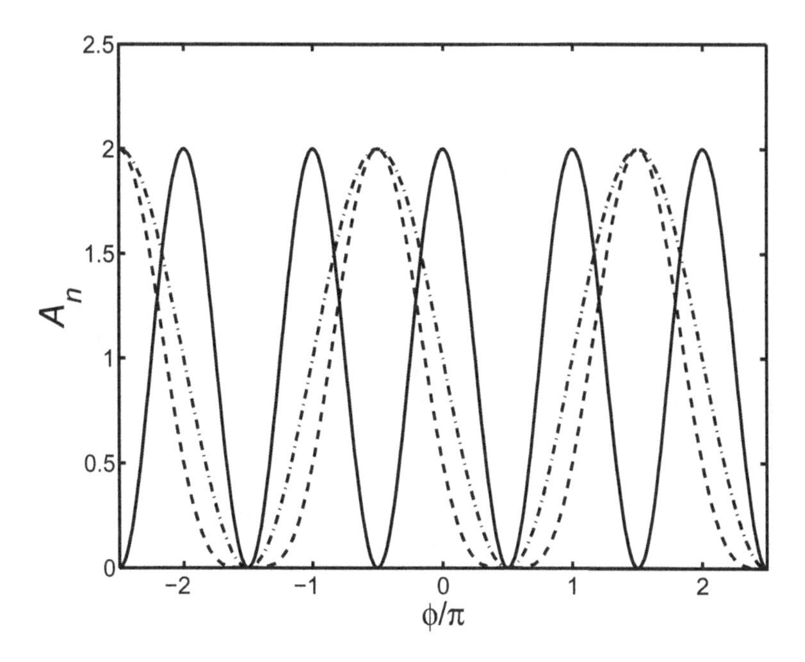

Figure 17.2 The absorption rates A_n as a function of the phase shift ϕ with one-photon excitation A_1 (dashed-dotted line), uncorrelated two-photon excitation $A_{2c}/2$ (dashed line) and quantum (entangled) two-photon excitation A_{2q} (solid line).

period half that of the corresponding classical interference pattern. A spatial interference pattern with the modulation period reduced by a factor of 2 has been observed experimentally by D'Angelo *et al.* [100]. This unusual property of the interference pattern is a consequence of the division of the two photons between the two channels in the input state $|\Psi_{2q}\rangle$.

The resolution can even be improved beyond the $\lambda/4$ limit if an n-photon correlated quantum field is used. It has been demonstrated experimentally by the Steinberg's group at University of Toronto who achieved resolution to the limit of $\lambda/6$ with a quantum three-photon correlated field [101], and the Zeilinger's group at Vienna University [102], who achieved resolution of order of $\lambda/8$ with a quantum four-photon correlated field.

17.4 Summary

Quantum lithography has a very good chance of becoming implemented in practice, especially as new sensitive multi-photon lithographic materials are being developed [103, 104]. We have also seen that classically simulated quantum lithography may be a realistic alternative approach which could be much more readily implemented [105, 106]. For example, Kiffner *et al.* [107] have proposed to study spectral resolution in terms of the coherent population trapping (CPT) rather than in terms of non-classical n-photon absorption. The advantage of the CPT is that the sub-wavelength resolution problem can be studied in terms of classical rather than quantum fields. Other interesting schemes of classically simulated quantum lithography include methods where sub-wavelength resolution was achieved by correlating wave vector and frequency in a narrow-band multi-photon detection process [108, 109], or inducing the multi-Rabi oscillation between energy levels of a two-level atom [110].

Exercises

17.1 The Rayleigh diffraction limit $\Delta x_{\min} = \lambda/2$ is proportional to the wavelength of light used. Explain, why it is not a common practice to increase the resolution simply by using light of a shorter wavelength?

17.2 Suppose, two plane waves characterized by wave vectors $\vec{k}_1$ and $\vec{k}_2$ fall on a surface at an angle θ from the normal to the surface. The wave vectors are given by

$$\vec{k}_1 = k(\sin\theta, \cos\theta), \quad \vec{k}_2 = k(-\sin\theta, \cos\theta).$$

(a) Find the interference pattern in the intensity $I(\vec{r})$ of the two plane waves at position $\vec{r}$ on the surface. Assume that the plane waves have the same amplitudes.

(b) Find the Rayleigh limit for the resolution of a feature size Δx on the surface oriented in the direction x.

(c) Show that the classical diffraction limit is $\Delta x = \lambda/4$ that occurs at grazing incidence of the waves, where λ is the wavelength of the incident waves.

17.3 Find the absorption rate for an input field being in the state

$$|\Psi_0\rangle = |n\rangle_a |m\rangle_b,$$

where n and m are the numbers of photons in the modes a and b, respectively. Determine the properties of the absorption rate in the cases of $n = m$ and $n \gg m$.

17.4 Evaluate the absorption rate of the output field when a pair of input beams in the quantum state $|\Psi_{2q}\rangle = |1\rangle_a |1\rangle_b$ is fed into an imperfect beam splitter with unequal reflection and transmission coefficients. Show that in this case the absorption rate contains both the slowly and rapidly oscillating terms, and explain what is the meaning of this result.

Chapter 18

Laser Model in the High-Q Limit

18.1 Introduction

In this chapter, we turn our considerations to the simplest model of a laser that includes all the essential features of any practical laser. The one-photon losses (due to coupling to the environment), linear gain (due to inverted atoms) and saturation (due to nonlinear loss), that are experienced by a single-mode (laser) cavity are all included in the calculations and their role in the laser properties is fully explored. We point out that all lasers must have these three elements to operate as a laser. If there is only a linear gain, then of course there can be no output, since an output causes a loss. With both gain and loss, but not saturation, the laser intensity is either near zero (well below threshold), or else it rises infinitely (well above threshold). Neither case is very useful as a model of a real laser, which has a finite output above threshold. This is why a saturation mechanism is needed, such as an intensity dependent nonlinear loss. This simple model used here is then obtainable in the limit of a high-Q laser operating not too far above threshold.

Quantum Optics for Beginners
Zbigniew Ficek and Mohamed Ridza Wahiddin

ISBN 978-981-4411-75-2 (Hardcover), 978-981-4411-76-9 (eBook)
www.panstanford.com

18.2 Master Equation

The laser model we consider here is based on the procedure of converting the master equation of the reduced density operator of the cavity mode into a Fokker–Planck equation (FPE) for the P representation [111, 112]. The equation is equivalent to a set of coupled stochastic differential equations (SDE) for the cavity field amplitude, which we solve for the steady-state and for two-time correlation functions using the phase diffusion model. These solutions will give information on the laser intensity and linewidth.

Consider a single-mode cavity containing an ensemble of atoms that decay to stationary states much faster than the cavity field. In this case, we may adiabatically eliminate the atomic variables and arrive to the reduced density operator of the cavity field alone. Including the linear loss, gain and nonlinear loss of the cavity mode, we can model the system (laser) by the following master equation

$$\begin{aligned}\frac{\partial}{\partial t}\rho = &-\frac{1}{2}\Gamma\left(\hat{a}^{\dagger}\hat{a}\rho + \rho\hat{a}^{\dagger}\hat{a} - 2\hat{a}\rho\hat{a}^{\dagger}\right)\\ &-\frac{1}{2}G\left(\hat{a}\hat{a}^{\dagger}\rho + \rho\hat{a}\hat{a}^{\dagger} - 2\hat{a}^{\dagger}\rho\hat{a}\right)\\ &-\frac{1}{2}\kappa\left(\hat{a}^{\dagger 2}\hat{a}^{2}\rho + \rho\hat{a}^{\dagger 2}\hat{a}^{2} - 2\hat{a}^{2}\rho\hat{a}^{\dagger 2}\right),\end{aligned} \tag{18.1}$$

where Γ is a linear loss due to spontaneous emission, G is linear gain and κ is nonlinear loss.

Transforming the master equation into the FPE for the Glauber–Sudarshan P representation, we obtain

$$\begin{aligned}\frac{\partial}{\partial t}P(\alpha, t) = &\left\{\frac{\partial}{\partial\alpha}\left[\frac{1}{2}(\Gamma - G)\alpha + \kappa|\alpha|^{2}\alpha\right]\right.\\ &\left.+\frac{\partial}{\partial\alpha^{*}}\left[\frac{1}{2}(\Gamma - G)\alpha^{*} + \kappa|\alpha|^{2}\alpha^{*}\right] + \frac{\partial^{2}}{\partial\alpha\partial\alpha^{*}}G\right\}P(\alpha, t).\end{aligned} \tag{18.2}$$

This FPE is equivalent to a set of coupled SDE for the field amplitudes.

18.3 Stochastic Differential Equations

The diffusion matrix D and the matrix B corresponding to the FPE, given in Eq. (18.2), are of the form

$$D = \begin{pmatrix} 0 & G \\ G & 0 \end{pmatrix}, \qquad B = \sqrt{\frac{G}{2}} \begin{pmatrix} 1 & i \\ 1 & -i \end{pmatrix}. \tag{18.3}$$

Using the Ito rule, we can write the following SDE

$$\begin{aligned} \frac{d\alpha}{dt} &= -\frac{1}{2}(\Gamma - G)\,\alpha - \kappa|\alpha|^2\alpha + F\,(t), \\ \frac{d\alpha^*}{dt} &= -\frac{1}{2}(\Gamma - G)\,\alpha^* - \kappa|\alpha|^2\alpha^* + F^*\,(t), \end{aligned} \tag{18.4}$$

where $F(t)$ and $F^*(t)$ are the noise terms corresponding to the complex conjugate variables α and α^*, respectively. The noise terms are given by

$$\begin{aligned} F\,(t) &= \sqrt{\frac{G}{2}}\,(\psi_1\,(t) + i\,\psi_2\,(t)), \\ F^*\,(t) &= \sqrt{\frac{G}{2}}\,(\psi_1\,(t) - i\,\psi_2\,(t)), \end{aligned} \tag{18.5}$$

where $\psi_1\,(t)$ and $\psi_2\,(t)$ are the independent real noise terms.

F^* is the complex conjugate of F, and therefore α and α^* remain complex conjugate. Thus, there is no the problem of employing the Glauber–Sudarshan P representation to this model. Note that this is only true because we have neglected in the FPE the diagonal diffusion coefficients due to the nonlinear loss.

Using the usual correlation properties (12.79) of the independent noise terms, we find that the only non-zero correlations of the noise terms are

$$\langle F(t)F^*(t')\rangle = \langle F^*(t)F(t')\rangle = G\delta(t - t'). \tag{18.6}$$

Following the above developments, one might ask a question: Why do we need the two-photon absorption term in the master equation?

From the SDE (18.4), when we put $\kappa = 0$ and ignore the noise terms, we obtain a simple equation of motion for α:

$$\frac{d\alpha}{dt} = -\frac{1}{2}(\Gamma - G)\,\alpha. \tag{18.7}$$

Its solution is

$$\alpha(t) = \alpha(0)\, e^{-\frac{1}{2}(\Gamma - G)t}. \tag{18.8}$$

We see that for $(\Gamma - G) > 0$ (below threshold)

$$\alpha(t) \to 0 \qquad \text{when} \qquad t \to \infty, \tag{18.9}$$

whereas for $(\Gamma - G) < 0$ (above threshold)

$$\alpha(t) \to \infty \qquad \text{when} \qquad t \to \infty. \tag{18.10}$$

Lasers operate in the steady-state regime and have a finite intensity, thus we need a nonlinear absorption process to get laser systems into a stable and finite steady-state intensity.

18.4 Semiclassical Steady-State Solution and Stability

The semiclassical steady-state solutions of the SDE follow immediately from Eq. (18.4). In the semiclassical limit the noise terms in the SDE are dropped ($F(t) = 0$) on the assumption that quantum fluctuations are small. Then, taking the steady-state limit ($d\alpha/dt = 0$), we readily find

$$\left[\kappa\, |\alpha_s|^2 + \frac{1}{2}(\Gamma - G)\right] \alpha_s = 0, \tag{18.11}$$

where α_s is the steady-state value of α. Equation (18.11) has two solutions

$$\begin{aligned} &1. \qquad \alpha_s = 0, \\ &2. \qquad |\alpha_s|^2 = \frac{G - \Gamma}{2\kappa}. \end{aligned} \tag{18.12}$$

The second solution is physically meaningful for $G - \Gamma > 0$ (above threshold), but contains no information about the steady-state phase ϕ_s of the complex amplitude $\alpha_s = |\alpha_s| \exp(i\phi_s)$.

Are the steady-state solutions stable?

We use the linearization technique to find whether the steady-state solutions are stable. This means, we take

$$\alpha(t) = \alpha_s + \delta\alpha(t), \tag{18.13}$$

and substitute into Eq. (18.4). We keep terms linear in $\delta\alpha$, and put the equations of motion for $\delta\alpha\,(t)$ and $\delta\alpha^*\,(t)$ in a matrix form

$$\frac{d}{dt}\begin{pmatrix}\delta\alpha\,(t)\\ \delta\alpha^*\,(t)\end{pmatrix} = -A'\begin{pmatrix}\delta\alpha\,(t)\\ \delta\alpha^*\,(t)\end{pmatrix} + \text{noise}. \qquad (18.14)$$

Next, we find the eigenvalues of the matrix A'.

The steady-state is stable if *all* the eigenvalues λ_i of the matrix A' have positive real parts. If there is a negative real part of λ_i or $\lambda_i = 0$, then the steady-state is unstable.

Consider the case 1. $\alpha_s = 0$.

Here, $\alpha\,(t) = \delta\alpha\,(t)$, and then Eq. (18.4) reduces to

$$\begin{aligned}\frac{d\delta\alpha}{dt} &= -\frac{1}{2}\,(\Gamma - G)\,\delta\alpha + \text{noise},\\ \frac{d\delta\alpha^*}{dt} &= -\frac{1}{2}\,(\Gamma - G)\,\delta\alpha^* + \text{noise}.\end{aligned} \qquad (18.15)$$

Thus, the matrix A' is of the form

$$A' = \begin{pmatrix}\frac{1}{2}\,(\Gamma - G) & 0\\ 0 & \frac{1}{2}\,(\Gamma - G)\end{pmatrix}, \qquad (18.16)$$

and its eigenvalues are

$$\lambda_{1,2} = \frac{1}{2}\,(\Gamma - G)\,. \qquad (18.17)$$

It is clear that the eigenvalues are positive for $\Gamma > G$. Thus, the steady state is stable for $\Gamma > G$ (below threshold).

Consider the case 2. $|\alpha_s|^2 = (G - \Gamma)/(2\kappa)$.

Here, Eq. (18.4) reduces to

$$\begin{aligned}\frac{d\delta\alpha}{dt} &= -\left[\frac{1}{2}\,(\Gamma - G) + 2\kappa\,|\alpha_s|^2\right]\delta\alpha - \kappa\alpha_s^2\delta\alpha^* + \text{noise},\\ \frac{d\delta\alpha^*}{dt} &= -\left[\frac{1}{2}\,(\Gamma - G) + 2\kappa\,|\alpha_s|^2\right]\delta\alpha^* - \kappa\alpha_s^{*2}\delta\alpha + \text{noise},\end{aligned} \qquad (18.18)$$

from which it can be found that the matrix A' is of the form

$$A' = \begin{pmatrix}\frac{1}{2}\,(\Gamma - G) + 2\kappa\,|\alpha_s|^2 & \kappa\alpha_s^2\\ \kappa\alpha_s^{*2} & \frac{1}{2}\,(\Gamma - G) + 2\kappa\,|\alpha_s|^2\end{pmatrix}. \qquad (18.19)$$

The eigenvalues of A' are

$$\begin{aligned}\lambda_1 &= 0,\\ \lambda_2 &= 2\kappa\,|\alpha_s|^2\,.\end{aligned} \qquad (18.20)$$

We see that one of the eigenvalues is equal to zero. This means that the steady-state solution is unstable. Thus, above threshold the system cannot be analysed correctly by the assumption of small fluctuations and using the linearization techniques. As we will show below, this instability is physically due to the phenomenon of phase diffusion.

18.5 Exact Steady-State Solution

In order to analyse the problem more carefully, we rewrite the FPE in terms of intensity and phase variables

$$I = \alpha\alpha^*, \qquad \phi = \frac{1}{2i}\ln\frac{\alpha}{\alpha^*}. \tag{18.21}$$

Using the polar coordinates ($\alpha = r\exp(i\phi)$), we get

$$\begin{aligned}\frac{\partial}{\partial\alpha} &= \mathrm{e}^{-i\phi}\frac{1}{2}\left(\frac{\partial}{\partial r} - i\frac{1}{r}\frac{\partial}{\partial\phi}\right),\\ \frac{\partial}{\partial\alpha^*} &= \mathrm{e}^{i\phi}\frac{1}{2}\left(\frac{\partial}{\partial r} + i\frac{1}{r}\frac{\partial}{\partial\phi}\right).\end{aligned} \tag{18.22}$$

Hence,

$$\begin{aligned}\frac{\partial}{\partial\alpha}\alpha &= \left(\frac{\partial}{\partial\alpha^*}\alpha^*\right)^* = \frac{\partial}{\partial I}I + \frac{1}{2i}\frac{\partial}{\partial\phi},\\ \frac{\partial^2}{\partial\alpha\partial\alpha^*} &= I\frac{\partial^2}{\partial I^2} + \frac{\partial}{\partial I} + \frac{1}{4I}\frac{\partial^2}{\partial\phi^2},\end{aligned} \tag{18.23}$$

and then the FPE takes the form

$$\begin{aligned}\frac{\partial}{\partial t}P(I,\phi) = \Bigg\{&\frac{\partial}{\partial I}\left[(\Gamma - G)I + 2\kappa I^2 - G\right]\\ &+ \frac{1}{2}\frac{\partial^2}{\partial I^2}(2GI) + \frac{1}{2}\frac{\partial^2}{\partial\phi^2}\left(\frac{G}{2I}\right)\Bigg\}P(I,\phi).\end{aligned} \tag{18.24}$$

The normalization condition will change to

$$\begin{aligned}\int d^2\alpha P(\alpha,\alpha^*) &= \int_0^{2\pi}d\phi\int_0^\infty dr rP\left(r\mathrm{e}^{i\phi}, r\mathrm{e}^{-i\phi}\right)\\ &= \frac{1}{2}\int_0^{2\pi}d\phi\int_0^\infty dI\,P(I,\phi) = 1,\end{aligned} \tag{18.25}$$

where $I = r^2$.

The above FPE is still exact and can be solved for the steady state. Since the I and ϕ terms are independent of each other, we look into the solution of the form

$$P_{ss}(I, \phi) = P_{ss}(I)\, P_{ss}(\phi)\,. \tag{18.26}$$

First, consider the part $P_{ss}(I) \equiv P_s$. This part satisfies a differential equation

$$\frac{\partial}{\partial t} P_s = \left\{ \frac{\partial}{\partial I}(-AP_s) + \frac{1}{2}\frac{\partial^2}{\partial I^2}(DP_s) \right\}, \tag{18.27}$$

where

$$\begin{aligned} A &= (G - \Gamma)\, I - 2\kappa I^2 + G, \\ D &= 2GI. \end{aligned} \tag{18.28}$$

In the steady state, $\partial P_s/\partial t = 0$, and then the FPE simplifies to

$$\frac{\partial}{\partial I}\left\{ -AP_s + \frac{1}{2}\frac{\partial}{\partial I}(DP_s) \right\} = 0. \tag{18.29}$$

First integration gives

$$\frac{\partial}{\partial I} DP_s = 2AP_s + \text{const}|_0^\infty. \tag{18.30}$$

If AP_s and $\frac{\partial}{\partial I} DP_s$ vanish at infinity, the constant is zero, and then we can write

$$\frac{1}{DP_s}\frac{d}{dI}(DP_s) = \frac{2A}{D}, \tag{18.31}$$

whose solution is

$$P_s = \mathcal{N}\frac{1}{D}\exp\left[2\int dI \frac{A}{D}\right]. \tag{18.32}$$

Since

$$\frac{A}{D} = \frac{1}{2}\frac{(G-\Gamma)}{G} - \frac{\kappa I}{G} + \frac{1}{2I}, \tag{18.33}$$

and assuming that $I \gg 1$, we obtain

$$\int \frac{A}{D} dI = \frac{1}{2}\frac{(G-\Gamma)}{G} I - \frac{1}{2}\frac{\kappa I^2}{G}. \tag{18.34}$$

Hence,

$$P_s = \mathcal{N}\exp\left[\frac{(G-\Gamma)}{G} I - \frac{\kappa}{G} I^2\right]. \tag{18.35}$$

Note that the phase part $P_{ss}(\phi)$ does not contain a drift term A. Thus, the solution for $P_{ss}(\phi)$ is a constant. As a consequence, the distribution $P_{ss}(I, \phi)$ depends only on the intensity I, not on the phase ϕ.

Example 18.1 (Exact value of the phase distribution) *The exact value of the phase distribution $P(\phi) = \Phi(\phi)$ can be found from the normalization condition*

$$\frac{1}{2}\int_0^{2\pi} d\phi \int_0^{\infty} dI\, P(I, \phi) = 1, \tag{18.36}$$

from which we have

$$\Phi(\phi) = \frac{1}{2}\int_0^{\infty} dI\, P(I, \phi), \tag{18.37}$$

and

$$\int_0^{2\pi} d\phi\, \Phi(\phi) = 1. \tag{18.38}$$

Since $\Phi(\phi)$ does not depend on ϕ, we obtain

$$\Phi(\phi) = \frac{1}{2\pi}. \tag{18.39}$$

This result reflects the fact that the phase of the cavity field is uniformly distributed over 2π and does not have a well-defined value, which is due to the phase diffusion.

The intensity distribution function (18.35) exhibits a peak corresponding to the most probable value of the intensity. The peak is located at $I = 0$ in the below-threshold regime ($\Gamma > G$), and at

$$I\left[\frac{(G - \Gamma)}{G} - \frac{\kappa}{G} I\right] \tag{18.40}$$

in the above-threshold regime, from which we find

$$I = \frac{G - \Gamma}{\kappa}. \tag{18.41}$$

The distribution function can be applied to calculate one-time correlation functions. For example, we can calculate the average number of photons

$$\langle \hat{a}^{\dagger}\hat{a}\rangle = \langle \alpha^{*}\alpha\rangle = \frac{1}{2}\int_0^{2\pi} d\phi \int_0^{\infty} dI\, I\, P_s(I, \phi) = 2\pi \int_0^{\infty} dI\, I\, P_1(I), \tag{18.42}$$

where $P_1(I) = P_s(I, \phi)/2$.

Integrating by parts, or approximately we can assume that $\langle \hat{a}^\dagger \hat{a} \rangle$ is maximal at the peak of $P_1(I)$, and obtain

$$\langle \hat{a}^\dagger \hat{a} \rangle \approx \frac{G - \Gamma}{2\kappa}. \tag{18.43}$$

Below the threshold, the term $(\Gamma - G)/G$ dominates in $P_1(I)$ (we can ignore I^2 term), and integrating by parts, we get

$$\langle \hat{a}^\dagger \hat{a} \rangle \approx \frac{G}{\Gamma - G}. \tag{18.44}$$

18.6 Laser Linewidth

The steady-state solution for $P(I, \phi)$ can provide exact results for one-time (stationary) correlation functions, but it does not solve the time-dependent problems, which are important to calculate the spectrum of the field. The spectrum provides the information about the laser linewidth. In order to evaluate the spectrum of the cavity field, we must consider the full spectrum, with the noise terms SDE. We will consider separately the below and above-threshold cases.

18.6.1 *Below Threshold*

In the below-threshold regime, the linearized approach to the fluctuations is valid, as the system is stable. Since

$$\alpha = \alpha_0 + \delta\alpha, \tag{18.45}$$

and below threshold $\alpha_0 = 0$, we find from the SDE the following equations of motion

$$\begin{aligned} \frac{d}{dt}\delta\alpha &= -\frac{1}{2}(\Gamma - G)\,\delta\alpha + F(t), \\ \frac{d}{dt}\delta\alpha^* &= -\frac{1}{2}(\Gamma - G)\,\delta\alpha^* + F^*(t), \end{aligned} \tag{18.46}$$

where the noise terms satisfy the relation

$$\langle F(t)F^*(t')\rangle = G\delta(t - t'). \tag{18.47}$$

Integrating the equation for $\delta\alpha$, we obtain

$$\delta\alpha(t) = \int_{-\infty}^{t} dt' e^{-\frac{1}{2}(\Gamma - G)(t-t')} F(t'). \tag{18.48}$$

This solution allows us to calculate the two-time correlation function

$$\langle\delta\alpha^*(t)\delta\alpha(t+\tau)\rangle. \tag{18.49}$$

Its Fourier transform is the spectrum of the fluctuations.

$$\begin{aligned}
\langle\delta\alpha^*(t)\delta\alpha(t+\tau)\rangle &= \int_{-\infty}^{t} dt' \int_{-\infty}^{t+\tau} dt'' \mathrm{e}^{-\frac{1}{2}(\Gamma-G)(t-t')}\mathrm{e}^{-\frac{1}{2}(\Gamma-G)(t+\tau-t'')} \\
&\quad \times \left\langle F^*(t'F(t'')\right\rangle \\
&= G\mathrm{e}^{-\frac{1}{2}(\Gamma-G)(2t+\tau)} \int_{-\infty}^{t} dt' \int_{-\infty}^{t+\tau} dt'' \mathrm{e}^{\frac{1}{2}(\Gamma-G)(t'+t'')}\delta(t'-t'') \\
&= G\mathrm{e}^{-\frac{1}{2}(\Gamma-G)(2t+\tau)} \int_{-\infty}^{t} dt' \mathrm{e}^{(\Gamma-G)t'} = G\mathrm{e}^{-\frac{1}{2}(\Gamma-G)(2t+\tau)} \frac{1}{\Gamma-G}\mathrm{e}^{(\Gamma-G)t} \\
&= \frac{G}{\Gamma-G}\mathrm{e}^{-\frac{1}{2}(\Gamma-G)|\tau|}.
\end{aligned} \tag{18.50}$$

Thus, the spectrum of the fluctuations (intensity spectrum) is given by

$$\begin{aligned}
I(\omega) &= \int_{-\infty}^{\infty} d\tau \left\langle \tilde{a}^\dagger(t)\tilde{a}(t+\tau)\right\rangle \mathrm{e}^{i(\omega-\omega_c)\tau} \\
&= \frac{G}{(\omega-\omega_c)^2 + \frac{1}{4}(\Gamma-G)^2},
\end{aligned} \tag{18.51}$$

where ω_c is the cavity central frequency, and

$$\tilde{\hat{a}}(t) = \hat{a}(t)\mathrm{e}^{i\omega_c t}, \quad \tilde{\hat{a}}^\dagger(t) = \hat{a}^\dagger(t)\mathrm{e}^{-i\omega_c t} \tag{18.52}$$

are slowly varying parts of the field operators. We see that the intensity spectrum is a Lorentzian with half width $(\Gamma - G)/2$. Since $\Gamma \gg G$, the with is approximately equal to the cavity linewidth.

18.6.2 *Above Threshold*

To analyse the spectrum in the above-threshold regime, we use the FPE for $P(I, \phi)$, from which, using the Ito rule, we get the following SDE

$$\frac{dI}{dt} = -(\Gamma - G)I - 2\kappa I^2 + G + \sqrt{2GI}\xi_I(t), \tag{18.53}$$

$$\frac{d\phi}{dt} = \sqrt{\frac{G}{2I}}\xi_\phi(t), \tag{18.54}$$

where

$$\left\langle\xi_I^*(t)\xi_I(t')\right\rangle = \left\langle\xi_\phi(t)\xi_\phi(t')\right\rangle = \delta(t-t'). \tag{18.55}$$

Note that the equations for I and ϕ are decoupled from each other.

We first check if we can use the linearized theory for I. For this, we look at the steady state of Eq. (18.53):

$$2\kappa I_0^2 - (G - \Gamma)\, I_0 - G = 0, \tag{18.56}$$

which has two solutions

$$I_{0(1,2)} = \frac{G - \Gamma}{4\kappa} \pm \frac{1}{4\kappa}\sqrt{(G - \Gamma)^2 + 8G\kappa}. \tag{18.57}$$

Since we can accept only the positive solution, we see that in the steady state, $I_0 > (G - \Gamma)/(4\kappa)$.

We now check if the steady state is stable. In terms of the linearized theory we can write $I = I_0 + \delta I$, and then we obtain from Eq. (18.53)

$$\frac{d\delta I}{dt} = (G - \Gamma)\, \delta I - 4\kappa I_0 \delta I + \text{noise}, \tag{18.58}$$

or equivalently

$$\frac{d\delta I}{dt} = -\left[4\kappa I_0 - (G - \Gamma)\right] \delta I + \text{noise}. \tag{18.59}$$

Since $I_0 > (G - \Gamma)/(4\kappa)$, we see that the coefficient at δI is positive, so the steady-state solution is stable.

We will treat the intensity equation within the linearized theory, but will solve the phase equation exactly.

Well above the threshold, we may assume that the fluctuations are small compared to I_0, and then replace I by I_0. We can write the phase equation as

$$\frac{d\phi}{dt} = F_\phi^0\,(t), \tag{18.60}$$

where

$$F_\phi^0\,(t) = \sqrt{G/(2I_0)}\xi_\phi(t) = \sqrt{D_\phi}\xi_\phi(t). \tag{18.61}$$

The phase equation is equivalent to the FPE of the form of a diffusion equation

$$\frac{\partial}{\partial t}\Phi\,(\phi, t) = \frac{1}{2} D_\phi \frac{\partial^2}{\partial \phi^2}\Phi\,(\phi, t), \tag{18.62}$$

where

$$D_\phi = \frac{G}{2I_0}. \tag{18.63}$$

We now solve the diffusion equation

$$\frac{\partial}{\partial t}\Phi = D\frac{\partial^2}{\partial \phi^2}\Phi, \tag{18.64}$$

where $D = D_\phi/2$. The equation is an analogue of the harmonic oscillator equation, and will find the solution of this equation in terms of harmonics of ϕ:

$$\Phi = \sum_m C_m(t)e^{im\phi}. \tag{18.65}$$

Substituting Eq. (18.65) into Eq. (18.64), we obtain

$$\dot{C}_m(t) = -Dm^2C_m(t). \tag{18.66}$$

Its solution is

$$C_m(t) = C_m(0)e^{-Dm^2t}. \tag{18.67}$$

We choose initial value for C_m as

$$C_m(0) = \frac{1}{2\pi}e^{-im\phi_0}. \tag{18.68}$$

This initial value ensures that the function Φ will be periodic such that $\Phi(\phi + 2\pi) = \Phi(\phi)$. Then,

$$\Phi(\phi, t) = \frac{1}{2\pi}\sum_m e^{im(\phi-\phi_0)}e^{-Dm^2t}. \tag{18.69}$$

Non-zero steady state is only at $m = 0$:

$$\Phi(\phi, t) = \frac{1}{2\pi}, \tag{18.70}$$

which agrees with the earlier exact result.

Using the approximation of the stable intensity ($I \approx I_0$), and Eq. (18.60), we can now calculate the two-time correlation function $\langle \hat{a}^\dagger(t)\hat{a}(t+\tau)\rangle$ and find the intensity spectrum of the laser operating well above threshold. Replacing I by I_0, we obtain for the correlation function

$$\begin{aligned}\langle \hat{a}^\dagger(t)\hat{a}(t+\tau)\rangle &= \left\langle \sqrt{I(t)I(t+\tau)}e^{-i[\phi(t+\tau)-\phi(t)]}\right\rangle \\ &\approx I_0\left\langle e^{-i[\phi(t+\tau)-\phi(t)]}\right\rangle.\end{aligned} \tag{18.71}$$

To evaluate the average exponent, we use the solution of the Ito equation (18.60) for the phase

$$\phi(t) = \phi(t_0) + \int_{t_0}^{t} dt' F_\phi^0(t'). \tag{18.72}$$

Since $F(t)$ has statistical properties of a Gaussian variable, we can use the relation for the Gaussian processes, that for a Gaussian variable x such that $\langle x \rangle = 0$, higher order moments factorize according to

$$\begin{aligned} \langle x^{2n+1} \rangle &= 0, \\ \langle x^{2n} \rangle &= \frac{(2n)!}{2^n n!} \langle x^2 \rangle^n. \end{aligned} \tag{18.73}$$

This allows us to write

$$\langle e^x \rangle = e^{\frac{1}{2}\langle x^2 \rangle}, \tag{18.74}$$

which leads to the following result for the correlation function

$$\langle \hat{a}^\dagger(t)\hat{a}(t+\tau) \rangle = I_0 \left\langle e^{-i[\phi(t+\tau)-\phi(t)]} \right\rangle = I_0 e^{-\frac{1}{2}\langle [\phi(t+\tau)-\phi(t)]^2 \rangle}. \tag{18.75}$$

Since

$$\phi(t+\tau) = \int_{t_0}^{t+\tau} dt' F_\phi^0(t'), \tag{18.76}$$

and

$$\phi(t) = \int_{t_0}^{t} dt' F_\phi^0(t'), \tag{18.77}$$

we obtain

$$\phi(t+\tau) - \phi(t) = \int_{t}^{t+\tau} dt' F_\phi^0(t'). \tag{18.78}$$

Hence

$$\begin{aligned} \left\langle [\phi(t+\tau) - \phi(t)]^2 \right\rangle &= \int dt' \int dt'' \left\langle F_\phi^0(t') F_\phi^0(t'') \right\rangle \\ &= D_\phi \int_t^{t+\tau} dt' = D_\phi \tau. \end{aligned} \tag{18.79}$$

Finally, we arrive at the following result

$$\langle \hat{a}^\dagger(t)\hat{a}(t+\tau) \rangle = I_0 e^{-\frac{1}{2}D_\phi \tau}. \tag{18.80}$$

The laser intensity spectrum above threshold is therefore

$$I(\omega) = \frac{I_0 D_\phi}{(\omega - \omega_c)^2 + \frac{1}{4}D_\phi^2}, \tag{18.81}$$

where the diffusion coefficient is

$$D_\phi = G/(2I_0). \tag{18.82}$$

The intensity spectrum is a Lorentzian centred at ω_c with half width $D_\phi/2$. The width is essentially due to the phase diffusion, and becomes narrowed with increasing I_0.

We can introduce the characteristic phase correlation time

$$t_c \sim \frac{1}{\text{linewidth}} = \frac{2}{D_\phi} = \frac{4I_0}{G}. \tag{18.83}$$

For large I_0, the coherence time is very long indicating that the laser operating well above threshold produces essentially coherent light. In other words, well above threshold the laser can maintain the value of its phase within a sufficiently long time period, since the phase diffusion is slow (D_ϕ is small).

18.7 Summary

We have the following important results for the high-Q laser model:

(1) Above threshold, the steady-state solutions are stable when the nonlinear losses are included.
(2) Below threshold, the intensity spectrum is a Lorentzian and its linewidth is approximately equal to the cavity bandwidth.
(3) Above threshold, the intensity spectrum is a Lorentzian and its linewidth is essentially due to the phase diffusion.
(4) Above threshold, the laser produces essentially coherent light.

Exercises

18.1 Two non-degenerate frequency and parametrically coupled field modes a and b satisfy the following Heisenberg equations of motion

$$\dot{a}^\dagger = -\gamma a^\dagger + \frac{1}{2}iGb + \sqrt{\gamma}\,\xi^\dagger(t),$$
$$\dot{b} = -(\gamma - i\Delta)\,b - \frac{1}{2}iGa^\dagger + \sqrt{\gamma}\,\xi(t),$$

where Δ is the detuning between the frequencies of the modes, $\xi(t)$ is the noise term, γ and G are (real) damping and mode coupling parameters, respectively.

(a) Under what conditions the modes decay to a stable steady state?

(b) What would be the stability conditions if the modes were degenerate in the frequency ($\Delta = 0$)?

18.2 Using the linearized theory and assuming that the noise satisfied the Gaussian statistics, calculate the normalized second-order correlation function $g^{(2)}(t) = \langle \hat{a}^\dagger(t)\hat{a}^\dagger(t)\hat{a}(t)\hat{a}(t)\rangle / \langle \hat{a}^\dagger(t)\hat{a}(t)\rangle^2$ below and above threshold.

Chapter 19

Input–Output Theory

19.1 Introduction

An obvious question is, how to relate the internal cavity mode to externally measured fields. Normally, the measuring instruments are external to the cavity. In this case, the input–output formalism is useful. Strictly speaking, the input–output results are approximations, and can be obtained from considering the boundary conditions at lossy cavity mirrors. In this chapter, we consider the input–output formalism for a single-sided cavity in which the cavity mode is coupled to an external multi-mode environment through a partially transparent broadband mirror.

19.2 Input–Output Relation

In the input–output formalism, we may consider the modes external to the cavity as a set of modes $\hat{b}_k$ separate from the internal cavity mode $\hat{a}$.

The external modes can be divided into two sets of the so-called input and output modes, as illustrated in Fig. 19.1. It is usual to take the infinite-volume limit of these modes, which therefore have

Quantum Optics for Beginners
Zbigniew Ficek and Mohamed Ridza Wahiddin

ISBN 978-981-4411-75-2 (Hardcover), 978-981-4411-76-9 (eBook)
www.panstanford.com

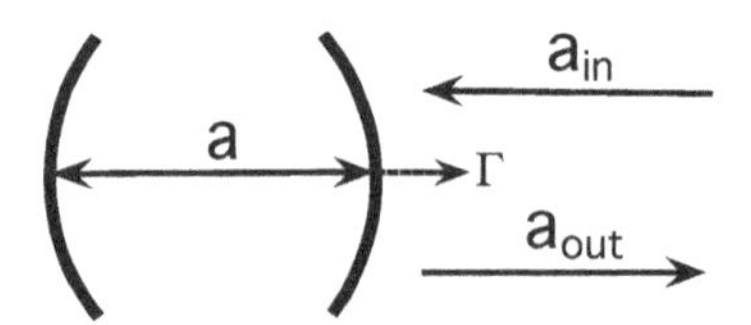

Figure 19.1 Schematic diagram illustrating the input–output formalism for a single-sided cavity damped with rate Γ to an external environment. The damping results from the coupling of the cavity mode to infinite number of the continuous modes of the environment.

commutation relation

$$\left[\hat{b}_k, \hat{b}^{\dagger}_{k'}\right] = \delta(k - k'). \tag{19.1}$$

Taking the Fourier transform of these modes over some finite bandwidth, we will take into account the fact that the output-measuring instrument has a finite bandwidth. This gives an external field for photons propagating in the $+x$ direction, defined as

$$\hat{c}_+(x) = \frac{c}{\sqrt{2\pi}} \int_{k_0-\Delta k}^{k_0+\Delta k} dk \hat{b}_k e^{ikx}. \tag{19.2}$$

This field has approximate equal-time commutators of the form

$$\left[\hat{c}_+(t, x), \hat{c}^{\dagger}_+(t, x')\right] = c^2 \delta(x - x'). \tag{19.3}$$

Next, if the mirror boundary is at $x = x_m$, the external photon flux operator can be readily defined as

$$\hat{a}_{\text{out}}(t) = \hat{c}_+(t, x_m). \tag{19.4}$$

From this it should be clear that there is also an input field, which corresponds to the field modes with $k = -k$, of the form

$$\hat{a}_{\text{in}}(t) = \hat{c}_-(t, x_m). \tag{19.5}$$

This input field is the source of quantum fluctuations inside the cavity that leads to the Γ-damping term in the master equation.

The following input–output relation occurs:

$$\hat{a}_{\text{out}}(t) = \sqrt{\Gamma}\hat{a} - \hat{a}_{\text{in}}(t). \tag{19.6}$$

19.3 Proof of the Input–Output Relation

Consider the Hamiltonian of the cavity field $\hat{a}$ interacting with the external modes $\hat{b}_k$:

$$\hat{H} = \hat{H}_c + \hat{H}_{ext} + \hat{H}_I, \tag{19.7}$$

where

$$\begin{aligned}
\hat{H}_c &= \hbar\omega_c \hat{a}^\dagger \hat{a}, \\
\hat{H}_{ext} &= \hbar \int d\omega_k \omega_k \hat{b}_k^\dagger \hat{b}_k, \\
\hat{H}_I &= i\hbar \int d\omega_k g(\omega_k) \left[\hat{b}_k \hat{a}^\dagger - \hat{a}\hat{b}_k^\dagger\right],
\end{aligned} \tag{19.8}$$

where $g(\omega_k)$ is a coupling constant between the cavity and the external modes.

The Heisenberg equation of motion for $\hat{b}_k$ is

$$\dot{\hat{b}}_k = -i\omega_k \hat{b}_k + g(\omega_k)\hat{a}, \tag{19.9}$$

and the cavity field operator $\hat{a}$ obeys the equation

$$\dot{\hat{a}} = -\frac{i}{\hbar}\left[\hat{a}, \hat{H}_c\right] - \int d\omega_k g(\omega_k)\hat{b}_k. \tag{19.10}$$

The solution of the equation of motion for $\hat{b}_k$ can be written in two ways depending on whether we choose to solve in terms of initial conditions at time $t_0 < t$ (the input), or in terms of the final conditions at times $t_1 > t$ (the output). The two solutions are

$$\begin{aligned}
\hat{b}_k(t) &= \hat{b}_k(0)e^{-i\omega_k(t-t_0)} + g(\omega_k)\int_{t_0}^{t} dt' \hat{a}(t')e^{-i\omega_k(t-t')}, \\
\hat{b}_k(t) &= \hat{b}_k(t_1)e^{-i\omega_k(t-t_1)} - g(\omega_k)\int_{t}^{t_1} dt' \hat{a}(t')e^{-i\omega_k(t-t')},
\end{aligned} \tag{19.11}$$

where $t_0 < t < t_1$, and $\hat{b}_k(0)$ is the value of $\hat{b}_k$ at $t = t_0$, while $\hat{b}_k(t_1)$ is the value of $\hat{b}_k$ at $t = t_1$.

Substituting the solution for $\hat{b}_k$ with $t_0 < t$ into the equation of motion for $\hat{a}$, we get

$$\begin{aligned}
\dot{\hat{a}} = &-\frac{i}{\hbar}\left[\hat{a}, \hat{H}_c\right] - \int d\omega_k g(\omega_k)\hat{b}_k(0)e^{-i\omega_k(t-t_0)} \\
&- \int d\omega_k g^2(\omega_k)\int_{t_0}^{t} dt' \hat{a}(t')e^{-i\omega_k(t-t')},
\end{aligned} \tag{19.12}$$

and in a similar manner, substituting the solution for $\hat{b}_k$ with $t < t_1$, we get

$$\dot{\hat{a}} = -\frac{i}{\hbar}\left[\hat{a}, \hat{H}_c\right] - \int d\omega_k g(\omega_k)\hat{b}_k(t_1)e^{-i\omega_k(t-t_1)} + \int d\omega_k g^2(\omega_k)\int_t^{t_1} dt'\hat{a}(t')e^{-i\omega_k(t-t')}. \quad (19.13)$$

We now make the Markov approximation, that the bandwidth of the external modes is large, so we can assume that the coupling constant $g(\omega_k)$ changes slowly with k, and replace $g(\omega_k) \approx g(\omega_c)$.

Next, we use the following properties of the delta function

$$\int_{t_0}^{t} dt' f(t')\delta(t-t') = \int_t^{t_1} dt' f(t')\delta(t-t') = \frac{1}{2}f(t), \quad (19.14)$$

where the factor $\frac{1}{2}$ arises from the fact that the peak of the delta function is at the end of the interval of integration.

We can also define the input and output operators, identical to $\hat{c}_+$ and $\hat{c}_-$, as

$$\hat{a}_{in}(t) = \frac{-1}{\sqrt{2\pi}}\int d\omega_k \hat{b}_k(0)e^{-i\omega_k(t-t_0)},$$
$$\hat{a}_{out}(t) = \frac{1}{\sqrt{2\pi}}\int d\omega_k \hat{b}_k(t_1)e^{-i\omega_k(t-t_1)}, \quad (19.15)$$

where the minus sign indicates that the input field propagates to the left in the opposite direction to the output field.

Using the relation

$$\int_{-\infty}^{\infty} d\omega_k e^{-i\omega_k(t-t')} = 2\pi\delta(t-t'), \quad (19.16)$$

and using the Markov approximation that $g(\omega_k) \approx g(\omega_c)$, we obtain

$$\dot{\hat{a}} = -\frac{i}{\hbar}\left[\hat{a}, \hat{H}_c\right] + \sqrt{2\pi}g(\omega_c)\hat{a}_{in}(t) - \frac{1}{2}\left(2\pi g^2(\omega_c)\right)\hat{a}(t), \ t_0 < t, \quad (19.17)$$

and

$$\dot{\hat{a}} = -\frac{i}{\hbar}\left[\hat{a}, \hat{H}_c\right] - \sqrt{2\pi}g(\omega_c)\hat{a}_{out}(t) + \frac{1}{2}\left(2\pi g^2(\omega_c)\right)\hat{a}(t), \ t < t_1. \quad (19.18)$$

Recognizing that $2\pi g^2(\omega_c) = \Gamma$ is the damping rate of the cavity mode, we have

$$\dot{\hat{a}} = -\frac{i}{\hbar}\left[\hat{a}, \hat{H}_c\right] + \sqrt{\Gamma}\hat{a}_{in}(t) - \frac{\Gamma}{2}\hat{a}(t), \qquad t_0 < t, \quad (19.19)$$

$$\dot{\hat{a}} = -\frac{i}{\hbar}\left[\hat{a}, \hat{H}_c\right] - \sqrt{\Gamma}\hat{a}_{out}(t) + \frac{\Gamma}{2}\hat{a}(t), \qquad t < t_1. \quad (19.20)$$

The above equations are Langevin equations for the damped amplitude $\hat{a}(t)$ in which the noise terms appear explicitly as the input (output) field. In Eq. (19.20), the output operator $\hat{a}_{out}(t)$ represents the coupling of the system to future external modes. Hence, this equation represents the backwards evolution of the system resulting in the negative damping term.

Finally, the relation between the input, output and cavity fields can be obtained by subtracting Eq. (19.20) from Eq. (19.19):

$$\hat{a}_{out}(t) = \sqrt{\Gamma}\hat{a} - \hat{a}_{in}(t). \quad (19.21)$$

This is the important result relating the input and output fields. It shows that the output field that is the *future* field outside the cavity is a sum of the input field and the field leaking the cavity with the rate Γ. The minus sign at the input field reflects the fact that the output and input fields have the opposite phase. If we consider normally ordered correlation functions, then the effect of the input field in this equation can be ignored, as long as the field is in the vacuum state (so that the field has vanishing normally ordered correlation functions). Otherwise, the measured functions include a contribution from the input field as well as from the internal cavity field. For a narrow-band external field, there is a finite correlation between the input and the cavity field, which makes this problem more complicated. Fortunately, it is generally possible to work with normally ordered correlation functions, where these complexities do not occur.

We have only considered a single-sided cavity. The input–output formalism can be extended to a two-sided cavity, where the couplings of the cavity mode to the external environment through partially transparent mirrors may not be the same [70].

Exercises

19.1 Consider a single-mode cavity described by the Hamiltonian $\hat{H}_c = \hbar\omega_c \hat{a}^\dagger \hat{a}$. Using Eqs. (19.19) and (19.21) and resolving the output, input and cavity field operators into the frequency components via the Fourier transformation

$$\hat{a}_i(\omega) = \frac{1}{\sqrt{2\pi}} \int_{-\infty}^{\infty} dt\, \hat{a}_i(t) e^{-i\omega(t-t_0)},$$

show that

$$\hat{a}_{\text{out}}(\omega) = \left(\frac{i(\omega - \omega_c) + \Gamma/2}{i(\omega - \omega_c) - \Gamma/2} \right) \hat{a}_{\text{in}}(\omega).$$

(a) What is the phase shift between the output and input fields at resonance ($\omega = \omega_c$)?

(b) What is the phase shift between the output and input fields at large detunings, $(\omega - \omega_c) \gg \Gamma$?

(c) What is the relation between the intensities $I_i(\omega) = \langle \hat{a}_i^\dagger(\omega) \hat{a}_i(\omega) \rangle$ of the output and input fields?

19.2 Consider the output–input relation (19.21). Under what conditions:

(a) The average number of the output photons equates to the average number of the cavity photons?

(b) The normally ordered variance of the output field equates to the normally ordered variance of the cavity field?

19.3 Using the output–input relation (19.21):

(a) Show that the output field commutator

$$\left[\hat{a}_{\text{out}}(t), \hat{a}_{\text{out}}^\dagger(t')\right] = \delta(t - t').$$

(b) Show that the commutator of the cavity field operators

$$\left[\hat{a}(t), \hat{a}^\dagger(t)\right]$$

is time-invariant despite the fact that the cavity field is damped in time.

(b) Assuming that the input field is in a vacuum state, show that the two-time correlation function of the output field

$$\langle \hat{a}_{\text{out}}^\dagger(t) \hat{a}_{\text{out}}(t') \rangle = \Gamma \langle \hat{a}^\dagger(t) \hat{a}(t') \rangle.$$

Chapter 20

Motion of Atoms in a Laser Field

20.1 Introduction

In quantum optical problems involving the interaction of atoms with laser fields, the atoms are usually considered being stationary during the interaction. Since every photon has a momentum, the atomic momentum (motion) can be changed in the process of absorption and emission of photons. This may force a stationary atom to move or to change momentum of an already moving atom. Therefore, the dynamical behaviour of atoms can be varied by the interaction with a radiation field. For a weak driving field, the influence of this field on the atomic motion can be neglected. However, the intensities of laser fields are generally very strong, and then the motion of the atoms can be considerably changed by the laser field. Here, we discuss this aspect, especially the effect of a driving field on the motion of atoms and the atomic momentum distribution. In particular, we will consider diffraction of atoms by a standing-wave laser field due to an exchange of momentum with the photons of the wave, and radiation force on atoms driven by a running- or standing-wave laser field.

Quantum Optics for Beginners
Zbigniew Ficek and Mohamed Ridza Wahiddin

ISBN 978-981-4411-75-2 (Hardcover), 978-981-4411-76-9 (eBook)
www.panstanford.com

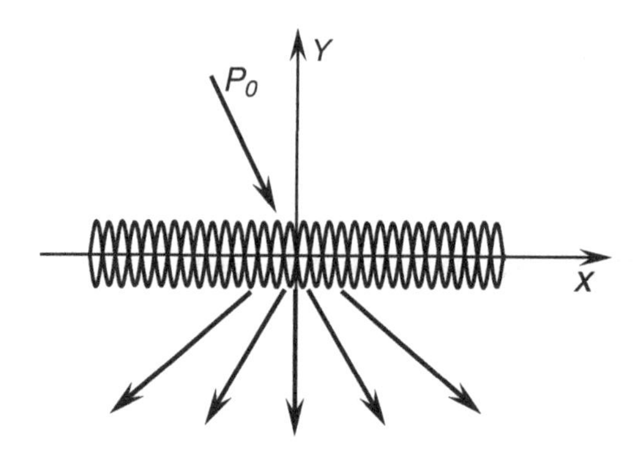

Figure 20.1 Illustration of the diffraction phenomenon of moving atoms. The atoms of an initial momentum $\vec{P}_0$ are sent to pass through a standing wave propagating in the horizontal x-direction.

20.2 Diffraction of Atoms on a Standing-Wave Laser Field

Consider the process of diffraction of a moving atom passing through a standing-wave laser field, as shown schematically in Fig. 20.1. Suppose, the laser field propagates along the x-axis, and the momentum of the atom, before entering into the laser field is in the direction $\vec{P}_0 = P_x\vec{i} + P_y\vec{j}$. The standing-wave laser field is equivalent to a superposition of two running-wave fields of the same amplitudes, but opposite propagation vectors

$$\begin{aligned}\vec{E}(x,t) &= 2\vec{E}_0\cos(kx)\cos(\omega t)\\ &= \left[\vec{E}_0\cos(\omega t - kx) + \vec{E}_0\cos(\omega t + kx)\right]. \end{aligned} \quad (20.1)$$

The Hamiltonian of the system of a moving atom interacting with a standing-wave laser field can be written in the standard form

$$\hat{H} = \hat{H}_{\mathrm{A}} + \hat{H}_{\mathrm{F}} + \hat{H}_{\mathrm{int}}, \quad (20.2)$$

where

$$\hat{H}_{\mathrm{A}} = \hbar\omega_0 S^z + \frac{|\vec{P}|^2}{2m} \quad (20.3)$$

is the Hamiltonian of the atom including the kinetic energy $|\vec{P}|^2/2m$,

$$\hat{H}_{\mathrm{F}} = \sum_{i=1}^{2} \hbar\omega_{\mathrm{L}}\hat{a}_i^{\dagger}\hat{a}_i \quad (20.4)$$

is the Hamiltonian of the laser field that is composed of two fields of the same frequency ω_{L} and wave-vectors $\vec{k}_i$ satisfying the relation

$$\vec{k}_1 = -\vec{k}_2 = \vec{k}. \quad (20.5)$$

The interaction Hamiltonian of the moving atom with the standing-wave laser field can be written as

$$\hat{H}_{\text{int}} = \frac{1}{2} i g \hbar \sum_{i=1}^{2} \left[S^{+} \hat{a}_i e^{i k_i x} - S^{-} \hat{a}_i^{\dagger} e^{-i k_i x} \right], \tag{20.6}$$

where x is the coordinate of the atom along the direction of the propagation of the laser field. Using the Hamiltonian (20.2), we find from the Schrödinger equation the state vector of the system, from which we can analyse the time evolution of the atomic momentum due to the interaction with the laser field.

Assume that initially the atom was in its ground state $|g\rangle$, and had a momentum $\vec{P}_0$. Moreover, assume that there were n_i photons in the ith field. Therefore, the initial state of the system can be written as

$$|\Phi_0\rangle = |n_1\rangle |n_2\rangle \left| g, \vec{P}_0 \right\rangle. \tag{20.7}$$

Moving atom enters to the field, and then can absorb photons from either of the two running-wave laser fields. When the atom absorbs a photon from the field '1', the state vector changes to

$$|\Phi_1\rangle = |n_1 - 1\rangle |n_2\rangle \left| e, \vec{P}_0 + \hbar \vec{k}_1 \right\rangle, \tag{20.8}$$

where $|e\rangle$ is the excited state of the atom.

When the atom absorbs a photon from the field '2', the state vector changes to

$$|\Phi_2\rangle = |n_1\rangle |n_2 - 1\rangle \left| e, \vec{P}_0 + \hbar \vec{k}_2 \right\rangle. \tag{20.9}$$

Suppose, the interaction of the atom with the laser fields is strong and the time of passing the field is short such that we can ignore spontaneous emission from the atom leaving only a possibility of stimulated emission to either of the two laser fields.

If the system is in the state $|\Phi_1\rangle$, the atom can be stimulated to emit a photon of the wave vector $\vec{k}_1$ or $\vec{k}_2$. If the wave-vector of the emitted photon is $\vec{k}_1$, the system returns to the initial state $|\Phi_0\rangle$. If the wave-vector of the emitted photon is $\vec{k}_2$, the state vector changes to

$$|\Phi_3\rangle = |n_1 - 1\rangle |n_2 + 1\rangle \left| g, \vec{P}_0 + \hbar \vec{k}_1 - \hbar \vec{k}_2 \right\rangle. \tag{20.10}$$

Similarly, if the system was in the state $|\Phi_2\rangle$ and the atom emits a photon of the momentum $\hbar \vec{k}_1$, the state vector changes to

$$|\Phi_4\rangle = |n_1 + 1\rangle |n_2 - 1\rangle \left| g, \vec{P}_0 + \hbar \vec{k}_2 - \hbar \vec{k}_1 \right\rangle. \tag{20.11}$$

Since the interaction of the atom with the laser fields is strong, there is a large number of absorption and emission processes during the time of passing through the field, which leads to the final state

$$|\Phi_n\rangle = \left|n_1 - \frac{n}{2}\right\rangle \left|n_2 + \frac{n}{2}\right\rangle \left|g, \vec{P}_0 + \frac{n}{2}\hbar\vec{k}_1 - \frac{n}{2}\hbar\vec{k}_2\right\rangle, \quad (20.12)$$

when n is an even number, and

$$|\Phi_n\rangle = \left|n_1 - \frac{n+1}{2}\right\rangle \left|n_2 + \frac{n-1}{2}\right\rangle \otimes \left|e, \vec{P}_0 + \frac{n+1}{2}\hbar\vec{k}_1 - \frac{n-1}{2}\hbar\vec{k}_2\right\rangle, \quad (20.13)$$

when n is an odd number. We will treat the states $|\Phi_n\rangle$ as complete basis states of the non-interacting system, and will find the state vector of the atom–field interacting system as a linear superposition

$$|\Phi(t)\rangle = \int d\vec{P} \sum_n C_n\left(\vec{P}, t\right) |\Phi_n\rangle . \quad (20.14)$$

Substituting the Hamiltonian (20.2) and Eq. (20.14) into the Schrödinger equation, we obtain the following set of coupled differential equations for the probability amplitudes

$$\begin{aligned} i\hbar\frac{d}{dt}C_n = &\left\{\left(n_1 - \frac{n}{2}\right)\hbar\omega_{\mathrm{L}} + \left(n_2 + \frac{n}{2}\right)\hbar\omega_{\mathrm{L}} \right. \\ &\left. + \frac{1}{2m}\left(\vec{P}_0 + \frac{n}{2}\hbar\vec{k}_1 - \frac{n}{2}\hbar\vec{k}_2\right)^2 - \frac{1}{2}\hbar\omega_0\right\} C_n \\ &+ i\frac{g}{2}\hbar\left(\sqrt{n_1 - \frac{n}{2}}C_{n-1} - \sqrt{n_2 + \frac{n}{2}}C_{n+1}\right), \end{aligned} \quad (20.15)$$

when n is an even number, and

$$\begin{aligned} i\hbar\frac{d}{dt}C_n = &\left\{\left(n_1 - \frac{n+1}{2}\right)\hbar\omega_{\mathrm{L}} + \left(n_2 + \frac{n-1}{2}\right)\hbar\omega_{\mathrm{L}} \right. \\ &\left. + \frac{1}{2m}\left(\vec{P}_0 + \frac{n+1}{2}\hbar\vec{k}_1 - \frac{n-1}{2}\hbar\vec{k}_2\right)^2 + \frac{1}{2}\hbar\omega_0\right\} C_n \\ &+ i\frac{g}{2}\hbar\left(\sqrt{n_1 - \frac{n-1}{2}}C_{n-1} - \sqrt{n_2 + \frac{n-1}{2}}C_{n+1}\right), \end{aligned} \quad (20.16)$$

when n is an odd number.

We can simplify the differential equations by introducing a notation

$$E_N = \left(n_1 + n_2 - \frac{1}{2}\right)\hbar\omega_{\mathrm{L}} + \frac{1}{2m}\left(P_x^2 + P_y^2\right), \qquad (20.17)$$

and a transformation

$$\tilde{C}_n = C_n \mathrm{e}^{iE_N t}. \qquad (20.18)$$

With these simplifications, the differential equations reduces to (n even)

$$i\frac{d}{dt}\tilde{C}_n = \frac{1}{2}\Delta\tilde{C}_n + \left[\frac{n^2\hbar}{8m}\left(\vec{k}_1 - \vec{k}_2\right)^2 + \frac{n\vec{P}_0}{2m}\left(\vec{k}_1 - \vec{k}_2\right)\right]\tilde{C}_n$$
$$+ i\frac{g}{2}\left(\sqrt{n_1 - \frac{n}{2}}\tilde{C}_{n-1} - \sqrt{n_2 + \frac{n}{2}}\tilde{C}_{n+1}\right), \qquad (20.19)$$

and (n odd)

$$i\frac{d}{dt}\tilde{C}_n = -\frac{1}{2}\Delta\tilde{C}_n + \left\{\frac{\hbar}{8m}\left[(n+1)\vec{k}_1 - (n-1)\vec{k}_2\right]^2\right.$$
$$\left. + \frac{\vec{P}_0}{2m}\left[(n+1)\vec{k}_1 - (n-1)\vec{k}_2\right]\right\}\tilde{C}_n$$
$$+ i\frac{g}{2}\left(\sqrt{n_1 - \frac{n-1}{2}}\tilde{C}_{n-1} - \sqrt{n_2 + \frac{n-1}{2}}\tilde{C}_{n+1}\right), \qquad (20.20)$$

where $\Delta = \omega_{\mathrm{L}} - \omega_0$ is the detuning between the laser frequency ω_{L} and the atomic transition frequency ω_0.

For an intense driving field, the number of photons in the field modes is large, $n_1, n_2 \gg 1$, which prompts us to make the following approximations

$$\sqrt{n_i - \frac{n}{2}} \approx \sqrt{n_i} \approx \sqrt{N},$$
$$\sqrt{n_i \pm \frac{n-1}{2}} \approx \sqrt{n_i} \approx \sqrt{N}, \qquad (20.21)$$

where $N = \langle n_i \rangle$ is the average number of photons in the laser fields.

Hence, the differential equations for $\tilde{C}_n$ simplify to

$$i\frac{d}{dt}\tilde{C}_n = \frac{1}{2}\Delta\tilde{C}_n + \left[\hbar\frac{n^2k^2}{2m} + \frac{nP_xk}{m}\right]\tilde{C}_n$$
$$+ \frac{i}{2}\Omega\left(\tilde{C}_{n-1} - \tilde{C}_{n+1}\right), \quad n \text{ even}, \qquad (20.22)$$

and

$$i\frac{d}{dt}\tilde{C}_n = -\frac{1}{2}\Delta\tilde{C}_n + \left[\hbar\frac{n^2k^2}{2m} + \frac{nP_xk}{m}\right]\tilde{C}_n$$
$$+\frac{i}{2}\Omega\left(\tilde{C}_{n-1} - \tilde{C}_{n+1}\right), \quad n \text{ odd}, \qquad (20.23)$$

where $\Omega = g\sqrt{N}$ is the Rabi frequency of the laser field.

We may introduce two parameters

$$\hbar b = \frac{\hbar^2k^2}{2m}, \qquad q = \frac{P_x}{\hbar k}. \qquad (20.24)$$

The parameter $\hbar b$ corresponds to the recoil kinetic energy of the atom after absorption or emission of a photon, and q is the ratio of the initial momentum of the atom in the x-direction to the momentum of photons.

With the parameters (20.24), the differential equations for $\tilde{C}_n$ reduce to

$$\frac{d}{dt}\tilde{C}_n = -i\left[\frac{1}{2}\Delta + b\left(n^2 + 2nq\right)\right]\tilde{C}_n$$
$$+\frac{1}{2}\Omega\left(\tilde{C}_{n-1} - \tilde{C}_{n+1}\right), \quad n \text{ even}, \qquad (20.25)$$

and

$$\frac{d}{dt}\tilde{C}_n = -i\left[-\frac{1}{2}\Delta + b\left(n^2 + 2nq\right)\right]\tilde{C}_n$$
$$+\frac{1}{2}\Omega\left(\tilde{C}_{n-1} - \tilde{C}_{n+1}\right), \quad n \text{ odd}. \qquad (20.26)$$

We now illustrate solutions of the above differential equations for $\tilde{C}_n$, from which we will find the time evolution of the atomic momentum under the interaction of the atom with a standing-wave laser field. We will discuss two cases:

1. $P_x = 0, \qquad \Delta = 0,$
2. $P_x \neq 0, \qquad \Delta = 0. \qquad (20.27)$

20.2.1 *The Case $P_x = 0$ and $\Delta = 0$*

In this case, the parameter $q = 0$, and assuming that the Rabi frequency is much larger than the recoil energy, we obtain

$$\frac{d}{dt}\tilde{C}_n = \frac{1}{2}\Omega\left(\tilde{C}_{n-1} - \tilde{C}_{n+1}\right). \qquad (20.28)$$

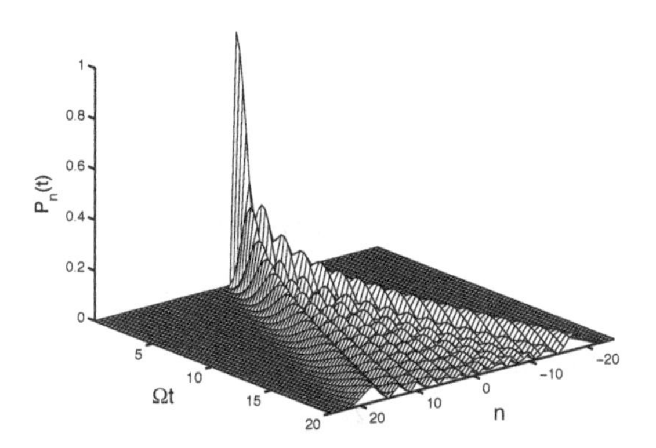

Figure 20.2 The atomic momentum distribution function for different values of Ωt.

Introducing a parameter $s = \Omega t$, we find that the coefficients $\tilde{C}_n$ satisfy a recurrence relation

$$2\frac{d}{ds}\tilde{C}_n(s) = \tilde{C}_{n-1}(s) - \tilde{C}_{n+1}(s). \tag{20.29}$$

Note, that the above recurrence relation is the same as that for the Bessel functions. Therefore, the probability amplitudes are

$$\tilde{C}_n(t) = J_n(\Omega t), \tag{20.30}$$

where J_n is the nth order Bessel function.

Hence, the atomic momentum $n\hbar k$ in the x-direction is

$$P_n(t) = \left|\tilde{C}_n(t)\right|^2 = J_n^2(\Omega t). \tag{20.31}$$

The momentum distribution is illustrated in Fig. 20.2. One can see that for $t = 0$, $P_n = 0$ for all n. As t increases, $P_n(t)$ increases and then the probability of finding atoms with momentum $n\hbar k$ increases.

20.2.2 *The Case $P_x \neq 0$ and $\Delta = 0$*

Consider now the second case in which $P_x \neq 0$. In this case, the term $2nbq$ in Eqs. (20.25) and (20.26) is different from zero. Then ignoring the recoil energy, we obtain

$$\frac{d}{dt}\tilde{C}_n = -2inbq\tilde{C}_n + \frac{1}{2}\Omega\left(\tilde{C}_{n-1} - \tilde{C}_{n+1}\right). \tag{20.32}$$

It is convenient to make a further transformation

$$\bar{C}_n = \tilde{C}_n e^{inbqt}. \tag{20.33}$$

Then

$$\begin{aligned}\frac{d}{dt}\bar{C}_n &= -inbq\bar{C}_n + \frac{1}{2}\Omega\left(\bar{C}_{n-1}e^{ibqt} - \bar{C}_{n+1}e^{-ibqt}\right) \\ &= -inbq\bar{C}_n + \frac{1}{2}\Omega\cos(bqt)\left(\bar{C}_{n-1} - \bar{C}_{n+1}\right) \\ &\quad + \frac{i}{2}\Omega\sin(bqt)\left(\bar{C}_{n-1} + \bar{C}_{n+1}\right).\end{aligned} \tag{20.34}$$

Introducing a parameter

$$z = \frac{\Omega}{bq}\sin(bqt), \tag{20.35}$$

we then can transform Eq. (20.34) into

$$\begin{aligned}2\frac{d}{dz}\bar{C}_n &= -i\frac{bq}{\Omega\cos(bqt)}\left[2n\bar{C}_n - z\left(\bar{C}_{n-1} + \bar{C}_{n+1}\right)\right] \\ &\quad + \left(\bar{C}_{n-1} - \bar{C}_{n+1}\right).\end{aligned} \tag{20.36}$$

Notice the recurrence relations for the Bessel functions

$$\begin{aligned}2nJ_n(x) &= x\left[J_{n-1}(x) + J_{n+1}(x)\right], \\ 2\frac{d}{dx}J_n(x) &= J_{n-1}(x) - J_{n+1}(x),\end{aligned} \tag{20.37}$$

and then, we find that

$$\bar{C}_n(t) = J_n\left(\frac{\Omega}{bq}\sin bqt\right). \tag{20.38}$$

Hence, the atomic momentum distribution function is

$$P_n = \left|\bar{C}_n(t)\right|^2 = J_n^2\left(\frac{\Omega}{bq}\sin bqt\right). \tag{20.39}$$

We see that the atomic distribution oscillates in time with the frequency bq.

In Fig. 20.3 we show the distribution for different values of bqt. As t increases from the initial value $t = 0$, the width of the distribution increases to its maximum value at $bqt = \pi/2$, and then the width decreases and reduces to zero for $bqt = \pi$. The maximum amplitude of the atomic distribution during each period of oscillation is equal to $\pm\Omega/(bq)$.

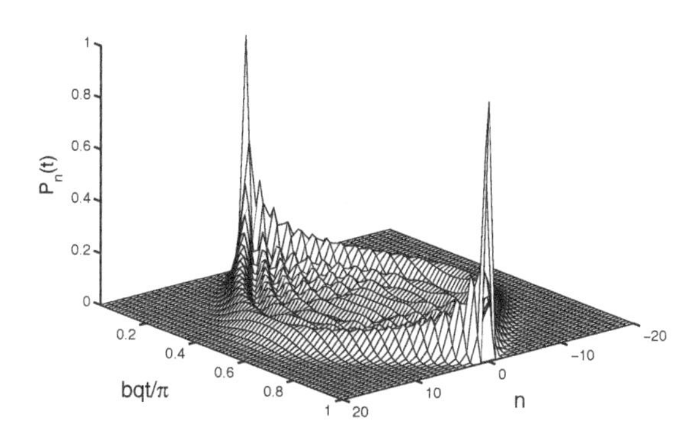

Figure 20.3 The atomic momentum distribution function for $\Omega/bq = 15$ and different values of bqt.

Thus, the diffraction of atoms with an initial momentum $P_x \neq 0$ is fundamentally different from that of $P_x = 0$. Instead of a continuous spreading of the atoms, the atomic distribution oscillates in time, periodically focusing and defocusing. This periodicity is observed in momentum space as an oscillation of the populations between the different momentum components.

In the derivation of Eqs. (20.31) and (20.39), we have ignored spontaneous emission from the atoms. It might be difficult to achieve this in actual experimental situations, in particular when the detuning $\Delta_{\rm L} = 0$. Alternatively, one can assume large detunings, $\Delta_{\rm L} \gg \Gamma, \Omega$. Then spontaneous emission can be neglected, as the atoms mostly reside in their ground states, with the upper atomic level adiabatically eliminated. To illustrate this, consider again the equations of motion (20.25) and (20.26), which in the Raman–Nath regime and with $P_x = 0$, reduce to

$$\frac{\partial}{\partial t}\tilde{C}_{2n} = -\frac{i}{2}\Delta_{\rm L}\tilde{C}_{2n} + \frac{1}{2}\Omega\left(\tilde{C}_{2n-1} - \tilde{C}_{2n+1}\right), \qquad (20.40)$$

and

$$\frac{\partial}{\partial t}\tilde{C}_{2n-1} = \frac{i}{2}\Delta_{\rm L}\tilde{C}_{2n-1} + \frac{1}{2}\Omega\left(\tilde{C}_{2n-2} - \tilde{C}_{2n}\right). \qquad (20.41)$$

According to Eq. (20.13), the equation of motion (20.41) for odd n corresponds to the time evolution of the probability amplitude of the excitation of the atomic upper state $|1\rangle$. When $\Delta_L \gg \Omega$, we can adiabatically eliminate $\tilde{C}_{2n-1}$ assuming that the amplitude does not change in time. Then, we put $\partial\tilde{C}_{2n-1}/\partial t = 0$ in Eq. (20.41), and obtain

$$\tilde{C}_{2n-1} = i\frac{\Omega}{\Delta}\left(\tilde{C}_{2n-2} - \tilde{C}_{2n}\right). \tag{20.42}$$

Substituting Eq. (20.42) into Eq. (20.40) and solving for $\tilde{C}_{2n}$, we find the probability amplitudes of the ground state

$$\tilde{C}_{2n} = \left[1 + 2\left(\frac{\Omega}{\Delta}\right)^2\right]^{-\frac{1}{2}} \exp\left[-i\left(\frac{1}{2}\Delta_L + \frac{\Omega^2}{\Delta_L}\right)t\right] J_n\left(\frac{\Omega^2 t}{\Delta_L}\right), \tag{20.43}$$

and then the probability distribution function is given by

$$P_{2n} = \left[1 + 2\left(\frac{\Omega}{\Delta}\right)^2\right]^{-1} J_n^2\left(\frac{\Omega^2 t}{\Delta_L}\right). \tag{20.44}$$

The above result holds for large detunings, but is in a form similar to Eq. (20.31), obtained for $\Delta_L = 0$. However, the result (20.44) is realistic experimentally, as for $\Delta_L \gg \Gamma$ spontaneous emission is negligible and can be ignored. The momentum distribution function (20.44) is an even function of n corresponding to the absorption of a photon from the $+\vec{k}$ component of the standing wave, followed by emission of a photon into the $-\vec{k}$ component. In this process, the atoms transfer photons from one component of the standing wave to the other, remaining in their ground states, but their momentum changes by $2n\hbar\vec{k}$.

20.3 Radiation Force on Atoms

In the previous section, we have shown that atoms can be diffracted by the interaction with a standing-wave laser field. The diffraction arises from a force acting on the atoms from the laser field. In other words, the force results from the transfer of momentum of laser photons to the atoms.

In the calculations, we have ignored spontaneous emission from the atoms. As we know, the atom being in its ground state can absorb

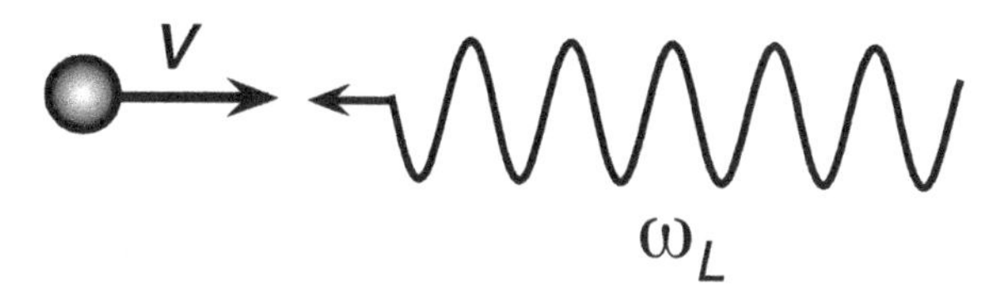

Figure 20.4 An atom moving with velocity $\vec{v}$ in the direction opposite to the direction of propagation of a running laser field.

one photon to make a transition to its excited state. In this transition process, the atom not only absorbs the photon energy, but also gains momentum (equal to the photon momentum) along the propagation direction of the laser field. After a short time (the lifetime of the atomic excited state), the atom spontaneously emits a photon and returns to its ground state. In the spontaneous emission process the atom also gains momentum (recoil momentum). Since the spontaneous emission is random and isotropic, the effective recoil momentum gained by the atom in many spontaneous emissions is zero. Thus, the momentum gained by the atom is only that along the propagation direction of the laser field, which results from the absorption of photons from the laser field.

Consider a two-level atom moving with a velocity $\vec{v}$ and interacting with a classical coherent field, as illustrated in Fig. 20.4. The Hamiltonian of the system, in the electric-dipole approximation, is given by

$$\hat{H} = \frac{|\vec{P}|^2}{2m} + \hbar\omega_0 S^z - \vec{\mu} \cdot \vec{E}(\vec{R}, t), \tag{20.45}$$

where $\vec{P}$ is the momentum of the atom, and the electric field is evaluated at the position $\vec{R}$ of the atom. In the case of many atoms, $\vec{P}$ and $\vec{R}$ correspond, respectively, to the momentum and position of the centre of mass.

We will use the full quantum mechanical picture of the atomic motion. Therefore, $\vec{P}$ and $\vec{R}$ will be treated as operators. In the Heisenberg picture, the operators $\hat{\vec{P}}$ and $\hat{\vec{R}}$ obey the following equations of motion

$$\frac{d}{dt}\hat{\vec{R}} = \frac{1}{i\hbar}\left[\hat{\vec{R}}, \hat{H}\right] = \frac{\hat{\vec{P}}}{m},$$
$$\frac{d}{dt}\hat{\vec{P}} = \frac{1}{i\hbar}\left[\hat{\vec{P}}, \hat{H}\right] = -\nabla\hat{H} = \nabla\left(\vec{\mu}\cdot\vec{E}(\hat{\vec{R}}, t)\right). \quad (20.46)$$

For simplicity, we assume that the spread of the atomic wave packet Δr is small compared with the laser wavelength, that is, $\Delta r \ll \lambda$. In this case, we can replace the atomic position operator $\hat{\vec{R}}$ by its expectation value $\vec{r} = \langle\hat{\vec{R}}\rangle$. Therefore, the average radiation force acting on the atom can be given by

$$\vec{F} = m\frac{d^2}{dt^2}\vec{r} = \langle\nabla(\vec{\mu}\cdot\vec{E})\rangle. \quad (20.47)$$

Suppose, the laser field is a single-mode plane wave of polarization $\vec{e}$ and amplitude $\mathcal{E}_0$:

$$\vec{E}(\vec{R}, t) = \frac{1}{2}\vec{e}\mathcal{E}_0 \exp[i(\omega_{\mathrm{L}}t - \vec{k}\cdot\vec{R})] + \mathrm{c.c.} \quad (20.48)$$

Then

$$\vec{F} = \langle\vec{\mu}\cdot\vec{e}\,\nabla\mathcal{E}(\vec{R}, t)\rangle = \langle\vec{\mu}\cdot\vec{e}\rangle\nabla\mathcal{E}(\vec{r}, t), \quad (20.49)$$

where

$$\mathcal{E}(\vec{R}, t) = \frac{1}{2}\mathcal{E}_0 \exp[i(\omega_{\mathrm{L}}t - \vec{k}\cdot\vec{R})] + \mathrm{c.c.} \quad (20.50)$$

Thus, in order to find the force acting on the atom, we have to calculate the time evolution of the atomic dipole moment. In order to do it, we express the atomic dipole moment in terms of the atomic dipole operators

$$\langle\,\vec{\mu}\cdot\vec{e}\,\rangle = \mu_{eg}\left(\langle S^+\rangle + \langle S^-\rangle\right), \quad (20.51)$$

and employ the Bloch equations to find the steady-state value of the average values of the atomic operators $\langle S^+\rangle$ and $\langle S^-\rangle$. For a two-level atom driven by a classical laser field, the Bloch equations are

$$\frac{d}{dt}\langle S^z\rangle = -\frac{1}{2}\Gamma - \Gamma\langle S^z\rangle - \frac{1}{2}\Omega\left(\langle\tilde{S}^+\rangle + \langle\tilde{S}^-\rangle\right),$$
$$\frac{d}{dt}\langle\tilde{S}^+\rangle = -\left[\frac{1}{2}\Gamma + i\left(\Delta + \frac{d}{dt}\theta(r)\right)\right]\langle\tilde{S}^+\rangle + \Omega\langle S^z\rangle,$$
$$\frac{d}{dt}\langle\tilde{S}^-\rangle = -\left[\frac{1}{2}\Gamma - i\left(\Delta + \frac{d}{dt}\theta(r)\right)\right]\langle\tilde{S}^-\rangle + \Omega\langle S^z\rangle, \quad (20.52)$$

where $\Delta = \omega_L - \omega_0$ is the detuning of the laser frequency from the atomic transition frequency, Γ is the spontaneous emission rate, Ω is the Rabi frequency of the laser field,

$$\langle \tilde{S}^{\pm} \rangle = \langle S^{\pm} \rangle \exp\left[\mp i \left(\omega_L t + \theta(t)\right)\right] \tag{20.53}$$

are the slowly varying parts of the atomic operators, and $\theta(r) = -\vec{k} \cdot \vec{r}$.

Substituting Eqs. (20.51) and (20.50) into Eq. (20.49), and neglecting terms rapidly oscillating with frequencies $\pm 2\omega_L$, we get

$$\vec{F} = \frac{1}{2}\hbar \left(U \nabla \Omega + V \Omega \nabla \theta\right), \tag{20.54}$$

where

$$U = \langle S^{+} \rangle + \langle S^{-} \rangle, \quad V = -i\left(\langle S^{+} \rangle - \langle S^{-} \rangle\right). \tag{20.55}$$

Note that

$$\begin{aligned} \frac{d}{dt}\theta(r) &= -\vec{k} \cdot \frac{d\vec{r}}{dt} = -\vec{k} \cdot \vec{v}, \\ \nabla\theta(r) &= -\vec{k}. \end{aligned} \tag{20.56}$$

Then, solving the Bloch equations for the steady state, we find

$$\begin{aligned} U &= -\frac{\Omega\left(\Delta - \vec{k} \cdot \vec{v}\right)}{4\left(\Delta - \vec{k} \cdot \vec{v}\right)^2 + 2\Omega^2 + \Gamma^2}, \\ V &= -\frac{2\Gamma\Omega}{4\left(\Delta - \vec{k} \cdot \vec{v}\right)^2 + 2\Omega^2 + \Gamma^2}. \end{aligned} \tag{20.57}$$

Hence, the force $\vec{F}$ on the atom exerted by the radiation field is given by

$$\vec{F} = -\hbar \frac{\left(\Delta - \vec{k} \cdot \vec{v}\right) \nabla \Omega^2 + \Gamma \Omega^2 \nabla \theta}{4\left(\Delta - \vec{k} \cdot \vec{v}\right)^2 + 2\Omega^2 + \Gamma^2}. \tag{20.58}$$

It is seen that the force depends on the gradient of the Rabi frequency, $\nabla \Omega^2$. If Ω is independent of r, which happens for a running-wave laser field, $\nabla \Omega^2 = 0$, and then the force reduces to

$$\vec{F} = \frac{\Gamma \Omega^2 \hbar \vec{k}}{4\left(\Delta - \vec{k} \cdot \vec{v}\right)^2 + 2\Omega^2 + \Gamma^2} = I(\Delta)\hbar\vec{k}. \tag{20.59}$$

The force is in the direction of the wave vector $\vec{k}$, and is equal to the Doppler shifted photon scattering rate

$$I(\Delta) = \frac{\Gamma\Omega^2}{4\left(\Delta - \vec{k}\cdot\vec{v}\right)^2 + 2\Omega^2 + \Gamma^2}, \tag{20.60}$$

multiplied by the photon recoil momentum $\hbar\vec{k}$.

20.3.1 *Slowing and Confining Atoms*

Let us discuss in more details the relationship between the detuning of the laser field from the atomic resonance and the magnitude of the radiation force. We shall distinguish between Doppler force useful for slowing down moving atoms and dipole force useful for confining atoms into a very small region.

Let us assume for a moment that the atom is stationary, $\vec{v} = 0$. We see that even in this case, the force is different from zero, proportional to $\hbar\vec{k}$. When we take into account the atom's motion, $\vec{v} \neq 0$, the direction of the force is still the same as the direction of the laser field.

Figure 20.5 shows the force as a function of Δ for two different directions of the moving atom. The force is a Lorentzian centred at $\Delta = \vec{k}\cdot\vec{v}$, where $\vec{k}\cdot\vec{v}$ is the Doppler frequency shift.

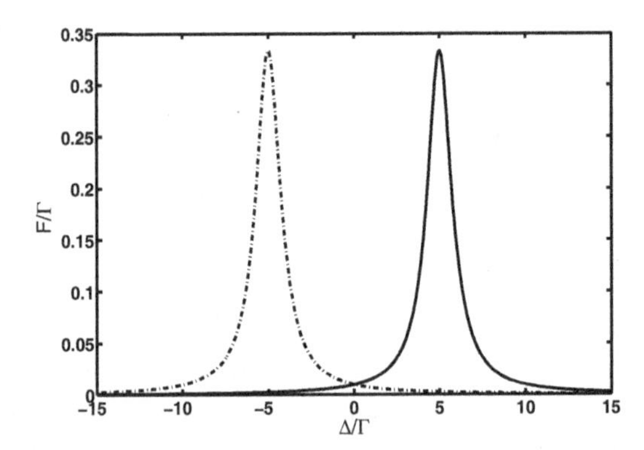

Figure 20.5 The force $\vec{F}$ per momentum $\hbar\vec{k}$ plotted as a function of the detuning Δ for $\Omega = \Gamma$ and two different directions of the moving atom, $\vec{k}\cdot\vec{v} = 5\Gamma$ (solid line) and $\vec{k}\cdot\vec{v} = -5\Gamma$ (dashed line).

Therefore, the force is called the Doppler force. For $\Delta > 0$, that is, when the laser frequency is *blue-detuned* from the atomic resonance, and the atom moving in the direction of the laser field, $\vec{k} \cdot \vec{v} = kv$, the force on the atom is large. However, when the atom is moving in the opposite direction to the direction of the laser field, $\vec{k} \cdot \vec{v} = -kv$, and then the force is small. The situation changes when $\Delta < 0$, that is, when the laser frequency is *red-detuned* from the atomic resonance. In this case, the force is large when the atom moves in the direction opposite to the laser field, and is small when the atom moves in the direction of the laser field. The case of $\Delta < 0$ is used in experiments to slow down (cool) trapped atoms.

Suppose now that Ω depends on r and θ is independent of r, which happens for a standing-wave laser field, the force $\vec{F}$ reduces to

$$\vec{F} = -\hbar \frac{\Delta}{4\Delta^2 + 2\Omega^2 + \Gamma^2} \nabla \Omega^2. \qquad (20.61)$$

The force exhibits a dispersive dependence on Δ, and its magnitude is proportional to the gradient of the field intensity. This force is called the dipole force. For $\Delta > 0$, the direction of the force is in direction of negative gradient of the field (decreasing intensity). For $\Delta < 0$, the direction of the force is in the direction of positive gradient of the field (increasing intensity). The force is zero for a resonant ($\Delta = 0$) driving field. This shows that the dipole force attracts the atom to regions of intense field when the laser is tuned below resonance, and repels the atom from these regions when tuned above resonance. This property of the dipole force is used in experiments to trap atoms in a very small area of a focused laser beam.

20.4 Summary

In this chapter, we have demonstrated the following:

(1) Atoms can be diffracted by the interaction with a standing-wave laser field. The diffraction arises from a force acting on the atoms from the laser field.

(2) The dipole force is different from zero only for a detuned laser field.
(3) The dipole force attracts the atom to regions of intense field when the laser is tuned below resonance, and repels the atom from these regions when tuned above resonance.

We conclude this chapter with a brief comment that laser-cooling techniques have developed to the level that led to create Bose–Einstein condensate [113–115] and to achieve ultra-low temperatures of optical lattices and single ions [116, 117]. A new area of science, ultra-cold physics have been developed [118] and a significant progress has been made towards not only to achieve ultra-low temperatures but also in cooling macroscopic objects, such as metallic or dielectric plates and biological samples (membranes) to temperatures of few microkelvins [119, 120].

Exercises

20.1 Using the Bloch equations (20.52), find the steady-state value of the average population inversion, $\langle S^z \rangle$. Then, show that Eq. (20.60) represents the steady-state photon scattering rate, defined as $I(\Delta) = \Gamma(\langle S^z \rangle - 1/2)$.

20.2 A driven two-level atoms undergoes dressing by the driving field.

(a) Starting with the expression from Eq. (20.54), write the force $\vec{F}$ in terms of the populations and coherences of the dressed states.

(b) Under what conditions the force would depend only on the populations of the dressed state?

20.3 The Doppler force is asymmetric with Δ that the case $\Delta < 0$ is better for slowing down moving atoms than $\Delta > 0$. Explain in your own words, why a red-detuned laser works better for slowing down moving atoms than a blue-detuned laser?

20.4 In practice, atoms may move with an uniform velocity. Assuming that the atomic velocities obey the Maxwell–

Boltzmann distribution

$$\rho(v) = \frac{1}{\sqrt{\pi}\sigma} \exp(-v^2/\sigma^2),$$

where σ is the width of the distribution, calculate the average force by integrating expression (20.58) over the velocity distribution. What is the average force in the limit $\sigma^2 \gg 2\Omega^2 + \Gamma^2$?

Final Remark

Although this book focuses on backgrounds of quantum optics, it is nevertheless appropriate to conclude by emphasizing the importance of quantum optics in the development of new areas in science and technology. The predictions of quantum optics have turned research and technology into new directions and have led to numerous technological innovations and the development of a new technology on the scale of single atoms and electrons, called quantum technology, or nanotechnology. The ability to manufacture tiny structures, such as quantum dots, and to control their dimensions allows us to engineer the unique properties of these structures and predict new devices such as quantum computers. A quantum computer can perform mathematical calculations much faster and store much more information than a classical computer by using the laws of quantum physics. The technology for creating a quantum computer is still in its infancy because it is extremely difficult to control quantum systems, but is developing very rapidly with little sign of the progress slowing.

We have seen in our journey through the backgrounds of quantum optics that despite its long history and the development toward quantum technology, quantum optics still challenges our understanding and continues to excite our imagination. We hope this book has provided a good guide toward current developments in quantum optics and has encouraged the reader to learn more.

References

1. J. D. Jackson, *Classical Electrodynamics*, 3rd ed. (Wiley, New York, 1999).
2. E. Merzbacher, *Quantum Mechanics* (Wiley, New York, 1998).
3. L. Allen and J. H. Eberly, *Optical Resonance and Two-Level Atoms* (Wiley, New York, 1975).
4. C. Gerry and P. L. Knight, *Introductory Quantum Optics* (Cambridge University Press, Cambridge, 2005).
5. P. W. Milonni, J. R. Ackerhalt and H. W. Galbraith, *Phys. Rev. Lett.* **50**, 966 (1983).
6. S. J. D. Phoenix, *J. Mod. Opt.* **38**, 695 (1991).
7. V. E. Lembessis, *Phys. Rev. A* **78**, 043423 (2008).
8. H. T. Ng and K. Burnett, *New J. Phys.* **10**, 123014 (2008).
9. L. Mandel, E. C. G. Sudarshan and E. Wolf, *Proc. Phys. Soc.* (London) **84**, 435 (1964).
10. J. Perina, *Coherence of Light* (Kluwer, Dordrecht, 1985).
11. B. T. Varcoe, S. Brattke, M. Weidinger and H. Walther, *Nature* (London) **403**, 743 (2000).
12. D. I. Schuster, A. A. Houck, J. A. Schreier, A. Wallraff, J. M. Gambetta, A. Blais, L. Frunzio, J. Majer, B. Johnson, M. H. Devoret, S. M. Girvin and R. J. Schoelkopf, *Nature* (London) **445**, 515 (2007).
13. M. Hofheinz *et al.*, *Nature* **454**, 310 (2008).
14. H. J. Carmichael and D. F. Walls, *J. Phys. B* **9**, 1199 (1976).
15. H. J. Kimble and L. Mandel, *Phys. Rev. A* **13**, 2123 (1976).
16. M. Dagenais, H. J. Kimble and L. Mandel, *Phys. Rev. Lett.* **39**, 691 (1977).
17. F. Diedrich and H. Walther, *Phys. Rev. Lett.* **58**, 203 (1987).
18. G. T. Foster, S. L. Mielke and L. A. Orozco, *Phys. Rev. A* **61**, 053821 (2000).
19. L. Mandel, *Opt. Lett.* **4**, 205 (1979).
20. R. J. Glauber, *Phys. Rev.* **130**, 2529 (1963); *Phys. Rev.* **131**, 2766 (1963).

21. P. A. M. Dirac, *Proc. R. Soc. A* **114**, 243 (1927).
22. L. Susskind and J. Glogower, *Physics* **1**, 49 (1964).
23. P. Carruthers and M. Nieto, *Rev. Mod. Phys.* **40**, 411 (1968).
24. D. T. Pegg and S. M. Barnett, *Phys. Rev. A* **41**, 3427 (1989).
25. S. M. Barnett and J. A. Vacaro (eds.) *The Quantum Phase Operator* (Taylor and Francis, London, 2007).
26. R. Tanaś, A. Miranowicz and Ts. Gantsog, Progress in Optics, vol. XXXV (Elsevier, Amsterdam, 1996), p. 355.
27. H. P. Yuen, *Phys. Rev. A* **13**, 2226 (1976).
28. C. M. Caves, *Phys. Rev. D* **23**, 1693 (1981).
29. Z. Ficek and P. D. Drummond, *Phys. Today* **50**, 34 (1997).
30. J. S. Peng and G. X. Li, *Introduction to Modern Quantum Optics* (World Scientific, Singapore, 1998).
31. P. D. Drummond and Z. Ficek (eds.) *Quantum Squeezing* (Springer, Berlin, 2004).
32. D. F. Walls and P. Zoller, *Phys. Rev. Lett.* **47**, 709 (1981).
33. K. Wódkiewicz and J. H. Eberly, *J. Opt. Soc. Am. B* **2**, 458 (1985).
34. M. Kitagawa and M. Ueda, *Phys. Rev. A* **47**, 5138 (1993).
35. D. J. Wineland, J. J. Bollinger, W. M. Itano and D. J. Heinzen, *Phys. Rev. A* **50**, 67 (1994).
36. A. S. Sorensen, L. M. Duan, J. I. Cirac and P. Zoller, *Nature* **409**, 63 (2001).
37. A. Messikh, Z. Ficek and M. R. B. Wahiddin, *Phys. Rev. A* **68**, 064301 (2003).
38. L. Lugiato and G. Strini, *Optics Commun.* **41**, 67 (1982).
39. G. J. Milburn and D. F. Walls, *Optics Commun.* **39**, 401 (1981).
40. H. P. Yuen and V. W. S. Chan, *Optics Lett.* **8**, 177 (1983).
41. B. L. Schumaker, *Optics Lett.* **9**, 189 (1984).
42. D. F. Walls and G. J. Milburn, *Quantum Optics* (Springer-Verlag, Berlin, 1994).
43. M. O. Scully and M. S. Zubairy, *Quantum Optics* (Cambridge University Press, Cambridge, 1997).
44. E. C. G. Sudarshan, *Phys. Rev. Lett.* **10**, 277 (1963).
45. P. D. Drummond and C. W. Gardiner, *J. Phys. A* **13**, 2353 (1980).
46. E. P. Wigner, *Phys. Rev.* **40**, 749 (1932).
47. W. P. Schleich, *Quantum Optics in Phase Space* (Wiley, New York, 2001).
48. E. T. Jaynes and F. W. Cummings, *IEEE Proc.* **51**, 89 (1963).

49. H. Paul, *Ann. der Phys.* **466**, 411 (1963).
50. J. H. Eberly, N. B. Narozhny and J. J. Sanchez-Mondragon, *Phys. Rev. Lett.* **44**, 1323 (1980).
51. S. Gleyzes, et al., *Nature* **446**, 297 (2007).
52. T. F. Gallagher, *Rydberg Atoms* (Cambridge University Press, Cambridge, 1994).
53. A. Wallraff, D. I. Schuster, A. Blais, L. Frunzio, R.-S. Huang, J. Majer, S. Kumar, S. M. Girvin and R. J. Schoelkopf, *Nature* (London) **431**, 162 (2004).
54. W. Louisell, *Quantum Statistical Properties of Radiation* (Wiley, New York, 1973).
55. H. A. Bethe, *Phys. Rev.* **72**, 339 (1947).
56. D. W. Wang, L. G. Wang, Z. H. Li and S. Y. Zhu, *Phys. Rev. A* **80**, 042101 (2009).
57. S. Yang, H. Zheng, R. Hong, S. Y. Zhu and M. S. Zubairy, *Phys. Rev. A* **81**, 052501 (2010).
58. S. F. Chien, M. R. B. Wahiddin and Z. Ficek, *Phys. Rev. A* **57**, 1295 (1998).
59. Z. Ficek and H. S. Freedhoff, *Prog. Optics* **40**, 389 (2000).
60. C. P. Slichter, *Principles of Magnetic Resonance* (Springer-Verlag, Berlin, 1980).
61. M. Lewenstein and T. W. Mossberg, *Phys. Rev. A* **37**, 2048 (1988).
62. C. Cohen-Tannoudji and S. Reynaud, *J. Phys. B* **10**, 345 (1977).
63. C. Cohen-Tannoudji, J. Dupont-Roc and G. Grynberg, *Atom–Photon Interactions* (Wiley, New York, 1992).
64. J. Zakrzewski, M. Lewenstein and T. W. Mossberg, *Phys. Rev. A* **44**, 7717 (1991); **44**, 7732 (1991); **44**, 7746 (1991).
65. N. Lu and P. R. Berman, *Phys. Rev. A* **44**, 5965 (1991).
66. Y. Zhu, Q. Wu, S. Morin and T. W. Mossberg, *Phys. Rev. Lett.* **65**, 1200 (1990).
67. D. J. Gauthier, Q. Wu, S. E. Morin and T. W. Mossberg, *Phys. Rev. Lett.* **68**, 464 (1992).
68. H. Risken, *The Fokker Planck Equation* (Springer-Verlag, Berlin, 1984).
69. H. J. Carmichael, *Statistical Methods in Quantum Optics* (Springer, Berlin, 1999).
70. C. W. Gardiner and P. Zoller, *Quantum Noise* (Springer, Berlin, 2000).
71. H. J. Carmichael, *Phys. Rev. Lett.* **70**, 2273 (1993).

72. H. J. Carmichael, *An Open System Approach to Quantum Optics*, Lecture Notes in Physics, Vol. 18 (Springer-Verlag, Berlin, 1993).
73. J. Dalibard, Y. Castin and K. Molmer, *Phys. Rev. Lett.* **68**, 580 (1992).
74. P. R. Rice and H. J. Carmichael, *Phys. Rev. A* **50**, 4318 (1994).
75. M. S. A. Hadi, M. R. Wahiddin and T. H. Hassan, *Phys. Rev. A* **68**, 063804 (2003).
76. G. X. Li, M. Luo and Z. Ficek, *Phys. Rev. A* **79**, 053847 (2009).
77. P. S. Epstein, *Am. J. Phys.* **13**, 127 (1945).
78. A. Elitzur and L. Vaidman, *Foundation Phys.* **23**, 987 (1993).
79. P. G. Kwiat, H. Weinfurter, T. Herzog, A. Zeillinger and M. A. Kasevich, *Phys. Rev. Lett.* **74**, 4763 (1995).
80. Z. Ficek and S. Swain, *J. Mod. Opt.* **49**, 3 (2002).
81. Y. Ou and L. Mandel, *Phys. Rev. Lett.* **62**, 2941 (1989).
82. A. Einstein, B. Podolsky and N. Rosen, *Phys. Rev.* **47**, 777 (1935).
83. A. Aspect, P. Grangier and G. Roger, *Phys. Rev. Lett.* **49**, 91 (1982).
84. J. S. Bell, *Science* **177**, 880 (1972).
85. M. D. Reid, P. D. Drummond, W. P. Bowen, E. G. Cavalcanti, P. K. Lam, H. A. Bachor, U. L. Andersen and G. Leuchs, *Rev. Mod. Phys.* **81**, 1727 (2009).
86. Q. Y. He, M. D. Reid, T. G. Vaughan, C. Gross, M. Oberthaler and P. D. Drummond, *Phys. Rev. Lett.* **106**, 120405 (2011).
87. C. K. Hong, Z. Y. Ou and L. Mandel, *Phys. Rev. Lett.* **59**, 2044 (1987).
88. L. Mandel and E. Wolf, *Optical Coherence and Quantum Optics* (Cambridge University Press, Cambridge, 1995).
89. Z. Ficek and S. Swain, *Quantum Interference and Coherence: Theory and Experiments* (Springer, New York, 2005).
90. P. Lambropoulos and D. Petrosyan, *Fundamentals of Quantum Optics and Quantum Information* (Springer-Verlag, Berlin, 2007).
91. M. A. Nielsen and I. J. Chuang, *Quantum Computation and Quantum Information* (Cambridge University Press, Cambridge, 2000).
92. R. H. Lehmberg, *Phys. Rev. A* **2**, 883 (1970); **2**, 889 (1970).
93. G. S. Agarwal, *Quantum Statistical Theories of Spontaneous Emission and Their Relation to Other Approaches*, Vol. 70 of Springer Tracts in Modern Physics (Springer-Verlag, Berlin, 1974).
94. Z. Ficek and R. Tanaś, *Phys. Rep.* **372**, 369 (2002).
95. R. H. Dicke, *Phys. Rev.* **93**, 99 (1954).
96. G. M. Palma and P. L. Knight, *Phys. Rev. A* **39**, 1962 (1989).

97. G. S. Agarwal and R. R. Puri, *Phys. Rev. A* **41**, 3782 (1990).
98. Z. Ficek, *Phys. Rev. A* **44**, 7759 (1991).
99. S. M. Barnett and M. A. Dupertuis, *J. Opt. Soc. Am. B* **4**, 505 (1987).
100. M. D'Angelo, M. V. Chekhova and Y. Shih, *Phys. Rev. Lett.* **87**, 013602 (2001).
101. M. W. Mitchell, J. S. Lundeen and A. M. Steinberg, *Nature* **429**, 161 (2004).
102. P. Walther, J.-W. Pan, M. Aspelmeyer, R. Ursin, S. Gasparoni and A. Zeilinger, *Nature* **429**, 158 (2004).
103. G. S. He, L. S. Tan, Q. Zheng and P. N. Prasad, *Chem. Rev.* **108**, 1245 (2008).
104. E. C. Spivey, E. T. Ritschdorff, J. L. Connell, C. A. McLennon, C. E. Schmidt and J. B. Shear, *Adv. Funct. Mater.* **22** (2012).
105. G. S. Agarwal, R. W. Boyd, E. M. Nagasako and S. J. Bentley, *Phys. Rev. Lett.* **86**, 1389 (2001).
106. R. W. Boyd and S. J. Bentley, *J. Mod. Opt.* **53**, 713 (2006).
107. M. Kiffner, J. Evers and M. S. Zubairy, *Phys. Rev. Lett.* **100**, 073602 (2008).
108. P. R. Hemmer, A. Muthukrishnan, M. O. Scully and M. S. Zubairy, *Phys. Rev. Lett.* **96**, 163603 (2006).
109. Q. Sun, P. R. Hemmer and M. S. Zubairy, *Phys. Rev. A* **75**, 065803 (2007).
110. Z. Liao, M. Al-Amri and M. S. Zubairy, *Phys. Rev. Lett.* **105**, 183601 (2010).
111. M. Lax and W. H. Louisell, *Phys. Rev.* **185**, 568 (1969).
112. H. Haken, *Laser Theory* (Springer, Berlin, 1984).
113. M. H. Anderson, J. R. Ensher, M. R. Matthews, C. E. Wieman and E. A. Cornell, *Science* **269**, 198 (1995).
114. D. J. Heinzen, *Int. J. Mod. Phys. B* **11**, 3297 (1997).
115. J.R. Anglin and W. Ketterle, *Nature* (London) **416**, 211 (2002).
116. C. Zipkes, S. Palzer, C. Sias and M. Köhl, *Nature* (London), **464**, 388 (2010).
117. G. K. Campbell, *Nature* (London) **480**, 463 (2011).
118. I. Bloch, *Science* **319**, 1202 (2008).
119. H. J. Kimble, *Nature* (London) **453**, 1023 (2008).
120. M. Aspelmeyer, S. Gröblacher, K. Hammerer and N. Kiesel, *J. Opt. Soc. Am. B* **27**, A189 (2010).

Index